Conscious Coastal Cities

Voula P. Mega

Conscious Coastal Cities

Sustainability, Blue Green Growth, and the Politics of Imagination

after a special authorisation by the European Commission

Springer

Voula P. Mega
PhD in City and Regional Planning
and Policy
Brussels, Belgium
Voula.Mega@post.Harvard.edu

ISBN 978-3-319-20217-4 ISBN 978-3-319-20218-1 (eBook)
DOI 10.1007/978-3-319-20218-1

Library of Congress Control Number: 2015953832

Springer Cham Heidelberg New York Dordrecht London

Printed on acid-free paper

Springer International Publishing AG Switzerland is part of Springer Science+Business Media (www.springer.com)

In memoriam Professor Sir Peter Hall

Imagination is more important than knowledge. For knowledge is limited to all we now know and understand, while imagination embraces the entire world, and all there ever will be to know and understand

Albert Einstein

Want of foresight, unwillingness to act when action would be simple and effective, lack of clear thinking, confusion of counsel until the emergency comes, until self-preservation strikes its jarring gong, these are the features which constitute the endless repetition of history

Winston Churchill

The greater danger for most of us lies not in setting our aim too high and falling short, but in setting our aim too low, and achieving our mark

Michelangelo

With every drop of water you drink every breath you take, you ‘reconnected to the sea. Health to the ocean means health for us.

Sylvia Earle

All things move and nothing remains still …

Heraclitus

Preface

Try to imagine the world without cities. It would be like a world without sea, without excitement. As humanity progressively and irreversibly becomes urban, the future of the planet is more and more decided in cities. It seems that, in the noblest adventures of the century, cities will lead and countries will follow. Most of the globe's economic activity is funnelled through cities and especially coastal and maritime cities. They have the opportunity and the responsibility to lead the world towards sustainable development.

The sheer size of cities is crucial, but also their potential to generate and capture positive synergies, their capacity for creativity and innovations and the interconnected local political leadership which heralded the era of urban geopolitics. Urban democracies, advocating for citizens to be at the heart of policies, could bring an extraordinary momentum in times of global democratic and trust deficit.

This book sheds particular light on coastal cities, places of exceptional opportunities and tensions at the junction of marine, land, air and human ecosystems. From seaside resorts and delta cities to maritime hubs, the spectrum of coastal cities includes very diverse typologies. They can all generate very diverse exceptional synergies, out of the interacting strengths and weaknesses of both cities and coasts. The book highlights particular threats and opportunities, and underlines effective imaginative responses that can inspire cities to address the challenges of the future in harmonious symbiosis with the sea.

Humanity owes much of its prosperity to seas and oceans. Oceans are the ultimate life support system. Water has not only been a precondition for life on earth but it has provided vital resources to society and endowed some cities with special assets, from fisheries and sea transport to recreational activities, raw materials and energy. Since the age of ancient metropolises on the sea, many port cities concentrated not only commercial and economic wealth, but also political and even cultural power. Privileged access to the sea has often invited economies to exchange, trade and expand. The proximity of the sea and, through maritime routes, of the shores of the world, makes coastal cities more open and inventive, rich in experiences that can be offered to both citizens and visitors.

Direct access to and symbiosis with the sea have been the generic criteria applied in this book which broadly focuses on coastal cities, without following a precise typology. The array of coastal cities includes places of potential conflicts between activities and ecosystem services and more vulnerable to open sea meteorological conditions and extreme phenomena that are also linked to advancing climate change. The escalating risks of flood and storm surge increasingly threaten cities, even if their citizens continue to wish to live by the sea. Authoritative reports analysing city-level data on population, assets, elevation and forecasts of rising sea levels, suggest that risks can be high and have to be collectively understood and addressed by all stakeholders. The spectre of sunken cities has reappeared. Images of Wall Street under water, in 2012, could be a twenty-first-century symbol for urban climate change risks.

Since Ulysses, the first navigator, humanity has been attracted to exploration of lands beyond the seas. Its interest in life below the waves probably is as recent as its interest in life beyond the planet Earth. Atlantis, the archetypal sunken city, lost the struggle with eternity, but underwater archaeology shed light on many wealthy and glorious urban times before being engulfed by the water. Many more lost cities could be found on the seabed, humanity's largest museum. At the other end of the spectrum, thriving cities reign on the seas and lead global maritime functions. Could present threats transform them into the sunken cities of the future?

Climate change, subsidence, and socio-economic change could cause more cities to sink. Given the populations migrating to coastal cities, their oversize economic impact, and the anchoring of global prosperity in maritime businesses, the eventual sinking of coastal cities is an economic, social and even ethical and cultural issue of the highest importance. Resilience of places and communities to withstand threats becomes decisive. Cities have to strike a balance between short-term action to address urgent extreme situations and, most important, long-term action for sustainable development.

The questions in the heart of the matter include: What could the journey of coastal cities into the world of the future be like? What auspicious and adverse winds and waves might there be to affect their unique assets and prospects? Why do citizens wish to aggregate by the sea? How does this proximity influence their quality of life? What are the risks of exposure of the world's port cities to coastal flooding, storm surges and high winds? What can the impact of climate change, subsidence, demographic developments and socio-economic activities be? What can cities do to protect their citizens and what can citizens do to make their cities more immune to shocks? How could cities best manage precious marine resources as valuable assets that must be used sustainably? What innovations can cities bring as strongholds of the emerging blue economy and the services provided by marine ecosystems? How could coastal cities be best eco-managed to grasp opportunities and manage risks while engaging with their stakeholders and the world and promoting ocean corporate responsibility?

* * *

Cities are major nodes of all energy fluxes in the complex, uncertain and interlinked urbanising world. The first chapter examines the opportunities and risks which could impact the future of cities as critical nexus points of social, economic and ecological change. It reviews global demographic, socio-economic and environmental trends, political developments and technological breakthroughs and presents an array of possible futures which invite cities to develop a vision on a preferred future. Last but not least, this part explores the potential of cities to lead the world in times of global democratic deficit.

The acceptable balance towards the civilisation of sustainability is a fundamental challenge for twenty-first century cities. Conscious cities have to address all interacting global societal challenges, striving for strong sustainability. They can head through the age of scarcity and towards a new world order offering better chances for all citizens, present and future, within the limits of the planet. In alliance with seas and the oceans, the supreme life support systems, cities can lead by example and through their magnitude reinforced by their multiple networks. They should care not only about their natural resources, including marine resources, but also lead coalitions of committed stakeholders and engage in protection of the oceans and the high seas, the ultimate world commons.

The second chapter reviews possible future trends, opportunities, tensions and risks for coastal cities. Backbones of both sea and land dynamics and functions, coastal and maritime cities can generate and enhance precious synergies among their diverse resources and interacting ecosystems and serve as bridges among the hinterland, and, through the maritime routes, faraway lands. To advance towards sustainable development, coastal cities have to enhance all aspects of their unique urban capital, natural and physical, on land and water, human, intellectual and social, cultural and political, financial and constructed, as strong sustainability asks for all forms of urban capital to be preserved, strengthened and transmitted to future generations. Climate change, subsidence and excessive population growth and socio-economic change are important risks especially for low-elevated densely populated areas, but coastal cities count on the very diverse assets which, if sustainably managed, can help them address the challenges of the future. The chapter invites us to a journey to some urban shores of the world, highlights strengths and vulnerabilities, and a multitude of possible actions, promises and perils.

The great wealth of coastal areas, whether in terms of access to the sea, maritime trade, tourism, fishing, or natural resources, attracts urban populations. However if this precious coastal capital is unsustainably managed, the seeds of its benefits can be easily undermined. Coastal urbanisation affects land use and cover of the shore, biodiversity, soil, water and air quality and global climate. Responsible coastal cities try to reduce their ecological debts on land and the sea. Urban coastal organs and functions have to boost the resilience of places and assist in transition to the civilisation of sustainability. Sustainable cities and oceans, among the priority areas of the Rio+20 conference and the dialogue for the post-2015 sustainable development goals, need ecological and environmentally-friendly cells and neurons to thrive in harmony with the planet and the seas. Together with their citizens and stakeholders, they can play a cardinal role for the exploration and protection of the precious marine resources.

Globalisation brings a closer interdependency among human settlements and places of extraction or transportation of resources. The autocratic balance has been modified. The third chapter focuses on coastal cities as vital and interdependent ecosystems able to manage crucial amounts of scarce resources and materials, ensure food security, offer sustainable goods and services and reduce greenhouse gas emissions and waste, especially preventing marine debris. Many coastal cities provide outstanding models of emission-neutral, waste-free eco-neighbourhoods and intelligently designed waterfronts to celebrate the sea which links them to the conscience of the world. Furthermore, they can send an impulse for maritime routes to become vectors of responsibility for sustainable development.

Responsible coastal cities have to ensure uninterrupted access for all to clean, secure, competitive and affordable energy and transport. Investments in energy efficiency and renewable energy sources can only be beneficial. Offshore energy generation can extend possibilities for cities, producers and consumers of green energy, to become energy autonomous. The fourth chapter offers an insight into the components and vectors of sustainable energy production and consumption in cities and the efforts made at the local level to overcome national and supranational energy policy targets. Blue green electricity from offshore wind, ocean thermal, tides and waves, is expected to be among the energy vectors of the future.

Coastal cities could be among the first post-carbon cities. Renewable energies have conquered Europe and the world. They become major players in the energy market and a significant generator of employment. The performance of the highly energy-intensive transport sector, the most intractable contributor to global warming, has to be radically improved. Sustainable mobility and accessibility are fundamental in coastal cities for linking the first and the last mile of all journeys to terrestrial and maritime transport networks and all parts of the city with the port and the seafront through integrated, smarter and cleaner services.

Coastal cities are privileged crossroads of exchanges on a globalising chessboard, the very places where many diverse value chains on land and the sea interact and economic flows can be converted and intensified to better satisfy the expectations of citizens, business and organisations. Since the great majority of the total volume of the world trade is done by sea, port cities by definition are the cities where most of the world wealth transits. But, they are not just 'passage' cities. Brainports and well-functioning harbours can reinforce global value chains and can generate sustainable value and employment. Their competitiveness depends on the strength of their functions, their maritime links and their hinterland connections. Sustainable infrastructures can help reconcile ports and cities with the oceans and the planet.

The fifth chapter highlights the capacity and achievements of coastal cities as strongholds of the coastal, blue and green economy. In times of uncertainty, multiple dividend innovations are essential for creating new assets, often out of liabilities, and capturing synergies. Industrial symbiosis in ports is critical for the creation of thriving maritime clusters and high value services. Eco-responsible blue green businesses have a key role to play in creating sustainable value out of values and leading corporate action for responsible cities and global ocean governance.

Concentration of very diverse people and activities are invaluable assets for cities, schools for respecting difference and living harmoniously together. The sixth chapter sheds light on the evolving social capital of coastal cities and their capacity for intergenerational and intercultural equity, social justice, public health, safety and solidarity. Coastal cities themselves are important direct and indirect sources of new employment, the first factor of social integration, followed by housing. The enhancement of urban social capital is of increasing importance in cities that face new forms of unemployment, poverty and exclusion, where more than three generations co-exist, and immigrants come also from the sea in search of better opportunities in a blue horizon. They should all be offered the conditions to enjoy a healthy and meaningful life.

Cities are the undisputed intellectual and cultural epicentres of countries, continents and the world. Coastal cities benefit from access to the extraordinary beauty, and unfathomable diversity of marine life. The bond between the city and the sea is often indissoluble and has a high impact on urban cultures and identities. The urban marine heritage, including also underwater heritage, is as important for humanity as the exceptional natural sites. The proximity of the sea has often inspired citizens to invest time and energy into new discoveries and artistic creation and transform urban environments into unique experiences. A sustainable city has to cultivate the seeds of freedom that science and the arts need and offer a space to all for expressing their creativity and enhancing their abilities and skills.

Coastal cities are gateways to both land and the seas. The seventh chapter examines the role of intellectual and cultural resources as key assets for sustainability. Marine cities, with their coastal and offshore dimensions, represent most fertile soil for the incubation of innovation and prosperity. Knowledge coastal cities invest in education and ingenuity, observation, surveillance and exploration of the seas, ocean literacy and international cooperation for excellence. The chapter presents a spectrum of inspiring actions to reinforce urban identity and make citizens proud of their terrestrial and marine environment.

Urban renaissance usually tends towards an ideal nurtured with the ideas of all citizens. Better co-designed policies should help citizens to live more fulfilling lives. They should start with a shared vision for the future of the place they live in and the planet. Strategic, holistic, transparent, ecosystem-based planning and management should strive to make the vision come true. Spatial planning in search of the sustainable regeneration of coastal cities, of their physical parts and of their extraordinary diversity, should not stop at the edge of the water but also consider marine resources. Urban sustainability agendas have to address many spatio-temporal patterns, both on land and the sea. Time is a scarce and precious resource and can serve as the litmus test of the well-being of societies. Local time and seasonal plans can enhance the capacities of cities and improve resource allocation and quality of life for inhabitants and visitors. Urban observatories and coastal sustainability indicators should take the pulse of coastal cities and their citizens and serve as compasses in the journey towards sustainable development.

The eighth chapter sheds light into the dynamics of coastal renaissance for the optimisation of urban coastal functions, in the context of multiple pressures such as climate changes, natural hazards, erosion, and increased maritime activity. Citizens can play a decisive role in shaping vital urban spaces and forging bonds out of degraded spaces, including harbour infrastructures and disused seafronts that turn their back to the sea and the world. Distressed peri-harbour areas should be transformed into vibrant inclusive communities which can withstand shocks and attract sustainable activities. Symbolic and public and cultural projects on the waterfronts can promote collective life and local democracy and bring more value to places.

Present and future citizens are the political stakeholders of cities in an increasingly multipolar and interconnected world. They have the right to consultation on the investments of the future and the duty to exercise democratic scrutiny of policies. Climate change and the path to sustainable development has encouraged the emergence of multiple forms of active citizenship. New governance architectures seek to enhance the potential of all invisible hands of urban societies and economies and build a social consensus on a future vision and the steps and means to make it become true. Citizen empowerment is a mobilising force and public-private partnerships are recognised drivers of transformational change.

The ninth chapter examines the emergence of new models of citizen participation in responsible coastal cities and links with world citizens. Institutional alliances, enriched with a variety of participatory leadership schemes, and innovative partnerships can maximise the potential of synergies, enhance the content and the methods of cooperation and serve as catalysts of change. A global solidarity bond involving coalitions and networks of cities, both from the emerging and the developed world, can play a major role in jointly addressing global common challenges, including the global ocean. Initiatives such as the Compact of Mayors, a ground-breaking agreement created during the 2014 Climate Week, can raise the bar of excellence and enhance resilience to climate change.

The European Union, world's first maritime power, is the continent with the densest network of conscious cities. Many coastal cities have developed a myriad of innovative flagship actions which can inspire world cities. But the way to sustainable development will be long and arduous. The final chapter builds on messages from pioneer cities and networks for increasing the resilience of vulnerable coastal cities and their potential to grasp opportunities offered by the blue green growth. Global strategic partnerships and alliances, such as the C40, the Asian Cities Climate Change Resilience Network and the 100 Resilient Cities, can greatly improve the capacity of coastal cities to withstand threats and seize opportunities for sustainable development. Promising emerging cities, such as Jakarta and Manila, can benefit from good cases and integrate no-regret multi-beneficial practices in their sustainability agenda to become more resilient and thrive together with their citizens.

* * *

This book intends to raise global awareness on the challenges and opportunities for coastal cities and the myriad of policies and stakeholders which impact upon them. Drawing on the most authoritative studies and asking further questions, the book presents a panoramic view of critical issues for cities and especially marine cities in symbiosis with seas. It embraces issues of smart, sustainable and inclusive growth, active social integration of all and everyone, environmental conscience and resilience and progress in education and science, culture and the arts, urban renaissance, accountable governance and ecosystem-based urban coastal planning and policy. It insists on the mobilising effect of consensual visions, explores preferable and desirable futures, reviews emerging strategies and responses, such as the blue green growth, and examines the forefront of exemplary policies and strategic alliances.

The book offers a systemic integrated view of key urban coastal sustainability issues. All chapters trace the horizon of the most prominent issues at the heart of the matter, and highlight, in a nutshell, the ways that they are best addressed by pioneer cities. The evolving forefront of innovations is illustrated by selected cases able to act as beacons, empower for action and inspire emulation. By adopting a bird eye's view, the book invites us to study more in depth the future possibilities and threats of coastal cities in specific urban and regional contexts. At the end of the book, the suggested references and Internet links offer the specialised reader the opportunity to dive deeper into the issues at stake and the best ways to address them.

This book should be seen as a spark for many new departures to go deeper into the multiple issues. Identifying and questioning a number of best upbeat examples but also bad practices, the book invites readers to a journey of excellence to coastal cities in times of uncertainties, shared questions and interlinked responses. Boston, New York, New Orleans, Panama, Rio de Janeiro, Valparaíso, Cartagena, San Francisco, Reykjavik, Oslo, Stockholm, Copenhagen, Helsinki, Amsterdam, Brest, Lisbon, Barcelona, Marseilles, Naples, Athens, Alexandria, Dhaka, Jakarta, Manila, Singapore, Shanghai, Tokyo and Sydney are just some parts of the journey into the world of the future. The COP 21 (Paris, 2015) and Habitat III (Quito, 2016) are major milestones ahead and this book has the noble ambition to inspire cities in their navigation forward.

The writing of the book was symbolically finalised on 1 March 2015, the Futures Day. Its ultimate ambition is to assist urban actors to co-create a better life, a better future.

Brussels, Belgium Voula Mega

Acknowledgements

This book has been an extraordinary journey to the coastal urban world of the future. It expands and enriches the journey reflected on the Trilogy on the Future Sustainable Cities, the three parts of which were published by Springer in 2005, 2010 and 2013. This time, I explored many ports, deltas and shores, from Rotterdam and Rio de Janeiro to Tokyo, Shanghai and Manila. From the Jakarta islands and the bay of Sydney to the Vancouver parks and the Thessaloniki seafront, I tried to capture something from the unique diversity and intensity of coastal cities.

Many colleagues and friends helped me to search the essentials in the symbiosis of cities with the sea. My most sincere thanks to all those who have inspired, nurtured and accompanied this exploration in blue horizons. Warmest thanks also to my editor and my known and unknown reviewers; their remarks truly helped me to improve the manuscript. I would also like to take this opportunity and thank, for the first time, my masters in painting and watercolours, who helped me to capture the light and, if possible, something from the soul of the places I have visited.

Last but not least, my thanks go to the European Commission which granted me permission to publish this book as an external activity.

This book is dedicated to the memory of a great city planner, professor and person, Sir Peter Hall, who passed away in July 2014. I had the privilege to cooperate with Peter, always ahead of his time. I benefited greatly from his wisdom and kindness. His prefaces to the two first parts of my Trilogy on the Future Sustainable Cities, published by Springer (2005–2013), bear witness of his enormous generosity. This dedication is a very modest demonstration of my infinite gratitude.

Contents

Acronyms

ACCCRN	Asian Cities Climate Change Resilience Network
ACRR	Association of Cities and Regions for Recycling
ASEAN	Association of Southeast Asian Nations
BASD	Business Action for Sustainable Development
BBNJ	Biodiversity Beyond National Jurisdiction
C40	Network of the World's Major Cities
CBD	Convention on Biological Diversity (UN)
CCI	Clinton Climate Initiative
cCCR	carbon Cities Climate Registry
CCS	Carbon Capture and Storage
CDP	Carbon Disclosure Programme
CEMR	Council of European Municipalities and Regions
CERES	Coalition of Environmentally Responsible Economies and Societies
CHP	Combined Heat and Power
CO_2	Carbon Dioxide
COP	Conference of the Parties (UNFCCC), COP 21 (Lima, 2014), COP 21 (Paris, 2015)
CoR	Committee of the Regions
CRPP	City Resilience Profiling Programme
CSO	Civil Society Organisation
CSR	Corporate Social Responsibility
EBSA	Ecologically and Biologically Significant Areas
ECSP	European Council of Spatial Planners
EEA	European Environment Agency
EEZ	Exclusive Economic Zone
EFTA	European Free Trade Association (Iceland, Liechtenstein, Norway, Switzerland)
EFUS	European Forum for Urban Safety
EMODNET	European Marine Observation and Date Network
EPA	Environmental Protection Agency (US)

EPBD	Energy Performance of Buildings Directive
ESCT	European Sustainable Cities and Towns
ESI	Environmental Ship Index
EU or EU28	European Union (28 Member States since 1.7.2013 Austria, Belgium, Bulgaria, Croatia, Cyprus, Czech Republic, Denmark, Estonia, Finland, France, Germany, Greece, Hungary, Ireland, Italy, Latvia, Lithuania, Luxembourg, Malta, Netherlands, Poland, Portugal, Romania, Slovakia, Slovenia, Spain, Sweden and United Kingdom)
FAO	Food and Agriculture Organization (UN)
G8 countries	Canada, France, Germany, Italy, Japan, United Kingdom, United States and Russia
G20 countries	Argentina, Australia, Brazil, Canada, China, France, Germany, India, Indonesia, Italy, Japan, Mexico, Russia, Saudi Arabia, South Africa, South Korea, Turkey, United Kingdom, United States and the European Union, represented by the Council and the European Central Bank
GDP	Gross Domestic Product
GEOSS	Global Earth Observing System of Systems
GHG	Greenhouse Gases (carbon dioxide (CO_2), methane (CH_4), nitrous oxide (N_2O) and the three main fluorinated gases)
GGBP	Greener, Greater Buildings Plan (New York)
GOC	Global Ocean Commission
GOOS	Global Ocean Observing System
GPA	Global Programme of Action for the Protection of the Marine Environment
GPC	Global Protocol for Community-Scale Greenhouse Gas Emission Inventories
GPO	Global Partnership for Oceans
HDI	Human Development Index
HIV/AIDS	Human Immunodeficiency Virus/Acquired Immune Deficiency Syndrome
HFA	Hyogo Framework for Action
IAPH	International Association of Ports and Harbours
ICLEI	International Council for Local Environmental Initiatives
ICPC	International Crime Prevention Centre
IUA	International Union of Architects
ICT	Information and Communications Technologies
IEA	International Energy Agency
ILO	International Labour Organization
IMCAM	Integrated Marine and Coastal Area Management
IMO	International Maritime Organization
IOC	Intergovernmental Oceanographic Commission (UNESCO)
IOM	International Organization for Migration
IPCC	Intergovernmental Panel on Climate Change

ISA	International Seabed Authority
ISOCARP	International Association of City and Regional Planners
ITF	International Transport Forum
JIUS	Joint Initiative on Urban Sustainability
IUUF	Illegal, Unreported, and Unregulated Fishing
LNG	Liquefied Natural Gas
MARPOL	International Convention for the Prevention of Pollution from Ships
MDG	Millennium Development Goals
MPA	Marine Protected Area
NGO	Non-Governmental Organisation
NOAA	National Oceanic and Atmospheric Administration (USA)
NPCC	New York City Panel on Climate Change
OAP	Ocean Action Plan (New York)
OBIS	Ocean Biogeographic Information System
OECD	Organisation for Economic Co-operation and Development
OWHC	Organization of the World Heritage Cities
PV	Photovoltaics
RES	Renewable Energy Systems
RES-E	Electricity from Renewable Energy Sources
RFSC	Reference Framework for Sustainable Cities
SDG	Sustainable Development Goal
SME	Small and Medium-sized Enterprise
STEM	Science, Technology, Engineering and Mathematics
STOA	Science and Technology Options Assessment (European Parliament)
UCLG	United Cities and Local Governments
UITP	Union Internationale des Transports Publics
UNCED	UN Conference on Environment and Development (Rio de Janeiro, 1992)
UNCLOS	UN Convention on the Law of the Sea
UNDP	UN Development Programme
UNEP	UN Environment Programme
UNESCO	UN Educational, Scientific and Cultural Organisation
UNFCCC	UN Framework Convention on Climate Change
UN-HABITAT	United Nations Human Settlements Programme
UNISDR	UN Office for Disaster Risk Reduction
UNPF	UN Population Fund
UNWTO	UN World Tourism Organisation
WBCSD	World Business Council for Sustainable Development
WEF	World Economic Forum
WHO	World Health Organisation
WMO	World Meteorological Organisation
WPCI	World Port Climate Initiative
WRF	World Resources Forum

WRI	World Resources Institute
WTO	World Trade Organisation
WWF	World Wild Fund
Y4PT	Youth for Public Transport

List of Watercolours

Watercolour 1
Tokyo, the top Coastal City of the World

Chapter 1
The Age of Cities: Urban Geopolitics and the Path Towards Sustainable Development

Abstract Cities are major nodes for all energy fluxes in the complex, uncertain and interlinked urbanising world. This chapter examines the opportunities and risks which could impact the future of cities as critical nexus points of social, economic and ecological change. It reviews demographic, socio-economic and environmental trends, political developments and technological breakthroughs, and presents an array of possible futures which invite cities to develop a vision for a preferred future. Last but not least, it explores the potential of cities to lead the world in times of global democratic deficit.

The acceptable balance enabling a civilisation of sustainability is a fundamental challenge for twenty-first century cities. Conscious cities have to address all interacting global societal challenges, striving for strong sustainability. They can head through the age of scarcity and towards a new world order offering better chances for all citizens, present and future, within the limits of the planet. In alliance with seas and the oceans, the supreme life support systems, cities can lead by example and through their magnitude, reinforced by their multiple networks. They should care not only about their natural resources, including marine resources, but also lead coalitions of committed stakeholders and engage in the protection of the oceans and the high seas, the ultimate world commons.

1.1 Urbanisation and the End of Demographic Explosion: Trends and Prospects

Are seas the future of the planet? Are cities the new countries? Signals from economics and politics suggest a positive response, especially to the second question. The world is witnessing the largest growth of urban population in human history, while rural communities continue to shrink. The progress of humanity towards urbanisation has been long and relentless. The 2014 UN World Urbanisation Prospects suggest that 54 % of the global population lives in urban areas, a number that is expected to increase to 66 % by 2050. Cities concentrated 3 % of the world population in 1800 and 14 % in 1900. They are expected to host more than six

---

V.P. Mega, *Conscious Coastal Cities*, DOI 10.1007/978-3-319-20218-1_1

billion urban citizens by 2050. The human condition becomes increasingly and irreversibly urban (UN 2014a).

Cities are those human organisations with the greatest longevity and also the most constantly changing ones. Authoritative thinkers and organisations suggest that urbanisation is among the definitive phenomena of the century. In a globalised world, city states are re-emerging as dominant economic and political players. Cities are, and will continue to be, the epicentres of distribution of essential goods and creation of wealth. They promise to play a crucial role in the transition to a post-carbon society and economy in an increasingly multi-speed planet. Last but not least, their agility in cooperating with their peers often makes cities more super-partners rather than superpowers. Engaged networks of committed cities seem to have a huge potential in the interconnected world of the future (Gore 2013).

The emergence of a more polycentric world and the move of the earth's economic centre of gravity eastwards appear to be influential global trends which could be accompanied by a shift of power from states towards private and civic actors. State fragility due to inequality and a massive increase in the urban population are expected to reinforce megacities and metropolitan areas, both in relation to their regions and as interconnected partners in a more diffused power landscape. By 2030, the 50 largest world cities could concentrate more resources than most small and medium-income states and play a more prominent international role.

Humanity took more than a century to grow from one billion in 1804 to two billion in 1927 and more than 30 years to reach three billion in 1960. Fifteen years later, in 1975, the world population had reached four billion, and by 1987 and 1999 had gained one and two additional billion, respectively. In 2011, the world hit the watershed of seven billion people. Annual growth rate reached its peak in the 1960s, with more than 2 % annual increase. Average density in developing countries is double those of Europe and Japan, which are double those of the United States, Canada, and Australia (UN 2013).

Massive urbanisation, accompanied by lower fertility rates, is a brake to population growth and brings an end to demographic explosion, expected to reach its apogee in the twenty-first century. Global population will have quadrupled during the period 1950–2100. The UN World Population Prospects foresee a global population of 9.3 billion people in 2050, and more than ten billion by the end of the century (UN 2013; UNFPA 2011).

Demographic dynamics are shifting from high mortality and high fertility to low mortality and low fertility (Millennium Project 2013). The global fertility rate has fallen from six children in 1900 to 2.5 in 2012. Life expectancy is projected to increase in developed and emerging countries. At the global level, it is projected to reach 76 years around 2045–2050 and 82 years in 2095–2100. By the end of the century, the average life expectancy in developed countries could be around 89 years, versus about 81 years in emerging regions. While the average number of children per woman has fallen rapidly in large emerging countries such as China, India, Indonesia, Brazil and South Africa, rapid growth is expected to continue over the subsequent decades in countries with high levels of fertility, such as Nigeria, Niger, the Democratic Republic of Congo, Ethiopia, Uganda and Afghanistan.

In the European Union, the fertility rate remains below the level of 2.1 children per woman which guarantees the replacement of generations and, in the future, positive net migration is expected to be the only demographic growth factor, delaying the rapid ageing of population (EC 2012b). As living and educational conditions improve for all world citizens and fertility rates continuously drop, demographic ageing is likely to diffuse gradually to more youthful parts of the world (UN 2013). By 2050, there could be as many citizens over 65 as under 15-years old.

The world landscape of demographic change seems very diverse. The population of developed regions is expected to slowly stabilise. In contrast, the Least Developed Countries, the most rural and/or rapidly urbanising countries, are projected to double in size from a population of around 900 million in 2013 to 1.8 billion in 2050. India is expected to become the world's largest country, overtaking China around 2028, when both countries will have populations of 1.45 billion. After 2028, India's population is expected to continue to grow and China's to start decreasing. Nigeria's population is expected to surpass that of the United States before 2050.

More than half of the demographic growth by 2050 is expected to take place in Africa. Asia and Africa together will account for close to 90 % of all increases in global urban population. Asia will still be the most populous continent by 2050. Its population is expected to peak over five billion around the middle of the century and then start a slow decline. Africa gains weight as its population more than triples. The African urban population is expected to increase to over 1.2 billion by 2050, while that of Asia could soar to 3.3 billion. The total population of all other major regions, the Americas, Europe and Oceania, amounted to 1.7 billion in 2011 and is projected to rise to nearly two billion by 2060 before stabilising and declining very slowly, remaining near two billion around the next turn of century (UN 2013, 2014a).

Demography is a capital force which impacts all human activities. World trends may signal the beginning of a demographic divide between countries with young and those with ageing populations, a fact which has important consequences for migration movements. The UN defines migrants as citizens residing in a foreign country for more than 1 year. According to the International Organization for Migration, the movement of people is likely to become even more significant in the future, as a result of continued globalisation, demographic dynamics and economic imbalances. Worldwide, migrants accounted for about 3.2 % of the world population in 2013. From asylum seekers to expats, and from freely chosen to compulsory itineraries, migration covers an increasingly diverse range of mobile people and patterns. One in ten new migrants is from China, and one in five from Asia. In the North, international migrants constituted 10.8 % of the total population in 2013, compared to 1.6 % in developing regions. Europe and Asia combined hosted nearly two thirds of all international migrants. Cities are the main magnet for migrants, as it is mainly in urban environments that migrant populations have increased access to economic, social, cultural and political conditions for the realisation of their potential (IOM 2011, 2013; OECD 2013, 2014b).

The majority of the world urban population lives in coastal cities having direct access to the sea. The two billion citizens expected before the global population peaks and begins declining will mostly be citizens of the coastal cities of the emerging world. The largest urban growth is expected to take place in India, China and Nigeria, with an additional 404, 292 and 212 million urban dwellers, respectively, followed by Indonesia and the United States. China and India will contribute more than one third of the global urban population increase between 2014 and 2050. Together with Nigeria, this will account for 37 % of the projected growth of the world's urban population up to 2050 (UN 2014a).

One in eight of the world's urban citizens lives in megacities, officially defined as urban agglomerations hosting more than ten million people. Their number has increased to 28 in 2014, versus 10 in 1990, and it is expected to further rise to 41 by 2030. Megacities, mainly coming from the global South, are home to 12 % of the world population, versus less than 7 % in 1990. Asia hosts 16 megacities, while four are located in Latin America, three each in Africa and Europe, and two in North America. Most of the largest cities in Asia, Africa and Latin America and the Caribbean are located on a coastline or along a river bank. Asia counts 12 coastal megacities (Tokyo, Shanghai, Mumbai, Osaka, Dhaka, Karachi, Kolkata, Manila, Guangzhou, Tianjin, Shenzhen and Jakarta), Europe two (London and Istanbul), Africa two (Lagos and Kinshasa), North America two (New York and Los Angeles) and Latin America two (Rio de Janeiro and Buenos Aires). More megacities could be considered coastal according to different definitions. Delhi, Mexico City, Moscow, Cairo and Chongqing are the megacities most distant from the seas. Concerning the shores of history, some of the megacities have existed for thousands years, while others, like the port cities of Guangzhou, Guangdong and Shenzhen on the Pearl River Delta, have sprung up overnight, and others, like Shanghai, underwent a dynamic reinvention to attract the opportunities of the future (UN 2014a).

Megacities have increasingly become networks of networks, representing the most complex human ecosystems. They dominate world development not only because of their sheer size, but also due to their ability and influence, often strengthened through networks and interactions. Preventing and combatting poverty and environmental degradation are prime concerns. The protection and enhancement of their marine environments have to be an organic part of their policies for sustainable development. The endowment and management of expanding infrastructures and the search for more inclusive, intelligent and effective metropolitan governance are critical issues for the future (Metropolis 1999, 2014; Sassen 2001).

As global population rises and life expectation increases, there are intensifying pressures on natural and physical capital and life support systems. These tensions pose serious challenges to resource, water, food and energy security, as climate change and the depletion of natural resources combine with loss of land to urbanisation and environmental degradation of terrestrial and marine ecosystems. The ocean, the single, contiguous body of water embracing land and the world's largest ecosystem, is under severe pressure. Risks increase for the ocean's rich diversity of plants, animals and microbes, from the largest animals on the planet, the blue whales, to plankton and bacteria. The aesthetic, cultural, religious and spiritual ecosystem services provided by the oceans are also under threat.

Urbanisation has to face a powerful set of grand societal challenges, including climate change and environmental quality, food, health and well-being, education and training, energy and transport, creation and distribution of wealth, social justice and accountable governance. Most of these issues intensify in coastal cities, being, at the crossroads of diverse ecosystems. Addressing these challenges will define the state of cities, which will determine quality of life for most citizens of the planet (ICLEI 2010).

Rapid urbanisation of coastal areas has many consequences, such as enlargement of natural inlets and dredging of waterways for navigation. There are critical links between increasing shoreline retreat and risk of flooding of coastal cities and the degradation of coastal ecosystems by human activity. The majority of marine pollution comes from cities, which have often overexploited the seas as a source of food and as a location for waste disposal. But to advance towards sustainability, waste can only be seen as a by-product, if not as a product, and environmental problems should never compromise the benefits offered by the seas nor threaten the viability and even the survival of coastal cities (UNEP 2012a, b).

The metamorphosis of many regions, countries and continents from rural lands into urban communities is stunning. China experienced a rapid transformation from a land of rural communities and regions into a gigantic urban country preparing for its first billion citizens. Cities expanded abruptly, especially on the East coast and around the major deltas. Chongqing, the third most populous Chinese city after Shanghai and Beijing and the sixteenth world megacity, developed as an inland port city in just a few years, serving as a global giant for laptop computers, producing one in four laptops sold in the world. Promoting human-centred urbanisation and balancing urban-rural development is capital for the second world economy, which, by 2025, could count 221 cities with more than one million inhabitants, including 23 cities with more than five million and eight megacities with over ten million people (MGI 2009; UN 2014a).

Africa, the most rapidly urbanising continent, has moved from economic stagnation to above 5 % GDP average annual growth and has weathered several global economic crises fairly well. The explosive urbanisation of the most rural continent should bring quantum leaps for education and health provision services, as concentrated populations are much easier to target, train and service. Rapid urbanisation is breeding enormous demand for natural resources, fresh water supplies, sanitation, energy and infrastructures, and can also generate social and environmental tensions (World Bank 2012b).

Africa seems a land of opportunity both for Africans and international investors. It hosts some of the most rapidly growing economies in the world, including Ethiopia, Ghana, Mozambique and Tanzania, which have all urbanised very quickly. As in Asia, this growth is leading to a burgeoning urban middle class, which creates new markets for goods and services and promises more political stability. Africa is likely to witness this rise by the third decade of the century, as millions are continuously lifted out of poverty (Brookings 2013).

In many parts of the emerging world, the demographic transition in urban areas, rampant poverty, threatened ecosystems, massive construction of infrastructure without concern for the environment and society, and competing stakeholder agendas have resulted in unbalanced and extremely vulnerable urban systems. Explosive urbanisation has led to the rise of mega-slums. Africa is home to nearly two-thirds of the global slum population. Over 70 % of the urban population suffers from inadequate housing, water supply or sanitation, and more than 60 % of sub-Saharan Africa's urban population lacks adequate access to clean water, nutritious and affordable food, sanitation, and energy. Megacities need to become better places to live in and develop in more harmonious symbiosis with the natural systems that support them. Offering true opportunities to the young and disadvantaged is a major challenge in emerging countries and cities (CSIRO 2011).

Human settlements have always tended to concentrate in coastal zones, given the economic benefits that accrue from access to ocean navigation, fisheries, offshore energy, and recreation. The exact magnitude of the world's urban coastal population is difficult to evaluate. Sweeping generalisations about the distribution of human population on earth suggest that half of humanity resides near sea coasts, rivers, and in low-lying areas, and that coastal areas are disproportionately urban, although the definitions of the terms "near coast" and "low-elevation" vary. Some organisations suggest that "near coast" means 60, 100 or 150 km from the sea, and that "low-elevation" means less than 10 m above mean sea level.

The Cities and Coastal Areas programme of the UN Environmental Programme (UNEP) suggests that half the world's population lives within 60 km of the sea, and three-quarters of all large cities are located in coastal areas (UNEP, 2011a, b). In the European Union, Eurostat informs us that nearly half of the population of EU countries with a sea border is located in coastal regions, a fact that was reinforced following the accession of Croatia to the EU in 2013. According to the official statistics, published for the first time in 2009, coastal regions are defined as standard statistical regions, which have at least half of their population within 50 km of the coast (Eurostat 2009; EEA 2014).

Archaeological evidence shows that many of the oldest human settlements were located along coastlines or on the banks of mighty rivers, lakes, and deltas. Oceans constitute the kidneys of Earth, a great biological pump at the heart of water and nutrient cycles and global atmospheric and thermal regulation. The proximity of the sea offered opportunities for fishing, which helped ensure a steady food supply. Coastal cities also served to link local economies to regional and global trade. Maritime transport provided vital economic links throughout history. As the volume of sea trade has steadily increased over the last decades, and is expected to continue to grow, port cities are likely to gain even more economic importance in the future (OECD 2013).

Proximity of and easy access to the sea has been the generic criterion followed in this book, which broadly focuses on coastal cities in which so many diverse dynamics interact, without respecting a precise typology. The aim is not to compare cities but to highlight common challenges and exemplary responses at the forefront of strategies and policies which can be shared by coastal cities, in an open innovation urbanising world.

Coastal areas are substantially more densely populated than other territories, and activity within them increased dramatically during the twentieth century, a trend that seems poised to continue through the twenty-first century. Coastal population growth in many of the world's deltas, islands and estuaries has led to widespread conversion of natural coastal landscapes to industrial and residential uses. The attractiveness of the coast has resulted in disproportionately rapid expansion of urban centres, economic activity, and tourist resorts. The United States' shoreline constitutes less than 10 % of its total land area, not including Alaska, but accounts for 39 % of its total population. From 1970 to 2010, the population of the coastal counties increased by almost 40 % and is projected to increase further (NOAA 2011).

Japan is also an overwhelmingly coastal urban country. Formerly a largely rural nation, it transformed into a prodigiously urban and coastal country within two decades. In 1950, Japan's inhabitants were dispersed throughout the country, with nearly half of these living in farming communities. By 1970, most Japanese were living in urban areas, the majority of which are located in the Pacific Coastal Belt. Despite the demographic decline of the third world economy and the extreme phenomena that damaged coastal regions, Japanese citizens prefer to live by the sea.

Rivers and delta regions have played an important role in the growth of European and Asian coastal cities and serve as gateways to inland areas. Coastal cities are also crucial for Australia. More than 80 % of the population lives within 50 km of the coast, a segment which has been growing faster than the population of the rest of the continent and is expected to continue to increase. Population growth in Australia seems to occur mainly around metropolitan areas and larger regional centres, through natural increase and internal and international migration.

Globalisation brings a closer interdependency among human settlements and places of extraction or transportation of resources. Competition for resources is likely to exacerbate tensions and trigger conflicts. Biodiversity losses and ecosystem crises may increase the sense that the world is entering an age of scarcity. But humanity has a choice. Visions for 2050 proposed by organisations and business associations underline that humankind can live well and within the limits of the planet, but it has to radically change its values and practices (EC 2013; WBCSD 2010).

Humanity stands at a juncture, a critical inflection point, probably one of no return. The multiplication of global actors and the diffusion of power could lead to a redefining point for urban futures. Many unpredictable events challenge human intelligence and demonstrate that the world has a high potential for surprises. Collective capacity to envision and shape the future and generate and distribute wealth, while reinventing and implementing good governance, has to be reinforced through open futures thinking. Many thinkers and organisations portraying different futures agree that the top three crucial world problems include inequality and instability, unsustainable development and climate change, and the global democratic governance deficit. Growing inequality, of opportunity or of income, is a critical economic, social and political challenge, and its significance has been amplified by the declining public confidence and the frail recovery (The Economist 2012).

Global change is intense and of unprecedented nature and speed. Forty years after the seminal 1972 report "Limits to growth", the Club of Rome suggested that humanity has 40 years to prevent the most negative consequences of resource over-consumption. Global population could peak at 8.1 billion people in 2042 because of rapid decline in urban fertility. Global wealth may grow, but much slower than generally expected because of slow productivity growth in mature economies. Global GDP may peak after 2052, and investment share of the GDP may grow as society is gradually forced to handle issues of depletion, pollution, biodiversity decline, climate change and inequality. Global energy use could reach a peak in 2040, because of continued increase in energy efficiency. CO_2 emissions could peak in 2030, because of a shift toward renewable energy sources. Nevertheless, CO_2 concentrations may grow, and the global average temperature could exceed the threshold of +2 °C by 2050. This could trigger self-reinforcing warming with a possible systemic collapse in the second half of the twenty-first century (Club of Rome 2012).

Some positive trends should not be unobserved. The UN suggests that, at the end of the first decade of the century and for the first time since trends began to be monitored, both the number of people living in extreme poverty and the poverty rates have fallen in every emerging region, including sub-Saharan Africa, where rates are the highest. The Millennium Development Goals (MDGs) on education, gender equality, child and maternal health, environmental stability, HIV/AIDS reduction, and a Global Partnership for Development find in cities the necessary infrastructure and political strength for significant progress. A failure would signal an extraordinary missed opportunity to transform global visions into tangible results and impacts (UN 2012).

Most global assessments confirm that humanity, even if it wins in relation to the basic conditions of health, education and wealth, as well as status and conditions of women, regresses in relation to key social and environmental conditions, with unemployment, income inequality, biodiversity decline, greenhouse gas (GHG) emissions and terrorism attacks. An interesting picture of the global situation and the prospects for 10 years ahead comes from the State of the Future Report by the Global Millennium Project. Most of the 15 global challenges, providing a framework for assessing the prospects for the future of humanity, are critically impacted by cities. Climate change, energy, water, resources, democracy, citizen participation and progress in science and technology are challenges which can best be addressed in cities (Millennium Project 2013).

The 2013 State of the Future Index, which provides a score card on humanity's performance in addressing the grand societal challenges, confirms that humanity wins in relation to human capital. Improvements are detected and further foreseen in relation to health and life expectancy at birth, undernourishment and infant mortality, literacy rate and secondary school enrolment, GDP per capita and Foreign Direct Investment, energy efficiency and share of renewables, access to clean water, and the empowerment and participation in elected parliaments of women.

However, political instability appears as a major global risk, and only a few ecological indicators, access to water, energy efficiency and renewable energies, mark progress at the global level. Income inequality is deepening, jobless growth persists, water stress is rising, terrorism incidents intensify, the environmental capital for life support is diminishing and climate change is advancing; all of these, against a lack of leadership and weak representative democracy, are highlighted as major longer-term trends (Millennium Project 2013; WEF 2015).

Economic globalisation could lead to the emergence of a tightly integrated, highly interactive economy of multinational corporations and long supply chains interacting in cities. It might progressively lead to the disruption of all three factors of production (capital, labour and natural resources), an increasing global wealth gap and two major changes: outsourcing and robot-sourcing. The advent of additive manufacturing could revolutionise manufacturing as profoundly as mass production did in the past (Gore 2013).

Globalisation, propelled by trade and rapid technological change, has brought prosperity and reduced poverty for millions of the world's citizens. However, in order that all reap the full benefits of globalisation, many elements have to be balanced, including access to quality education, social cohesion, efficient markets, sound policies for research, innovation and development and good public and corporate governance. Inequality is not inevitable and globalisation can and must be equitable. Ideally, globalisation is about people sharing the planet in freedom of movement and exchange, and provides opportunities for new ideas and wealth creation and individual and collective fulfilment. Adverse effects of loss of diversity, autonomy and sovereignty can and must be prevented (OECD 2007, 2010a, b).

Political instability and corruption are crucial hindrances to social inclusion and sustainable development. Responsible globalisation should not only be equitable but also reinforce ethical principles. Without enhanced ethical values and democratic governance, the global crisis of 2008 could have an enduring impact. Issues of trust are of crucial importance for navigating out of the crisis and into a more sustainable trajectory.

Climate change is the most evident manifestation that the life support systems are seriously threatened. Climate change is already happening in all continents and across all oceans (IPCC 2014). Most studies suggest that the universally accepted limit of a 2 °C increase will be very difficult to achieve (IEA 2014). The focus of scientists and decision makers has increasingly expanded from mitigation to the embrace of adaptation. More than adaptation, the resilience of cities, regions, and coastal areas represents a strong goal reinforcing the immunity and robustness of places, communities and functions.

Earth's resilience is being tested by rising global temperatures and more frequent and severe extreme weather events. The complementary green, clean and resilient agendas require balanced portfolios of policies combined to provide multiple benefits for all world citizens (World Bank 2012a). The long-term policies should go hand in hand with immediate measures to halt environmental degradation leading to extreme disasters, including desertification, floods and storms, hurricanes, typhoons and cyclones of increasing magnitude and frequency. Humanitarian crises, mainly due to the scarcity of water and energy resources and high food pressures and health emergencies, could become recurrent.

Food and energy resources are geostrategic. One sixth of the world's population is undernourished and lacks access to modern energy services. Urban food security is increasingly highlighted as a critical twenty-first century policy issue.

Access to sufficient, nutritious and affordable food is not only a basic human right but provides the foundation for citizens' ability to grow, learn, prosper, and reach their full potential. The global community has witnessed a revival of efforts after the 2007–2008 food price spikes. Cities can be decisive actors in global and local food distribution networks and promote food and nutritional security, also enhancing the potential of fisheries and aquaculture.

Global energy demand could rise by more than one third to 2035, especially driven by emerging economies. Energy efficiency is highlighted as being just as important as unconstrained energy supply. Global demand for electricity grows almost twice as fast as energy consumption, and huge investments are needed to satisfy rising demand in developing cities and to replace ageing energy infrastructure. Of the generation capacity that is to be built up to 2035, around one-third is needed to replace old plants. Half of the new capacity is expected to come from renewable sources of energy. Renewables have become the world's second-largest source of power generation but fossil fuels still dominate the global energy mix. Especially after the hype of shale gas in the USA, a better understanding of the opportunity costs of each energy option is crucial (IEA 2012).

Transport emissions are the most intractable anthropogenic emissions. The International Transport Forum informs us that, despite uncertainty, passenger transport could double and freight transport could quadruple by 2050. Reducing the sector's global carbon footprint and the heavy cost of congestion are key priorities (ITF 2013).

The changing global resource landscape suggests continued growth in demand for food, energy, minerals and timber resources, until at least 2030. Tensions over scarce resources and raw materials could cause conflict and require new forms of crisis management. The economic and environmental interconnections are frequently analysed in international forums which suggest that the world is more at risk as persistent economic weakness saps collective ability to tackle environmental challenges. Many studies underline that having to choose between preserving the environment and developing the economy is a continuous false dilemma. In the long term, economic and social development is weak and fragile without the sustainable management of resources (EEA 2013; OECD 2012).

Energy security, the extreme phenomena linked to climate change, irreversible biodiversity loss, terrorism, pandemics, crimes, nuclear disasters and nanotechnology threats are among the shocks regularly detected by world organisations. The array of interrelated risks, economic, technological, environmental, geopolitical and societal, and their perceived likelihood, impact and interconnectivity are subjects of important analyses which highlight the role of cities as incubators of innovative responses to risk governance (WEF 2015).

Cities have to improve their ability to engage with the future uncertainty and address major global risks involving significant and sudden, potentially high-impact events, the timing and magnitude of which are very hard to foresee. The OECD recommendations for the governance of critical risks underline the importance of creating an institutional environment that commits all stakeholders and allows for building collective resilience against future shocks (OECD 2014a).

1.2 The Crucial Role of Cities in the Journey to Strong Sustainability

In 2015, sustainable development issues are receiving increasing attention, against a backdrop of environmental degradation and persistent poverty, compounded by the consequences of climate change. The spectre of strategic resource insecurity has come back as the world undergoes a period of uncertainty and intensified resource stress, driven in part by the scale and speed of demand growth from emerging cities and markets. The horizon seems full of supply disruptions, volatile prices, environmental degradation and political tensions over resource sufficiency and access.

But no hard times in human history were devoid of opportunity. Cities are the very places where people concentrate and interact to create and seize opportunities. By the seventh century BC, Alcaeus had already suggested that "cities are not made from houses, stone walls, bridges and canals, but from people able to enhance their chances and make the most out of them". For cities can generate synergies from the cross-fertilisation of very diverse people and resources. They constitute laboratories of the future, places where growth patterns and lifestyles are shaped before they are disseminated into the world (Mega 2013).

In times of mistrust in public policies, it is probably worth remembering that "the aim of a good city is to lead citizens towards a happy life" (Plato). Each city is a public good, made out of tangible and intangible assets that help citizens in their pursuit of happiness. Cities are human nests and havens, places of social interactions, theatres of civilisation, schools of abilities and values, and temples of culture and democracy. Cities, especially the coastal ones, are our most complex and dynamic ecosystems. As a biological species which concentrate in mutually supportive environments, humans come together to optimise their chances for prosperity. Proximity and diversity contribute to the generation of precious complementarities and synergies, which, when captured and enhanced, can bring about important benefits.

Synergies are a strong urban characteristic, closely linked to the density of diverse people and assets. Populations have always come together to optimise very diverse activities, ranging from defence and religion to trade and politics. Once the ancient metropolises became strong, they initiated processes of colonisation and created the first urban networks, mainly linked by maritime routes. Cities like Athens considerably expanded their radius of influence, creating not just satellite cities but true allies. Most colonial cities were built in close proximity but a safe distance from the sea and could rely on cooperation from the mother city in case of attacks and hostilities.

In the twenty-first century, cities face complex, interlinked and rapidly evolving global challenges, such as climate change, migration flows, transnational governance demands, financial volatility, social inequalities and environmental deterioration. And some of these problems have only been aggravated as many cities have been hit hard by the crisis, in particular, by unemployment, notably youth unemployment and exclusion. Restoring trust, addressing challenges and seizing opportunities requires ingenuity and innovation, whether in education and research, policy-making, investment decisions or everyday actions.

Cities have been defined as human establishments which generate, in an endless way, economic growth from their indigenous resources, including those from the sea, and through the "disordered order" of human interactions orchestrating all activities and stimulating synergies. A sustainable city is, first and foremost, a city which carries the seeds of progress in balance with nature and the sea, and sustains its ability to reinvent itself and grasp opportunities for a brighter future for the present and future generations. Innovation is a sine qua non condition for continuously reinventing and enhancing precious assets (Jacobs 1969).

Urgency for cities is increasing, as the combined effects of various crises have to be addressed on the ground. Humanity is realising that the flawed economic compass of recent decades is leading to unsustainable growth, pollution and emissions, and the depletion of not only non-renewable resources, but also renewable resources which are regenerated at a pace far slower than their depletion rate. However, the outgrowth of humanity, colliding with approaching limits on the supply of critical natural resources, can be a major driver of change, from the local to global levels.

Cities are among the main actors in the unrelenting global commitment towards the advancement of sustainable development. Their engagement was renewed at the UN Conference on Sustainable Development Rio+20, which was intended to create an impulse and turn sustainable development from aspiration and unsystematic progress in unchartered waters into a genuine path to prosperity for all, in both the current and future generations. The Rio+20 highlighted that rapid urbanisation is breeding enormous concentrated demand for natural resources, water supplies, sanitation, energy and transport infrastructures and services. But the high density and cross-fertilised ingenuity of cities can bring efficiency gains and breed multi-beneficial innovations, each one of which can bring a cascade of possible improvements (UN Conference on Sustainable Development Rio+20 2012).

In 2012, twenty years after the ground-breaking UN Conference on Environment and Development (UNCED), the Rio+20 conference highlighted that cities occupy just 2 % of the Earth's land cover, but account for 60–80 % of energy consumption and greenhouse gas emissions. The urban explosion in the emerging world brings dramatic changes to the composition of the global population and further strain on limited natural resources. A vicious circle brings more urban poverty and environmental degradation to an unprecedented extent. In Least Developed Countries, 62 % of carbon emissions come from changes in land uses (Angel et al. 2005, 2011).

The UN Conference Rio+20 also shed light on the world's oceans and seas, vital conduits for health, food, energy, transport and trade. Over three billion people depend on marine and coastal biodiversity for their livelihoods. Marine fisheries directly or indirectly employ over 200 million world citizens, especially in the developing world. Oceans serve as the world's largest source of protein, with more than 2.6 billion people depending on them for their primary source of protein. Globally, the market value of marine and coastal resources and industries is estimated at about 5 % of the total wealth. A significant portion of the world oceans is heavily affected by human activities, including sea pollution, illegally depleted fisheries, decline of marine and coastal habitats and ecosystems, and climate change and acidification. In 2002, the World Summit on Sustainable Development was already recommending that the United Nations undertake a Regular Process for reporting and assessing the state of the marine environment, including socioeconomic aspects.

An integrated system of observations and analysis is crucial for the sustainable management of coastal resources. The Panel for Integrated Coastal Observation responsible for the design of the coastal module of the Global Ocean Observing System underlined that ecosystem goods and services are more concentrated in coastal zones than in any other part of the globe (GOOS 2008). The GOOS is the oceanographic component of the Global Earth Observing System of Systems (GEOSS) being built since 2005 on the basis of a 10-year implementation plan aimed at enhancing the relevance of Earth observations for global governance. Cities should avail themselves of all physical, chemical, biological and geological information on their surrounding ecosystems, share them with their constituencies and make informed decisions concerning coastal assets. Global observation infrastructures, vessels and platforms help integrate in situ and remote sensing observations and inform citizens and policies (UNEP, 2011a; UNESCO/IOC 2003).

Advancing climate change and the contribution of glaciers and ice-sheets to sea-level rise represents an additional threat to many coastal areas. Other problems also have to be high on the list of priorities. The prevention of marine debris is a noble cause for which coastal cities can fight. The amount of plastic going into the oceans is increasing at an alarming rate, keeping pace with global plastic production and mismanagement, especially in the developing world. Ocean acidification is one of the most critical aspects of the impact of increasing CO_2 emissions on ocean ecosystems. Ocean acidity has increased by about 30 % globally since the industrial revolution and is projected to increase by 170 % by the end of the twenty-first century, as compared to the pre-industrial level, unless global carbon emissions are significantly curtailed. This is "the other CO_2 problem" and could cause irreversible damage to marine species. Long-term observations within ocean, coastal, and coral reef environments to better assess the impacts and models are essential for scientific and policy communities (IPCC 2014).

The massive urbanisation of the coasts of the emerging world should be seen as a golden opportunity to integrate sustainability into the design of all urban functions, including housing and services, industrial spaces and energy and transport systems. The advancing radical disruption of the relationship between human beings and the earth's ecosystems could signal the beginning of a revolutionary transition marked by the transformation of production and consumption patterns. As energy from renewable energy sources becomes cheaper, and offshore activities expand, the sustainable use of marine resources is crucial and their contribution to this transition could be critical (EEA 2014).

Sustainable development can only be promoted through active citizen participation and local democratic processes. Democracy is the very essence of citizenship and collective life in cities. There is urgency to restoring the healthy functioning of representative democracy at all levels and making political systems sustainable. Democratic governance is critical in directing the flows of power and shaping the interrelated global and local patterns. It can prevent power imbalances and social inequalities interacting with environmental distress and resulting in even more worrying situations. Global governance has often weakened the voices of emerging world cities and excluded marginalised groups. Approaches emboldening cities to bring about socio-economic and political change hold enormous promise for human-centred policies at the local and global levels.

Enlightened citizens can play a great role in cities getting into sustainable trajectories and reinventing a new world order, nurtured with the growing sense of belonging to a single human community. Citizens are increasingly interconnected by extensive networks and interpersonal transnational flows, and this could lead to a new age of convergence of values. The emergence of a global conscience is increasing momentum for sustainable development, and the shift of power away from states towards local, civic and private actors are often presented as compelling drivers of change. Greater awareness of the global nature of citizens' most important concerns can bring a higher conviction for the potential of shared solutions and a change in direction, which seems unlikely to be realised within the existing global governance frameworks (EUISS 2012; National Intelligence Council 2013).

Near-universal access to education, the energising effects of information and communications technology, and the importance of life-long education, together with the evolution in the status of women in most countries, are promising elements of a new world. The process toward gender political-economic equality seems irreversible. Even if it takes another 80 plus years for humanity to achieve gender parity in the workplace, women are becoming empowered throughout the world and are increasingly engaged in professional life and representative democracy at all governance levels. The Global Gender Gap Report, by the World Economic Forum, offers, since 2006, a framework for capturing the magnitude and scope of gender-based disparities in key areas such as access to healthcare, education, political participation and economic equality. The Forum suggests that there is a strong correlation between those countries that are most successful at closing the gender gap and the most competitive economies. The top performing countries, according to the latest Global Competitiveness Index, also rank best according to their Global Gender Gap Index (WEF 2014a).

Ports often attract more newcomers than the other areas of cities. Access to the sea holds multiple opportunities and an urban state of mind more open to diverse cultures and values. Global change has already caused the first wave of climate migrants, assessed to be around 32 million citizens in 2014, versus 50 million asylum seekers in the world. The cost of inaction is likely to be higher than the cost of policy measures, especially if reducing the likelihood of knotty population displacement. Giving urgent policy attention to migration in the context of environmental change would prevent a much worse and far more costly situation in the future (UK Parliament 2011).

The strength of local networks and their interconnection with world networks could be decisive for coastal cities in welcoming and integrating migrants and weaving precious ties with their places of origin. Migration towards lower risk places could pose stimulating questions on notions of citizenship and identity and generate societal tensions, which often fuel populist political movements. The conception and deployment of sustainable infrastructure would require partnerships among civil society organisations, the private sector, governments, and international institutions. Systemic approaches ultimately help to upscale micro-solutions for macro-urban environments (National Intelligence Council 2012).

Coastal cities integrate the strengths and weaknesses of urban and coastal areas and can provide chances and choices for all, while trying to reduce their disadvantages and address threats. Cities generally have higher Human Development Indexes, composed out of life expectancy, literacy, education and standards of living. The Human Development Index shifted the focus from income to well-being and people-centred policies, such as education and health services, which can better reach populations concentrated in cities. This concentration, however, intensifies new challenges for generating urban jobs and providing the services for the collective quality of life. The 2013 Human Development Report highlighted the profound shift in global dynamics driven by the fast-rising powers of the emerging world and its long-term implications for human development (UNDP 2013). The 2014 Report recognises that, besides effective policies for reducing inherent vulnerabilities, cities have to build capacities for disaster preparedness and recovery from shocks (UNDP 2014).

Cities should expand access to public goods and consolidate rights to the commons for all, including migrants. In cities, human beings search for satisfaction of basic needs in essential goods and services, but also seek fulfilment and happiness. They are the places where the prospects of individual and collective well-being can be fulfilled. The UN report on the State of the World's Cities 2012/2013 highlighted that the city is the "home of prosperity", but prosperity for all has been compromised by a narrow focus on economic growth. A fresh approach to prosperity should duly take into account the quality of life, adequate infrastructures, equity and environmental sustainability. The yardstick for measuring progress should be reinvented. The City Prosperity Index, together with a conceptual matrix, the Wheel of Prosperity, could assist decision makers in designing comprehensive policy interventions (UN-HABITAT 2012).

Cities could be inspired to rate their degree of happiness by the Happy Planet Index, which takes into account the natural and physical capital and reveals the performance of governments in supporting citizens to live long lives and experience well-being, while ensuring that their ecological footprint does not prevent others from doing the same in the future. Introduced by the New Economics Foundation, founded in 1986 by the leaders of The Other Economic Summit in search of a different model of wealth creation based on equality, diversity and economic stability, the index reveals that the planet Earth is largely still an unhappy planet. Since 2006, the Happy Planet Index provides a tool to ensure that some fundamental issues are taken into account in crucial policy decisions. It also indicates the limits of the indicative indicators, which make countries with very different levels of development appear to be "equally" happy (New Economics Foundation 2012).

The right of the city should be enjoyed by everybody, exactly as the duties to the city apply to everyone. Since 2002, the biennial World Urban Forum, established by the United Nations, has brought together government leaders, mayors, members of national, regional and international associations of local governments, non-governmental and community organisations, experts, academics, women's and youth's organisations, and slum dwellers groups as partners working for better conditions in cities. The forums focus on burning urban issues, involve formal and informal dialogues and conclude with action-oriented proposals. The fifth forum (Rio de Janeiro, 2012) focused on "The right to the city: Bridging the urban divide" (UN-HABITAT 2011).

The UN Human Settlements Programme suggests that urbanisation, if well harnessed, can greatly support social equality, economic prosperity, ecological security and cultural vitality. Naples hosted the sixth World Urban Forum which included the "One Minute for Change" campaign. Just after the Rio+20 conference, the sixth Forum's manifesto for Cities called upon public, private and social partners to commit to a Global Urban Change Agenda. As committed partners striving to become "change makers", united by shared goals and a common vision for the city of the twenty-first century, they were invited to overcome the current models of socially, economically and environmentally unsustainable urbanisation. The "Urban Future we want" advocated for a world in which cities generate justice, knowledge and happiness. The manifesto set key principles, proposed essential paths for realising the Global Urban Agenda, and initiated momentum towards the Habitat III Conference (UN-HABITAT 2012).

The World Habitat Day, designated by the United Nations to be celebrated every year on the first Monday of October, reminds the world of its collective responsibility for the future of cities and the importance of the recognition of exceptional action. The purpose is also to reflect on the state of cities and the fundamental right of access to adequate shelter. The World Habitat Scroll of Honour, attributed on the same day since 1989, already acknowledges initiatives which have made outstanding contributions in improving human settlements and the quality of urban life.

The month of October 2014 was definitely an urban October. On 31 October 2014, cities and countries around the world celebrated the first World Cities Day. The day was proposed by the UN-Habitat to raise awareness of the global role of cities in humanity's future. World Cities Day promotes advocacy, outreach and cooperation activities for sustainable urban development. The theme of the first World Cities Day was "Leading Urban Transformations". The City of Shanghai hosted the main festivities and many cities of the world celebrated in unison. On the way to Habitat III, the theme offers the opportunity to redefine the urban paradigm for future generations.

Urban and ocean issues are part of the major 2015 UN process, a global dialogue for the post-2015 development agenda and Sustainable Development Goals (SDGs). The framework is being negotiated and the final agenda will be launched in September 2015, which is the target date for urbanising the MDGs. The process, led by Member States, enjoys broad participation from civil society stakeholders. The contributions to the agenda include a set of SDGs proposed by an open working group of the General Assembly, and the report of an intergovernmental committee of experts on sustainable development financing (UN 2014a). Furthermore, the UN Informal Working Group on Biodiversity Beyond National Jurisdiction (BBNJ), the technical group mandated by Rio+20 in 2012 to address the high seas, is conducting negotiations for an international agreement for the sustainable use and conservation of marine biodiversity in the high seas.

Habitat III, which will take place in Quito in October 2016, in line with the bi-decennial cycle which was begun in Stockholm in 1976, and continued in Istanbul in 1996, offers a generational opportunity to reinvigorate the global commitment to sustainable urbanisation and a "New Urban Agenda". The objective of the Conference is to secure renewed political commitment for sustainable development, assess accomplishments versus objectives, tackle urban poverty, and identify and address new and emerging challenges. Habitat III will be the first UN global summit after the adoption of the Post-2015 Sustainable Development Agenda. It offers a unique opportunity to discuss the ways that cities, including coastal cities, are preparing to fulfil their role in the climate change era and the implementation of new goals for sustainable development.

Sustainability involves a continuous renaissance and rebalance, resilience to threats, and a permanent reinvention of new and better opportunities. Qualitative leaps and paradigm shifts are crucial to maximise and optimise investments in capital, labour, and chance. Innovation and sustainability share a desire for permanent youth and the grasping of new opportunities. Against the backdrop of global crisis, cities have to become smarter and regenerate their body and all its vital organs, but also their mind, with its ideas and concepts, and their soul, with its values and emotions. They have to reinforce their capacity to co-create the masterpieces of the future. The synthesis report by the UN Secretary-General focused on six major elements: dignity, people, prosperity, planet, justice and partnership, which find in cities the most concentrated opportunities for accomplishment (UN 2014b).

The progress of a city can be evaluated through the increase or decrease in all forms of capital forming the aggregate capital stock. A key question concerns the extent to which the various components of wealth can be substituted for each other. For weak sustainability, the heart of the matter is not whether a particular form of capital, such as fisheries, will be available infinitely, but whether human ingenuity can preserve and increase the global capital (Mega 2005). An "integrated minimum" of the total urban capital has often been proposed as a non-negotiable political line, allowing for flexibility of substitution among the various forms of capital.

Strong sustainability implies that substitution of social and environmental capital is impossible, and that irreversible loss of any form of precious urban capital should absolutely be prevented. This is the plea of Agenda 21 and the Charter of European Cities and Towns towards Sustainability, which state that cities should be founded on the carrying capacity of nature and advance towards social justice, prosperous economies and environmental health. Equity is considered to be a precondition to sustainability, as the inequitable distribution of wealth causes both unsustainable patterns and resistance to cultural change (ICLEI 1995).

Embarking on the journey towards urban sustainability demands multidimensional efforts in all domains and at all levels, local, regional, national and global, against the depletion of the capital stock. Cities have to manage their assets to address the specific challenges linked to each type of capital, without losing the overall sustainability perspective. Effectively managing the co-evolving dynamic capitals requires a holistic approach, since all forms of urban capital are interconnected and interdependent. Advancing towards strong sustainability through the simultaneous reinforcement of all urban capitals demands strong political capital made out of robust leadership and citizen participation, so that the aspirations of the citizens are understood and visions for the future get inspired. Beyond converging bottom-up and top-down approaches, nurturing "mutual trust" dynamics is essential for enduring change. Ethical values are a critical element of good leadership, which engages in action founded on collective and shared principles that are communicated clearly and can mobilise citizens (PricewaterhouseCoopers 2014).

Public policy for sustainable development has to orchestrate responses to the demands of many different groups and manage the allocation of resources between various, and often competing, claims. Developing the capabilities to ensure that cities manage their capitals effectively is fundamental for responding to citizen expectations. Cities need to make sure that they can measure the performance at all policy fronts and create a culture of continuous improvement. As with any programme of substantial change, effective risk management is an essential tool. City governments have to identify and understand challenges, gauge their capacity to respond and link their policies to integrated frameworks of governance and compliance.

The total urban capital is composed out of human and social, natural and environmental resource capital, cultural and political, technical and financial capital. All forms of capital are highly interconnected and interdependent, yet each one is to be developed according to its distinct identity. Human capital includes personal health and well-being assets, personal skills and education, ethical values and lifestyles. Human and intellectual capital is identified by city leaders as the most important asset that their cities possess, which can make a critical difference towards sustainable development. Often, city leaders state that they are trying to develop policies to attract skilled and entrepreneurial people, a key factor for success in the twenty-first century (PricewaterhouseCoopers 2011).

Social capital is a definitive feature of a city as a society. It expresses the richness of informal and formal relationships, the extent to which people feel connected, the dynamism of local networks, the strength of the social bond. It comprises community traditions and patterns. Engaging all citizens is crucial for cities that wish to prevent crime, segregation and exclusion.

Cities can boost human capital, as they offer conditions for better education and training, information and awareness, health and well-being. The right to housing can enable the poor to create wealth. In the developing world, the vast majority of slum dwellers live in extreme poverty. However, the social capital is often high. Squatter cities are vibrant self-organised and self-constructed places with support communities and highly active groups. The multifactorial context, together with their predominant youth, makes slum dwellers one of the most vulnerable groups in terms of HIV/AIDS, malaria and tuberculosis.

Natural capital is described as the physical endowments, assets typically transacted in markets, while physical or environmental resource capital includes the life support systems, offering their services on land and sea. The aquatic capital of cities includes not only water but life, the rich marine biodiversity. The natural and physical capital of cities is being affected by accumulating stress on ecological systems, terrestrial and marine, massive loss of biodiversity and vulnerability to natural disasters, soil erosion, water stress and desertification. Tensions over scarce resources and raw materials may also cause conflict and aggravate crisis.

Financial capital is vital for investments, and depends on pertinent framework conditions and the ingenuity of the private sector. Technical capital includes the hardware of cities, infrastructures and buildings. It is harnessing the evolving possibilities of smart grids and new technologies to improve services such as housing, water and sanitation, energy and transport services, and access to world networks.

Cultural and leisure capital depends on a complex array of attributes defining the unique identity, the distinctiveness of a city. They include all cultural assets which form the visible essentials of a city and differentiate it from other cities on the world scene. Many cities try to develop a brand that encapsulates their exceptional qualities and generates powerful and memorable positive associations. Strategies designed to develop cultural capital need to understand the public perception of cities by citizens, businesses and visitors, and develop the steps to project cities forward into the experience that they wish to offer.

The "city and the sea experience" is a precious element of urban life, an increasingly vital component of the cultural capital capturing all possible activities to be experienced in a given place, and links to the aesthetic, religious and spiritual ecosystem services provided by the oceans. It also relates to iconic values and scenic experiences by the sea, for example, world monument sites during high tides or temples in fascinating canal mazes, and the search for the ultimate discovery experience. The aesthetic and cultural capitals of many European cities make the urban experience unparalleled and constitute a key factor for them as the world's prime tourist destinations. The most visited cities are also leading advocates for halting the depletion of natural resources and improving quality of life. They can provide lessons for preserving and enhancing the urban capital, with its tangible and intangible aspects (Beatley 2000).

Last but not least, the political capital of a city is very important. It depends on institutions and citizen participation, transparency and accountability, in order to achieve the shared commitment of a city on its journey into a sustainable future. City leaders stress the fundamental issues of engagement and participation. In times of general distrust, many governments try to re-invigorate the participation of citizens in civic activities end develop ways to make themselves more accountable, increase the transparency of decision making and engage civil society in the conception, implementation and evaluation of policies in the full sustainable development agenda (PricewaterhouseCoopers 2011).

Meaningful communication with citizens is at the heart of political capital. Channels for interaction are expanding and social media can further enrich the dialogue between cities and citizens. Developing accountability is a key responsibility for cities wishing to engage their stakeholders. It involves the true implication of citizens in discussing possible futures and visions, defining objectives and fixing targets and disclosing their performance against them, creating collaborative committees and forums with stakeholders and specific interest groups, and organising innovative public services responding to the true needs of the communities.

Preserving and enhancing each part of the composite urban capital, and at the same time enhancing it towards the future, requires innovations exploiting new and different paths and destroying old and outdated patterns. From the initial invention to the final transformation, a chain of interactive processes could bring a thorough change that opens up the range of opportunities and creates sustainable value. Trial and error are important, and failures often carry the seeds of future successes. Efficient but non-creative use of human or financial urban capital or technology can weaken the promising collective seeds of cities (Mega 2010).

1.3 Smart Cities, Network Societies and Possible Futures

In times of crisis and frail recovery, cities have to run just to remain still. To lead the future world, which is expected to be a dual world, with a physical component and a digital counterpart, cities have to become smarter. The potential of information and communication technologies has to be further harnessed to dramatically improve the ways that cities function. In the future, citizens are expected to be linked by intelligent devises, robots, sensors, databases and an infinite number of artefacts and interpersonal networks. Interconnection can be a force, together with education and empowerment. If the twentieth century was marked by the rise of a middle class in developed countries, the twenty-first century is expected to see the rise of a burgeoning middle class in many emerging countries. A wave of one billion additional consumers can boost competition for an array of products and services, which enter their sphere of affordability. Market growth takes off at a rate that can substantially exceed income growth until it reaches a saturation point (MGI 2012a). By 2009, Asia Pacific already counted for 28 % of the world's middle class, against the 36 % of Europe and the 18 % of North America (OECD 2010c).

The interconnection of humanity progressed rapidly and nearly 40 % of world citizens were estimated to be connected via the Internet in 2013 (Millennium Project 2013). By 2030, the digital revolution is expected to have covered the entire planet. A global society is a powerful driver of universal change. It is emerging through worldwide digital communications. It connects intelligent machines, robots, ubiquitous sensors, and databases in a planet-wide network of networks and the "Internet of Everything". The integration of the real and digital world provides for the formation of online cultures and lifestyles that transcend physical boundaries, and facilitates the integration of totally different spaces and forms of exchanges. The policy implications in relation to ethical values, privacy and security, and norms are important (Gore 2013).

Diversity thrives in cities and can help them envision many possible futures and address many scarcities, in energy, water, land, species, labour and strategic natural resources. Important tensions have yet to be tackled in regard to the current methods of production and consumption, along with the scarcity of non-renewable resources and the overconsumption of renewable ones. Cities also have to address social exclusion and inequality, the extreme phenomena linked to climate change, interdependent systemic risks, new pandemics and forms of organised crime, and terrorism.

Open knowledge is an inexhaustible resource which can be enhanced and enriched through the density of urban networks and transactions. Urban possibilities for integrating functions, including product, process and service innovations, social and public, and addressing threats are attracting increasing attention from scientists and decision makers. Cities can serve as the nexus for large scale replicable innovations and support global innovation ecosystems. Open innovations offer the possibility for each city to build on other best urban achievements and capitalise on improvements as a global network partner. Both incremental innovation, mainly bringing optimisation, and disruptive innovation bringing substitution can compose pathways for the future. Innovations, however, have to be responsible, as, in a highly interlinked world with more value created and migrating on line, important ethical questions arise.

Science and Technology are vital in the search for substitutes for scarce and exhaustible resources and the adoption of new models of sustainable production and consumption. Green technologies and eco-innovations have to prevent the destruction of habitats and ecosystems and the depletion of natural resources. Anthropogenic degradation often comes through intensive use of resources and excessive waste production, such as overfishing and marine debris. Environmental degradation has strong effects on health and on the availability of resources, especially on land, water and biological resources. As strings of extreme natural phenomena are developing into major catastrophes, affecting citizens and infrastructures, human livelihoods and economic systems, understanding and measuring the impacts of climate change through real time data gathering and global networks of sensors is crucial.

The development of circular or spiral economy models can foster the sustainable use of renewable resources, reduce primary demand for materials and limit waste to the bare minimum, thanks to prevention, reuse and continuous upcycling. As mass production is being replaced by on-demand, custom manufacturing, new opportunities arise. Additive manufacturing, also known as 3D-printing, and more recently 4D-printing, could revolutionise the productive web of cities.

Technology can democratise access to knowledge for all citizens. The Internet has been a great transformer of lives and mobile services have been integrated into the shifting lifestyles of citizens. The global network of networks is being proposed as the guardian of a free society and the proliferation of social networks has spurred public expression and communication. The Internet of things is advancing with the connection of real world objects embedded with sensors. The resulting networks hold promises for new and better business models and processes (MGI 2010, 2011, 2013).

The quantity of information available to citizens, unprecedented in human history, continues to grow exponentially, but it does not necessarily lead to illumination and wisdom. The sheer volume of information is in danger of creating more noise than value. Big data are expected to accelerate multiplication of knowledge, foster new technological advances, and enable management of increasing complexity and diversity. Transparent innovations in data structuring, mining and processing could help in filtering out the noise and enhancing the wealth of strategic knowledge so as to decisively address societal challenges (MGI 2010, 2011).

Smart grids enable cities to enhance Information and Communication Technologies as drivers of sustainable development and key enablers of change. Cities should harness scientific and industrial capabilities and take advantage of transformations in manufacturing, infrastructure, and services. Future smart cities may have to strive to balance sustainable infrastructure with the provision of smarter services to citizens. They may advance towards an integrated supply and demand structure that provides food, energy, materials, security, mobility and communication more effectively and efficiently (EC 2012a).

The most promising technologies to appear at the top of trustworthy world rankings shed light on key enabling technologies adding value to information, communication and participation, synthetic biology and metabolic engineering, green technologies for increased food and biomass, nanotechnologies, systems biology and simulation of chemical and biological systems, energy storage and wireless power, personalised medicine, nutrition and disease prevention and enhanced education technologies (MGI 2013).

Social and professional interactions are increasingly independent from the urban context and can, in virtual form, take place instantly on-line, but these relationships seem unable to replace the real world's personal connections and their underlying formation, maintaining and breaking trust and cooperation. To seek identity and connectedness with other individuals or groups that share the same interests, beliefs or ideals is a fundamental aspect of human conduct. At virtually zero marginal cost, citizens can interact with a much larger range of people across the globe, but direct eye to eye contact maintains its cardinal importance for social bonds (MGI 2012b).

In a networked society, traditional institutions can apply openness through digital participatory methods, such as crowdsourcing. The Committee for the Future of the Parliament of Finland suggests that crowdsourcing offers new possibilities for citizen involvement in sharing insights and impacting policy-making. Digital participatory methods are additional means towards an equal democratic society. They should never be perceived as "political window-dressing", for this could damage the credibility of the process and the motivation of citizens to participate in the future (Parliament of Finland 2011).

Helsinki is a strong city in the Smart Government arena. More than 1000 open data sets actively promote citizen engagement. The city played host to the first global Open Knowledge Festival in 2012 and launched the Forum Virium Smart City Project to provide ubiquitous data to citizens for improving quality of life. Electronic participation brings one more option for citizens to exercise democracy but it does not necessarily make a society more democratic. Digital identity futures face tensions between high levels of security and operability (STOA 2011).

Public acceptance is capital for the further development of digital services. Studies conducted by the European Parliament concluded that some digital services, like electronic procurement, met with early-stage uncertainties and resistance. The interest of public authorities in conducting cross-border, fully electronic procurements did not appear to be sufficiently proven. On-line health applications met with problems of public acceptance. Last but not least, the build-up of a comprehensive system for electronic voting in Europe could not be recommended for the near future (STOA 2011, 2012, 2013).

The electronic participation of citizens in local public decisions can increase transparency and help fight corruption. The Cracow Declaration on the Local Agenda i2010, adopted by the European Information Society Conference of local and regional governments, focused on digital solidarity among the cities of the world. It promoted the UN's Digital Solidarity Initiative and highlighted the need to fully acknowledge the role of local and regional governments in bridging the divide (European Information Society Conference of Local and Regional Governments 2005).

From eGovernance and eBusiness to eLearning or eHealth, the most challenging goal is to build digital ecosystems for knowledge sharing and upgraded public services, as well as the creation of smart small and medium enterprises. Each eGovernment must advance from the simple reproduction in digital format of procedures administered during office hours to the effective and efficient reengineering of constant and uninterrupted provision of on-line services. This implies a true shift in culture from a use of technologies to support public sector operations to the integration of digital technologies in shaping agendas for public sector reform.

Governments, enterprises and citizens continuously ask for more powerful devices and applications and more cost-effective technology. The supply chains to satisfy these demands lead to explosive growth in data and analytics and to new competition in almost every scientific and industrial domain. Genomic, biotechnology, neuroscience, and life sciences revolutions are radically transforming the fields of medicine, agriculture, and molecular science, and are putting control of evolution in human hands. The possibilities for artificial land and even artificial life multiply and the era of post-humanity seems close.

The convergence of digital technologies and the life sciences is a potent driver of universal change which opens the way to the digitalisation of life, complex ethical equations and the large-scale maps of neural wiring. Just as the discovery of DNA led to the mapping of human genome, the discovery of the neurons' connection in the brain could lead to the mapping of the connectome (Gore 2013). A great brain race has begun, with the EU, US, Japan and China promoting programmes to better understand the functioning of the brain and the creation of artificial brains (Millennium Project 2013).

Interactions between the physical and the virtual world are being massively facilitated by advances in materials and biological sciences. Human bodies could further benefit from artificial parts. New materials will allow the production of soft robots with organic tissues. Artificial and human intelligence may compete or coexist and reinforce each other harmoniously, leading to a change in human nature. Cyborgs could perform complex tasks like humans and in all economic activities, and may be particularly interesting for working in open seas conditions.

Public health is a critical urban system constantly challenged and evolving. Antibacterial and antimicrobial compounds (antibiotics) may no longer be readily available in the near future. Every dose of antibiotics creates selective evolutionary pressures. New antibiotics have been continuously developed to replace older, increasingly ineffective ones. However, human innovation may no longer be outpacing bacterial mutation. None of the new drugs currently under development may be effective against certain new mutations of killer bacteria that could turn into a pandemic, which could reach extreme dimensions in urban environments (WEF 2015).

Well-being has been significantly affected by excessive lifestyles and their resulting contradictions. Obesity is a prominent example. One out of eight world citizens are undernourished, yet more than one in three adults are overweight and almost one third of the produced food gets lost or wasted (UNEP 2014). Accelerating progress in prevention of excess weight is extremely important since this is often a major cause of chronic disease, disability, and death. Obesity risks are often disproportionate among minority, low-income, and less educated populations, due to inequitable distribution of health resources and community risk factors that contribute to disparities in obesity prevalence. Schools are uniquely positioned to be focal points for prevention and wellness. Many countries issue guidelines for school food, ranging from voluntary measures, to complete bans, including on marketing (Institute of Medicine 2012).

Ageing, as a critical lifetime process, creates new demands for food supply, with older adults not only being more susceptible to certain foodborne illnesses or related health complications but also likely to experience significant changes in dietary needs. Research on optimal nutrition for older adults provides insights into new food technologies and consumer messaging that can help cities better care for their older citizens.

The natural world, including the largely unexplored marine world, reflects the immense potential inherent in the genetic code of all living organisms. Progress in synthetic biology and metabolic engineering capitalises on this potential in unprecedented ways, enabling the development of new biological processes and organisms to serve specific purposes of personalised medicine, nutrition and disease prevention. Developments are laying the foundation for a revolution in healthcare for less resource intensive and more effective well-being. Systems biology and computational modelling and simulation are playing increasingly important roles in designing better therapeutics, materials and processes. Realising the potential offered by information and communication technologies in education and training could allow learner-centred education to thrive and lead to better lives.

Advanced automation can provide multiple improvements in care and contribute to the general adaptation of technologies and infrastructures to supporting ageing populations, including performance enhancing technologies and treatments. Migration can generate needs for innovations to support the governance of the process, including control and integration technologies, but also to enhance the chances of immigrants, for instance, through education and intercultural dialogue practices.

Public transparency and trust in the ability of technological systems to prevent negative surprises are important issues for society, the ultimate frontier. This is particularly important in a time in which humanity needs huge infrastructural investments. Urban infrastructures in emerging countries require the creation of whole functional districts. In industrialised cities, infrastructures are ageing and need renewal to deliver services such as water and energy distribution and public transport. Widespread sensor networks are needed to help manage the new infrastructures and allow the plethora of applications. The development of personalised services may be followed by another wave of innovations aimed at further tailoring and targeting services, from manufacturing to leisure.

The substitution of scarce and depleted resources is a key issue for the Green Revolution 2.0, which has been stimulated by the rising global demand from more empowered citizens and consumers for healthy environments, nutritious food and fresh water for all. Progress across the biological and physical sciences and innovation hold the promise of further reinforcing optimisation of resources, as in achieving maximum sustainable yields in agriculture and fisheries, minimising impact on biodiversity, and decreasing the carbon and environmental footprints of economic activities.

The escalating demand on natural resources over the last two decades has largely offset the benefits achieved through eco-efficiency technologies and practices. Research and development of new materials and energy sources and technologies are of particular interest for cities striving to achieve emission reductions. Strong, light structural materials reducing energy costs of buildings and transport promise to increase urban resilience. The energy transition is expected to stimulate technology developments in renewable electricity generation, and new technologies like fuel cells, the smart grid, and carbon capture and storage. Large-scale use of hydrogen as a fuel may complement the progressive electrification of everything, which also relies also on offshore resources.

The electricity sector is going through a major unprecedented transition. The penetration of renewable energies and the technological improvements combined with efficiency measures provide the opportunity to advance towards the post-carbon society and economy, while reducing dependence on imported fuels. The captured CO_2 after combustion could be turned into an important resource. Novel catalysts, based on nanostructured materials, can potentially transform carbon dioxide into high value hydrocarbons and other carbon-containing molecules, to be used as more sustainable alternatives by the chemical industry.

Cross-industry cooperation and dialogue with policy-makers are needed to define legal frameworks, reach international standards and set up public-private partnerships to address global challenges. Successful players need to take advantage of new opportunities associated with hyperconnectivity, such as big data analytics, while managing risks through data privacy rules and cyber security measures (WEF 2014b).

In the context of scarce resources and the search for new spaces of opportunities, marine and maritime technologies achieved significant progress over the last decade. There has been rapid technological progress in working in offshore locations in ever-deeper waters through robotics, submersible technologies and multi-purpose platforms.

The Apollo missions offered the first images of Earth from space, bringing attention to the other 70 % of the overwhelmingly planet (UNEP 2011a, b). Since then, ocean observation has progressed significantly, through more advanced observation satellites and in situ sea-borne sensors, mapping and forecasting. The exploration of the unfathomable diversity of ocean resources asks for submersible technologies, robotics, and video-surveillance and high performance equipment for marine operations.

Smart grids also rely on ocean infrastructure which may face conflicts among diverse uses. Submarine cables play an important role in all forms of communication through the internet, and increasing demand is expected to result in more capacity and more submarine cables. Although submarine cables, and the protective corridors around them, cover only very narrow strips of seabed, they introduce a dividing line across the open ocean seabed which may create tensions with other activities.

Submarine pipelines also have great importance for transporting oil and gas through coastal zones and between continents and regions. The growth of the offshore oil and gas technology and industry has mounted the demand for access to the ocean space within national jurisdictions. In the United States, about 550,000 km^2 of the whole EEZ is subject to oil and gas leases, which represent, in the Gulf of Mexico, 66 % of the United States EEZ in that area. Such dominant uses of ocean areas could create ecological disruptions and conflicts with other activities.

Offshore renewable energy is progressing fast and it must be proven that these methods do not cause marine environmental problems. In the EU, the world leader in offshore wind energy, average offshore wind farm size was 485 MW in 2013, 78 % more than the previous year. Continued effort to reduce the cost of offshore wind technology is expected to accelerate this growth. There is also a challenge to accelerate the development of ocean geothermal, wave and tidal energy through reductions in technology costs. Different combinations of geographic and oceanographic conditions require different technologies, providing a more predictable base-load supply of electricity which compensates for fluctuations in supplying electricity from wind (EWEA 2014).

1.4 Cities Rising on the World Scene: Leading Coalitions of the Committed

Urban democracy is a key conquest of cities and a major asset for rising in the world scene and leading sustainability-oriented action. Anticipatory democracy is vital for building momentum towards desired futures. Globalisation is at the crossroads of trust and rebalance, as, in an increasingly polycentric world, wealth and power shift from west to east and north to south, from local to global and back, emerging G20 economies drive more growth and south–south exchanges get intensified. Although poverty is decreasing, debt and economic insecurity are mounting, and income disparity is widening. Government indebtedness increased much in OECD countries where the total debt climbed from 74.2 % of total GDP in 2007 to 112.5 % in 2014.

A more conscious search of sustainable development could bring a better future for all (UNDP 2013; UNFPA 2011; Millennium Project 2013).

In the multi-partner, multi-speed world of the next decades, no single power is expected to play a hegemonic role. China and India ("Chindia") are expected to be the most influential players, while Russia and Japan could lose the great power status that they have enjoyed in the past. The impact of all powers will also depend crucially on their ability to act as inspiring models for future economic, political and societal developments (EC 2009; EUISS 2012; Boillot and Dembinski 2013).

The evolving world order could generate greater opportunities for smaller powers, at national and subnational levels. Traditional middle OECD counties, such as Canada and Australia, are expected to sustain their level of influence while dynamic cities with one million inhabitants and less could also become middle range players, with a significant regional impact. Europe, the continent of medium-sized cities, could benefit from its cities' leading roles in international forums and networks for the sharing of knowledge and technologies.

Improving living conditions in the world's megacities will be a major challenge for humanity. Megacities attract huge investments for infrastructures, roadways and rail structures, ports, power grids and telecommunications which have to be efficiently and effectively managed. Business and civil society networks are also expected to play a critical role. Their influence could become greater than that of many states, and lead to new forms of governance and civic action.

Coastal cities could play strategic roles in building a sustainability-oriented consensus among the many diverse players in a multi-layered global network of urban communities. The true influence of coastal megacities is hard to evaluate. Despite the knowledge about megacities and urban processes, and about coasts and their physical processes, there is relatively little work on the intersection between large-scale urbanisation and the coast (Pelling and Blackburn 2013).

The scale and speed of growth have profound implications for multi-level, multi-stakeholder governance, since the emerging profile of future cities confronts national structures. Strong and interconnected local leaders play an increasing role in international politics. The rising power of intermediate cities could signal important growth and innovation coming from cities at a faster pace than present. Furthermore, intense networking, alliances and coalitions of cities can be much more agile, responsive and effective than internationally established organisations.

The global democratic governance deficit may further weaken the legitimacy of national governments and international organisations and open the opportunities for leadership by cities. As no single power will be able to play a leading role in the search for shared solutions to global problems, strengthened links between local, regional, national, and global governance, and between state and non-state actors or among cities and civil society, seem to be decisive for the future of humanity. Cities could instil a new sustainability ethos in the global debate, as leading interconnected democratic actors in an era of citizen disengagement and mistrust. The transition from the nation-state system to a world with a powerful multipolar set of actors outside traditional government structures will require strategic coalitions consisting of governments, corporations, NGOs, and academic institutions. Democratic cities, networks of networks, could lead in networking with their peers.

Dynamic cities could become the world powerhouses and the most vibrant neuralgic centres (ESPAS 2013; National Intelligence Council 2013). Stellar megacities could be emulated on a global scale, reducing their ecological impact, while networking has a clear emulation effect. However, anthropogenic global warming, intensified land use, significant sea-level rise, and extreme weather phenomena could render many human settlements unsustainable, and eventually uninhabitable (EEA 2015a, b).

A powerful world city integrates an array of magnetic powers that can attract creative people and top companies from all over the world. This was the definition proposed by the first Asian ranking by the Institute for Urban Strategies in Tokyo, which evaluates and ranks the Comprehensive Power of the major world cities on the basis of 70 indicators making up six major urban functions defining the strength of a city: Economy, Research and Development, Cultural Interaction, Liveability, Environment, and Accessibility. Evaluation is carried out through the lenses of four global actors leading urban activities, including Managers, Researchers, Artists, and Visitors and, of course, local Residents. A matrix of city actors and city functions indicators helps in appraising and ranking cities (Mori Memoriam Foundation 2010a, b, 2011, 2012, 2013, 2014).

The 2014 ranking according to the Global Power City index reconfirmed the position of London, New York, Paris and Tokyo as the four most powerful cities on the world chessboard. The same cities form the league of what the World Economic Forum calls "super cities" for real estate investment. The future of London as an economically competitive and influential force in the world seems well established. The city can build on its soft power and use its international cultural radiance to best effect. It seems that the combination of systems and empathy lies at the heart of London's success. But the city's future is far from secure. Although London often appears on the verge of a new golden age, it also shows increasing signs of falling systems (Leadbeater 2014).

The Asian ranking exercise is keyed to the relative performance of Tokyo, which remained unchanged from previous years. As with other cities, there was considerable change in the city's relative ranking across the six criteria. Tokyo's competitiveness was strengthened by a decrease in local taxes, enhanced cultural interaction, and improvements in liveability. The score difference between Tokyo and Singapore, ranked at fifth place, has shrunk, while, in sixth place, Seoul has also largely closed the gap on Singapore. From the perspective of actors, London and Singapore appeared to be most attractive to managers. The island city-state, ranked as the most transparent and competitive Asian country by Transparency International and the World Economic Forum, according to the corruption perception and competitiveness indexes, respectively, is positioning itself high in many global city rankings. Researchers ranked New York and Tokyo first and second, respectively. In terms of appeal to residents, Paris is the incontestable top city, with a considerable lead over London in second position (Mori Memoriam Foundation 2014).

Although command of the sea is not among the criteria for cities to rule the planet, four out of the five cities with the highest world comprehensive power are coastal cities. Tokyo still ranks first as a population megacity and it is expected to remain first by 2030, despite many past UN predictions that it would have already been dethroned by Mumbai. The historical cityscape was seriously damaged during World War II, yet the fundamental structure of Tokyo, with its unique urban form surrounding the Imperial Palace, persists, often hidden behind or beneath heteroclite modern additions. Tokyo's strengths lie in market size, economic vitality and human capital, as well as in research and development and the environment. Its weaknesses lie in market attractiveness and regulations and risks, cultural resources, cost of living and international transportation network and traffic convenience. It also seems that the Japanese capital represents a mature megacity fostering fewer new opportunities for attracting people and capital and caring more for its shrinking and ageing population (UN 2014a).

Tokyo is linked to many coastal cities through constellations of urban networks. The top world coastal city is the seat of the Headquarters of the International Association of Ports and Harbours (IAPH). Over the past six decades, IAPH has developed into a global alliance of ports, representing some 200 ports in 85 countries, collectively handling well over 60 % of the global sea-borne trade and nearly 80 % of the world container traffic. The alliance strives for industry engagement and excellence, promoting the interests of ports and advancing sustainable practices.

Most high-promising Asian cities have close links with the sea. Singapore, a microcosm of Asia, is ranked the continent's top City of Opportunity. It is a major commercial hub, the fourth-world financial centre and one of the five busiest ports. Its globalised and diversified economy depends heavily on trade, especially manufacturing, representing more than one fourth of Singapore's GDP. In terms of purchasing power parity, Singapore has the third highest income per capita in the world. It also ranks high with regard to education, healthcare, water management and economic competitiveness (PricewaterhouseCoopers 2014).

In 2011, Shanghai became the largest global port according to its handling capacity. The city is emblematic of the progress marked by a colonial town to become a metamodern metropolis within less than 20 years. Founded in the thirteenth century as an administrative centre, a traditional circular city with canals and narrow streets, Shanghai has always been open to international movements of architecture and city planning, integrating classical European models with local and national ones.

As a major port on the east coast of China, on a unique geostrategic location, Shanghai grew into a modern metropolis in the 1920s and 1930s. From the early 1950s to the 1990s, political circumstances dethroned Shanghai as a major centre of development, but satellite cities, heavy industry, and harbour zones and residential quarters continued expanding. All public buildings respected Chinese architectural forms of monumentality and dignity.

In the 1980s, Shanghai was one of the first Chinese cities to achieve universal primary and junior secondary education. The Chinese government's decision to embrace a more open policy in the early 1990s gave the city an opportunity to become, once again, a world centre of finance, trade and shipping. The restructuring of the Pudong area has led to a new urban front with extraordinary high-tech buildings for housing, offices and other activities. Shanghai again became an open city, ready to seize opportunities and continue a competitive tradition underlying its dynamic nature and its entrepreneurial spirit (LSE 2005).

In the second decade of the third millennium, Shanghai is the undisputed largest and wealthiest city in China. It hosts more than 18 million inhabitants in search of harmonious coexistence, in balance among local and post-modern realities in the port, the old city and Pudong. The highly populated area within its traditional city boundaries has 6.5 million people. The Shanghai economy is expected to continue expanding rapidly. The riverscape has been given much attention, with factories, shipyards and old warehouses being gradually replaced by high-quality public open spaces and cultural and recreational services.

Cities are the most powerful and agile links in the chains between local and global dynamics. Shanghai welcomed the 2010 Expo "Better City, Better Life" hosted for the first time by a metropolis of the emerging world, which envisioned and proposed excellent cases, mosaics of practices and intercultural exchanges. The Shanghai declaration highlights that people's understanding and pursuit of a better life are both the foundations and the engines of urban development. It suggests that it is necessary to re-examine the relationship between people, cities and the planet and advance towards "Cities of Harmony".

Cities of harmony can lead by example and by network emulation to prevent, in extremis, the "tragedy of the public commons", resulting from the overexploitation of precious environmental resources. The interconnected local democratic leadership can be instrumental for global awareness and radical change in direction. Local short-term profits should not prevail but serve global collective long-term oriented efforts. The joint consequences could be highlighted through the well-known metaphor of the "prisoner's dilemma" which advocates for the art of cooperation and concerted action.

Advancing climate change gave a new impetus to coalitions and alliances of cities and multi-stakeholders, including citizens. Political short-termism and human greed have been identified as the main causes for the outgrowth of humanity. Global leaders call for a radical shake-up in politics and business to advance on climate change, reduce economic inequality, improve corporate practices and address the chronic burden of disease. The report "Now for the Long Term" by the Oxford Martin School, resulting from a process of research and debate undertaken by a group of eminent leaders, recommended, *inter alia*, the creation of a world coalition made up of G20 countries, 30 companies, and 40 cities to counteract climate change.

More and stronger coalitions could accelerate action on climate change, through initiatives for energy-efficient buildings, faster market penetration of efficient technologies and reduction of emissions. Cities, especially through "Fit Cities" networks, could play a major role in public health and help reduce the burden on health systems. Cities can also help end discrimination against future generations by revising discounting methods and adjusting them to take account of the uncertainties, risks and ethical implications for the long term. Social protection measures, such as conditional cash transfer programmes, can be used to break the intergenerational cycle of poverty, and reduce threats of long-term unemployment and deprivation (Oxford Martin School 2013).

Cities can lead coalitions of the committed for sustainable resource security. Such coalitions should include key actors, such as the world's principal resource producing and consuming countries and third countries of systemic significance and cities that impact resource distribution. Cities can work together to change the fundamental conditions that gave rise to the tight resources markets in the past years. A series of critical interventions could involve new informal dialogues among committed leaders to tackle resource price volatility and to improve confidence and coordination in increasingly integrated global markets (Chatham House 2012).

Coastal cities can play a dynamic leading role in coalitions for a sustainable blue economy, and the health of the marine resources which underpin it. The blue growth comprises a broad array of activities related to food, jobs and opportunities provided by ocean and coastal assets. The concept emphasises conservation of ecosystems and habitats and sustainable management of aquatic resources and equitable benefits to the coastal communities that directly rely on them. If marine ecosystems are in good environmental status, sharing a space as immense as the seas would bring multiple benefits to humanity (Beatley 2011).

More than 30 years after the signature of the UN Convention of the Law of the Sea (UNCLOS), known as the Constitution for the Oceans, which came into force in 1994, ocean governance is recognised as a major issue. Advanced technologies like robotics, marine biotechnologies or underwater systems call for the regulation of new activities, especially in deep sea waters, in areas beyond national jurisdiction. Better coordination is needed for marine protected areas (MPAs), environmental impact assessments, capacity building and rules on the transfer of marine technology, genetic resources and benefit sharing. The private sector, operating in transport, oil and gas, wind and ocean energy, fisheries, aquaculture or coastal tourism, is an important player and marine social society organisations are active drivers of change (Blue Ribbon Panel 2013).

A coalition of committed coastal cities could go far beyond their immediate blue horizons and more globally address the issue of the oceans. About one third of the global ocean is controlled and managed by governments. But on the high seas, i.e. the international waters that lie beyond national jurisdiction, which make up nearly half of the Earth's surface, there is little care and monitoring. High seas are considered to be Earth's last global commons. They are no longer protected by their inaccessibility. Their complex ecosystems are subject to negative impacts from human activities ranging from shipping and deep sea overfishing, all compounded by a lack of comprehensive knowledge, effective legal instruments and coherent governance for environmental protection (GOC 2014). United in their engagement to help reverse ocean degradation, cities could advocate both for the value and the need for the sustainable use of the global ocean.

On the eve of important steps to be taken by the UN, the independent Global Ocean Commission brought the debate about the future of the high seas and the value of this immense and fragile ecosystem out from the margins of political debate and much closer to the mainstream. The Commission concluded that a vicious circle has been fuelled by the indifference and benign neglect by the majority, and active abuse by the minority, while global governance deficit led to an absence of accountability (GOC 2014).

A sustainably managed healthier ocean can provide more food and energy, wealth and employment. Combatting illegal fishing would improve prospects for nature, ecosystem services and responsible businesses. It could also ensure that the benefits from the exploitation of ocean resources can be equitably shared. At the Rio+20 conference, governments raised issues of concern, including the threats from climate change and ocean acidification, overfishing, illegal fishing, and subsidies that drive unsustainable consumption. They have also debated the need to conserve and protect marine ecosystems to both restock the ocean and build its resilience to change.

Powerful coastal cities could lead ocean governance and protection of the high seas. Although more than 12 % of the Earth's land surface is protected, the figure for the global ocean is about 2.3 %, and only 1 % in the high seas. MPAs in the ocean are currently increasing, but only limited parts are truly preserved reserves, implying a complete ban of extractive or polluting activities. The 2002 Johannesburg World Sustainable Development Summit called for the implementation of an international network of MPAs. Although a MPA does not necessarily imply an area in which all human activities are excluded, in many cases it does imply that some, or most, such activities will be at least controlled.

Accountable leadership by conscious global cities can bring a major contribution to the protection and enhancement of the world blue commons. In the 2010 UN Convention on Biological Diversity in Nagoya, the international community agreed to protect 10 % of the oceans by 2020 and identify ecologically and biologically significant areas (EBSA) of ocean beyond national jurisdiction. However, as of yet, no legal mechanism currently exists for establishing high seas marine reserves. There is a plethora of organisations with overlapping competencies and mandates, but none with overall responsibility (GOC 2014).

Coastal cities could play an influential role in a multi-stakeholder scheme to enhance global ocean accountability and demonstrate that they do care not only about their immediate waters but the interlinked global ocean and the planet. Together with their marine industries, universities and civil society associations, they can engage in the protection of watersheds and ocean sheds, including care and recovery for the high seas, the ending of overfishing and harmful high sea subsidies, the closing of ports and markets to illegally harvested seafood, preventing it from entering the commercial streams, and the prevention and removal, and if possible transformation into energy, of marine litter, especially plastics. It seems that protecting the high seas could have a huge environmental impact, a small environmental cost and a large economic benefit to be shared by all.

References

Angel S et al (2005) The dynamics of global urban expansion. Lincoln Institute of Land Policy, Cambridge, MA

Angel S et al (2011) Making room for a planet of cities. Lincoln Institute of Land Policy, Cambridge, MA

Beatley T (2000) Green urbanism: learning from European cities. Islands Press, Washington, DC

Beatley T (2011) Blue urbanism: the city and the ocean. Islands Press, Washington, DC

Blue Ribbon Panel (2013) Indispensable oceans. Aligning ocean health and human. Wellbeing, Washington, DC

Boillot J-J, Dembinski S (2013) Chindiafrique. Odile Jacob, Paris

Brookings Institution (2013) Foresight Africa Report. Brookings Institution, Washington, DC

Chatham House (2012) Resources futures. Chatham House, London

Club of Rome (2012) EU Chapter. 2012–2052: a global forecast for the next forty years. Club of Rome, Rome

Commonwealth Scientific and Industrial Research Organisation (CSIRO) (2011) Our future world. An analysis of global trends, shocks and scenarios. CSIRO, Canberra

Economist (The) (2012) The world economy. 13 Oct 2012

European Commission (EC) (2009) The world in 2025. Rising Asia and socio-political transitions. EC, Brussels

EC (2012a) Smart cities and communities. European innovation partnership. EC, Brussels

EC (2012b) Figures for the future. 20 Years of sustainable development in Europe. EC, Brussels

EC (2013) 7th Environment action programme. EC, Brussels

European Environment Agency (2013) Signals—Building the future we want. European Environment Agency, Copenhagen

European Environment Agency (EEA) (2014) Marine messages: our seas, our future. EEA, Copenhagen

EEA (2015a) The European environment. State and outlook 2015. Synthesis report. EEA, Copenhagen

EEA (2015b) The European environment. Assessment of global megatrends. EEA, Copenhagen

European Information Society Conference of Local and Regional Governments (2005) The Cracow declaration on Local Agenda i2010 in Europe and the promotion of digital solidarity among the cities of the world. European Information Society Conference of Local and Regional Governments, Cracow

European Strategy and Policy Analysis System (ESPAS) (2013) The world in 2030 highlights, Brussels

European Union Institute for Security Studies (EUISS) (2012) Global trends 2030 citizens in an interconnected and polycentric world. EUISS, Paris

European Wind Energy Association (EWEA) (2014) The European offshore wind industry—key trends and statistics 2013. EWEA, Brussels

Eurostat (2009) Statistics in focus agriculture and fisheries 47/2009. Eurostat, Luxemburg

Global Ocean Commission (GOC) (2014) From decline to recovery. A rescue package for the global ocean. GOC, Oxford

Global Ocean Observing System (GOOS) (The) (2008) Panel for integrated coastal observation (PICO). GOOS, Paris, Report No. 172

Gore A (2013) The future: six drivers of global change. Random House, London

International Council for Local Environmental Initiatives (ICLEI) (1995) Towards sustainable cities and towns. Report of the first European conference on sustainable cities and towns. Aalborg, May 1994. Freiburg

ICLEI (2010) Future of cities. 20th anniversary conference. Incheon

International Energy Agency (IEA) (2012) World energy outlook 2012. IEA, Paris

IEA (2014) World energy outlook. IEA, Paris

Institute of Medicine (The) (2012) Accelerating progress in obesity prevention: solving the weight of the nation. Institute of Medicine (The), Washington, DC

International Organization for Migration (IOM) (2011) World Migration Report. Communicating effectively about migration. IOM, Geneva

IOM (2013) World Migration Report. Migrant well-being and development. IOM, Geneva

International Transport Forum (2013) ITF transport outlook 2013: funding transport. ITF, Paris

International Panel of Climate Change (IPCC) (2014) Climate change fifth assessment report. IPCC, Bonn

Jacobs J (1969) The economy of cities. Penguin, London
Leadbeater C (2014) The London recipe: how systems and empathy make the city. Centre for London, London
London School of Economics (LSE) Cities (2005) Shanghai: the fastest city? LSE Cities, London
McKinsey Global Institute (MGI) (2009) Preparing for China's Urban Billion
MGI (2010) The Internet of things
MGI (2011) The great transformer: the impact of the Internet on economic growth and prosperity
MGI (2012a) Urban world: cities and the rise of the consuming class
MGI (2012b) Global cities of the future: an interactive map
MGI (2013) Disruptive technologies: Advances that will transform life, business, and the global economy
Mega V (2005) Sustainable development, energy and the city. Springer, New York
Mega V (2010) Sustainable cities for the third millennium. The odyssey of urban excellence. Springer, New York
Mega V (2013) Quintessential cities, accountable to the future. Innovation, sustainability and leadership. Springer, New York
Metropolis (1999) A network of cities for world citizens. VI world congress documents. Metropolis, Barcelona
Metropolis (2014) Cities for all. XI world congress documents. Metropolis, Hyderabad
Millennium Project (The) (2013) 2013–14 State of the future report. Millennium Project (The), Washington, DC
Mori Memorial Foundation (The). Institute for Urban Strategies (2010a) Global power city index yearbook 2011. Mori Memorial Foundation (The). Institute for Urban Strategies, Tokyo
Mori Memorial Foundation (The). Institute for Urban Strategies (2010b) Tokyo future scenario 2035. Let's design the future of Tokyo. Mori Memorial Foundation (The). Institute for Urban Strategies, Tokyo
Mori Memorial Foundation (The). Institute for Urban Strategies (2011) Global power city index yearbook 2011. Mori Memorial Foundation (The). Institute for Urban Strategies, Tokyo
Mori Memorial Foundation (The). Institute for Urban Strategies (2012) Global power city index yearbook 2012. Mori Memorial Foundation (The). Institute for Urban Strategies, Tokyo
Mori Memorial Foundation (The). Institute for Urban Strategies (2013) Global power city index yearbook 2013. Mori Memorial Foundation (The). Institute for Urban Strategies, Tokyo
Mori Memorial Foundation (The). Institute for Urban Strategies (2014) Global power city index yearbook 2014. Mori Memorial Foundation (The). Institute for Urban Strategies, Tokyo
National Intelligence Council (2012) Alternative worlds: global trends 2030. National Intelligence Council, Washington, DC
National Intelligence Council (2013) Alternative worlds: global trends 2030. National Intelligence Council, Washington, DC
New Economics Foundation (2012) The happy planet 2012 report. New Economics Foundation, London
National Oceanic and Atmospheric Administration (NOAA) (2011) State of the coasts. NOAA, Washington, DC
Organisation for Economic Co-operation and Development (OECD) (2007) OECD forum on innovation, growth and equity. OECD, Paris
Organization for Migration (IOM) (2011) Migration report 2011. IOM, Geneva
OECD (2010a) OECD forum. Road to recovery: innovation, jobs and clean growth. OECD, Paris
OECD (2010b) Cities and climate change. OECD, Paris
OECD (2010c) The emerging middle class in developing countries. OECD working paper No 285. OECD, Paris
OECD (2012) OECD environmental outlook to 2050: the consequences to inaction. OECD, Paris
OECD (2013) The competitiveness of global port cities. Synthesis report. OECD, Paris
OECD (2014a) Boosting resilience through innovative risk governance. OECD, Paris
OECD (2014b) International migration outlook. OECD, Paris
Oxford Martin School (2013) Now for the long term. Oxford Martin School, Oxford

Parliament of Finland (2011) Crowd-sourcing for democracy. A new era in Policy-Making. Parliament of Finland, Helsinki

Pelling M, Blackburn S (eds) (2013) Megacities and the coast. Risk, resilience and transformation. Routledge, London

PricewaterhouseCoopers (2011) Cities of opportunity. PricewaterhouseCoopers, New York

PricewaterhouseCoopers (2014) Cities of opportunity. PricewaterhouseCoopers, New York

Sassen S (2001) The global city. Princeton University Press, Princeton

STOA, European Parliament (2011) E-public, e-participation and e-voting in Europe—prospects and challenges. STOA/European Parliament, Brussels

STOA, European Parliament (2012) Urban transport technology options. STOA/European Parliament, Brussels

STOA, European Parliament (2013) Security of e-Government: E-passport, E-procurement, E-Health. STOA/European Parliament, Brussels

UK Parliament (2011) Migration and global environmental change: future challenges and opportunities, London

UN Conference on Sustainable Development Rio+20 (2012) Future perfect. UN Conference on Sustainable Development Rio+20, Rio de Janeiro

UN Convention on Biological Diversity, Stockholm Resilience Institute and ICLEI (2012) Cities and Biodiversity Outlook. Action and Policy. UNEP, Paris

UN Development Programme (UNDP) (2013) Human development report 2013. The rise of the South. UNDP, New York

UNDP (2014) Human development report 2014. Sustaining human progress: reducing vulnerabilities and building resilience. UNDP, New York

UN Environmental Programme (UNEP) (2011a) The other 70%. UNEP, Paris

UNEP (2011b) Taking steps toward marine and coastal ecosystem-based management—An introductory guide. UNEP, Paris

UNEP (2012a) Global environmental outlook, 5th edn. UNEP, Paris

UNEP (2012b) Sustainable, resource efficient cities—Making it happen! UNEP, Paris

UNEP (2014) Annual report highlights 2013 achievements. UNEP, Paris

UNESCO, IOC (2003) The integrated, strategic design plan for the coastal ocean observations module of the global ocean observing system. UNESCO/IOC, Paris

UNFPA (2011) State of world population 2011. People and possibilities in a world of 7 million. UNFPA, New York

UN-HABITAT (2011) State of the world's cities 2010–2011. Bridging the urban divide. UN-HABITAT, Nairobi

UN-HABITAT (2012) State of the world's cities 2012–2013. Prosperity for all. UN-HABITAT, Nairobi

United Nations (UN) (2012) Millennium development goals 2012 report. UN, New York

UN (2013) World population prospects: the 2012 revision. UN, New York

UN (2014a) World urbanization prospects. The 2013 revision. UN, New York

UN (2014b) The road to dignity by 2030: ending poverty, transforming all lives and protecting the planet. Synthesis report of the Secretary General on the post-2015 sustainable development agenda. UN, New York

World Business Council on Sustainable Development (WBCSD) (2010) Vision 2050. WBCSD, Geneva

World Bank (2012a) Toward a green, clean, and resilient world for all. A world bank group environment strategy 2012–2022. World Bank, Washington, DC

World Bank (2012b) Cities at the crossroads: unlocking the potential for green urban growth. World Bank, Washington, DC

World Economic Forum (WEF) (2014a) The global gender gap report 2014, 9th edn. WEF, Geneva

WEF (2014b) Connected world transforming travel, transportation and supply chains. WEF, Geneva

WEF (2015) The global risks report 2015, 10th edn. WEF, Geneva

Selected Internet Links

https://www.gov.uk/government/collections/future-of-cities
www.c40cities.org
www.deltacities.com
www.citymayors.com
www.csiro.au
www.eolss.net
www.lincolninst.edu
www.metropolis.org
www.millennium-project.org
www.oecd.org
www.oceansatlas.org
www.un.org/esa/sustdev
www.uncsd2012.org
www.unhabitat.org
www.undp.org
www.unep.org
www.unfpa.org
www.urbact.eu
www.whitehouse.gov/issues/urban_policy
www.wef.org
www.wff.org
http://wisdom.unu.edu/en/ageing-societies
www.worldwatchinstitute.org
www.zeri.org

Watercolour 2
New York, an Unrivalled World City

Chapter 2
Opportunities, Tensions and Risks for Coastal Cities

Abstract This chapter reviews possible future trends, opportunities, tensions and risks for coastal cities. Backbones of both sea and land dynamics and functions, marine and maritime cities can generate and enhance precious synergies among their diverse resources and interacting ecosystems and serve as bridges among the hinterland, and, through the maritime routes, faraway lands. To advance towards sustainable development, coastal cities have to enhance all aspects of their unique urban capital, natural and physical, on land and water, human, intellectual and social, cultural and political, financial and constructed, as strong sustainability requires all forms of urban capital to be preserved, strengthened and transmitted to future generations. Climate change, subsidence and excessive population growth and socio-economic change are important risks, especially for low-elevated zones, but coastal cities count on the very diverse assets which, if sustainably managed, can help them address the challenges of the future. The chapter invites the reader on a journey to urban shores of some of the world seas, highlights strengths and vulnerabilities, and a multitude of possible actions, promises and perils.

2.1 Tales of Sunken Atlantises and the Treasured Vulnerable Cities

Atlantis could serve as an allegory for all cities in their struggle for eternity. First mentioned in Plato's dialogue Critias (c. 360 BC), Atlantis, allegedly a naval power "in front of the Pillars of Hercules", had conquered many parts of Europe and Africa in the tenth millennium BC, and sank into the ocean "in a single day and night of misfortune". Some scholars argue that records of the Thira volcano eruption or the Trojan War might have influenced Plato. The legend was taken up by Humanists in utopian works, such as Bacon's New Atlantis and More's Utopia. In the United States, "Atlantis: The Antediluvian World", by I. Donnelly, published in 1882, suggested that Atlantis was the island where humanity first rose from a state of barbarism to civilisation and the true antediluvian world, a universal memory of a mighty land (Zangger 1993).

V.P. Mega, *Conscious Coastal Cities*, DOI 10.1007/978-3-319-20218-1_2

The discovery of the coveted Atlantis has been a dream for generations of archaeologists. Seas often conceal fascinating histories of ancient worlds drowned by disaster. Archaeologists trawl the seas for traces of submerged cities of universal significance. The seafloor is waiting for the identification of its many treasures. Drowned stone gods and sphinxes often provide a snapshot of ancient civilisation at its height, frozen in time. In 2013, a discovery off Egypt's coastline revealed captivating relics of Thonis-Heraklion, the combined Egyptian and Greek names of the city. Papyrological and epigraphical evidence suggest that the city, endowed with a temple to Heracles, was a wealthy, blooming city, the obligatory port of entry to Egypt before the foundation of the major trading post of Alexandria (Goddio 2013; Zangger 1993).

Alexandria was founded around 331 BC by Alexander the Great, who commissioned its planning to Dinocrates of Rhodes. Capital of Hellenistic, Roman and Byzantine Egypt for almost 1000 years, Alexandria prospered as the most glorious Mediterranean city under the last Pharaohs of the Ptolemaic Dynasty. The city, best known for its Lighthouse, one of the seven wonders of the ancient world, and the Great Library, the largest in ancient times, was also an ancient knowledge powerhouse.

Harbour cities were often built to be discovered from the sea. Strabo described the principal buildings of Alexandria as seen from a ship entering the Great Harbour during the first century BC. The Royal Palaces at the Northeast angle of the town have been engulfed by the sea. Next is the Great Theatre which was used by Julius Caesar as a fortress. Maritime activities along the seafront were represented by the Emporium (Exchange) and the Navalia (Docks). A bustling promenade hugged the crescent-shaped Eastern Harbour. Cultural activities were represented by the Temple of the Sea God, close to the theatre, and the Great Caesareum, which later became the Patriarchal Church. The city's outstanding treasure, the Lighthouse, lies beneath the waters, together with other awe-inspiring ruins, undisturbed for centuries.

Alexandria's supremacy declined in the fourth century AD, while earthquakes and tidal waves brought a final blow to the city. After a long eclipse, passionate explorers started dredging up the illustrious past, until a UNESCO survey confirmed findings and stirred up interest. It seems that a great part of the royal and civic quarters sank beneath the harbour due to a tidal wave and subsidence in 365 AD. Ongoing maritime archaeological activities have already disclosed unknown facets of the history of Alexandria. The city is witnessing the beginnings of a recreational diving industry, along with plans to exploit its submerged treasures for maritime tourism.

Thonis-Heraklion was the result of a chance discovery. In search of warships, French archaeologists stumbled upon ruins and artefacts remarkably preserved. Their unearthing divulged the beauty and glory of a mighty city, one of the great port cities of the world, 2300 years ago. Underwater ruins revealed 64 ships, 700 anchors, a treasure trove of gold coins, statues and a temple to the god Amun. The city was criss-crossed with a network of canals and its islands were home to small sanctuaries. The ancient landmarks of the city, such as the temple of Amun and the harbours that controlled trade into Egypt, came into light. Many of the finds illustrate the importance of the city to the ancient world economy, also witnessed by the huge amount of gold coins and bronze, lead and stone weights used to measure the value of goods. The importance of the city has also been demonstrated by the discovery of the ships, which is the largest number of ancient vessels ever found in a single place.

The discovery of Thonis-Heraklion is the last in a chain of achievements of underwater archaeology. In 2001, after years of searching, archaeologists came upon the site of the lost Classical Greek city of Helike, buried under silt on the Southwest shore of the Gulf of Corinth after an earthquake struck the region in 373 BC. Beyond the classical city, excavators have found the remains of an Early Bronze Age site, as well as evidence of a road that, during the Roman period, linked Helike with the cities of Patras and Corinth. Although preserved for more than 2000 years, the site, most of which has yet to be excavated, was threatened by a rail project and the World Monument Fund inscribed it in its watch list (World Monuments Fund 2006).

Other awe-striking cases include Pavlopetri, discovered in 1967, which claims the title of the oldest submerged town in the world. Nearly 3000 years ago, the Bronze Age port of the city located on the Southern coast of mainland Greece, slipped beneath the waves as a result of gradual erosion or a storm rise. Although eroded over the centuries, the layout of the city is in good condition. The site faces modern threats of damage by the tourism industry and especially boats dragging anchors. In 2009, underwater archaeologists produced a complete digital map of Pavlopetri's ruins, indicating streets, public buildings, residences, courtyards, and graves.

Pavlopetri provided a field for taking underwater archaeology into the twenty-first century. It is the first submerged town digitally surveyed in three dimensions with the help of sonar mapping techniques developed by military and oil prospecting organisations. The site has been surveyed by the Australian Centre for Field Robotics which developed several unique robots for this purpose. Research concluded that the town was the centre of a thriving textile industry, given the many unearthed loom weights. The jars and pots which were excavated also indicate a major trading port.

The chorus of sunken cities is long and captivating. Baiae, once an attractive seaside resort on the Bay of Naples, endowed with thermal springs and a pleasant climate, was abandoned during a malaria outbreak in the sixteenth century. Volcanic activity caused the ancient city's ground level to sink, plunging its ruins under water. In 2002, the underwater Archaeology Park of Baiae was instituted as a protected marine area. The principal objectives of the underwater park include the environmental and archaeological enhancement, dissemination of knowledge about marine biology and ancient heritage, the promotion of educational and scientific research programmes and endogenous socio-economic development.

Many more cities remain buried under water. In the Americas, the underwater city of Port Royal, a shipping hub in Southeast Jamaica, has a turbulent history. It was an affluent city, compared to the sin cities of Sodom and Gomorrah. The bustling city of Port Royal was stricken by a disaster of biblical proportions in 1692 and preserved in situ. Unlike most archaeological locations, Port Royal sank into the harbour in a matter of minutes, remaining perfectly preserved as it was on the day of the earthquake. Buildings and streets slid into the sea, creating one of the western hemisphere's most significant underwater archaeological sites.

Common essentials in all sunken cities include the accumulated wealth and the sudden collapse due to extreme physical phenomena, especially earthquakes and tsunamis. More recently, the world has learned lessons from its experiences after major losses caused by hurricanes like Andrew, in 1992, and Ivan, in 2004. In 2005, one of the most distressing recent cases, hurricane Katrina swept across New Orleans and the Gulf Coast region, devastating local communities and their remarkable heritage. The failure of the flood protection system left 80 % of New Orleans underwater and caused the mandatory evacuation of the city.

The history of New Orleans is intrinsically linked to struggles with water. Located at the confluence of the Mississippi river and the Gulf of Mexico, Greater New Orleans has always been shaped through the constant flows of people and resources in close interaction with the aquatic environment. Extensive levees were built to mitigate flood risk, and surrounding wetlands were drained to fight disease such as yellow fever and to prepare the ground for further urbanisation. As a consequence, peaty soils compressed, subsided and sank below sea level. The levees prevented sediment-rich waters of the Mississippi River from adequately replenishing the floodplains and wetlands, and intensive engineering has changed the course of vast quantities of water. This rapid deterioration of coastal wetlands has undermined the region's resilience to storm rise.

The disastrous flooding caused by hurricane Katrina in 2005 triggered mechanisms for the restoration of green infrastructure to protect against future storms and rising sea levels (UN Convention of Biological Diversity et al. 2012). Post-Katrina New Orleans faces ongoing urban and environmental risks from subsidence and flooding. Despite massive water pumping capacity, heavy downpours overwhelm drainage systems, causing frequent flooding. Increasing subsidence is mainly due to inadequate water management and excessive pumping methods, which dry out organic soils. The city also continues to grapple with issues resulting from the frequent interruption of economic activity due to flooding, and also the legacies of Katrina, such as unemployment, and a devalued property market, especially for the low-income population, which is constantly the worst hit by all disasters.

In 2010, a comprehensive, integrated and sustainable water management strategy was conceived, followed by the Greater New Orleans Urban Water Plan in 2013. This was the first regional urban water plan of its kind in the United States, devised to concretise a long-term vision for urban water management. The plan provides a roadmap for mitigating flood risk, limiting subsidence, and improving the quality of water. Rather than pumping away the water and thereby increasing subsidence, this approach slows down the water using bio-retention and infiltration techniques. This involves, for example, rain gardens and storage of water in the landscape. The plan includes three phases, comprising first the implementation of smart retrofits, interventions to reshape legacy infrastructure or ongoing projects to incorporate sustainable water management techniques (up to 2020), improving water flows and connectivity (2020–2030), and the longer-term diffusion of the strategies throughout Southeast Louisiana (Greater New Orleans 2013).

In 2012, hurricane Sandy, the largest hurricane to ever form in the Atlantic Basin, hit New York. Water levels at Battery Park rose to a record 13.88 ft, surpassing the previous record of 10.02 ft more than 50 years earlier. Wave heights reached a record of 32.5 ft in New York harbour. During Sandy, numerous cities such as Atlantic City, Baltimore and Philadelphia set records for the all-time low pressure ever recorded. About 85 % of Atlantic City streets were under water at high tide. For the first time since 1888, the New York Stock Exchange was forced to close for two consecutive days due to weather. The US Department of Energy stated that more than 8.5 million customers reported blackouts in 21 states (WRI 2012).

Some places, sadly, have become synonymous with disaster risk; now, they strive to become related to humanity's efforts to overcome and address risk. Kobe, the capital of Japan's Hyogo Prefecture, in the wake of the devastating 2004 Indian Ocean tsunami which claimed more than 227,000 human lives, hosted the conference which crafted the Hyogo Framework for Action (HFA), designed to reduce losses in lives and assets. In March 2015, UN Member States will gather for the third World Conference for Disaster Risk Reduction in another Japanese city, Sendai, the heart of the region that bore the brunt of the 2011 earthquake and tsunami that led to the Fukushima nuclear disaster. Sendai must seal a revised HFA, to scale up efforts for a safer world, effectively addressing the risks posed by rapid population growth, exploding urbanisation, global warming and rising sea levels.

In the last 50 years, disasters caused by weather, climate, and water-related hazards have led to more than 3.5 million deaths. In Asia, where 80 % of the world's disasters are concentrated, the number of citizens directly affected has dropped by almost one billion, thanks to measures like the Indian Ocean tsunami early-warning system. China has worked hard to keep economic losses below 1.5 % of GDP, while the Philippines and India managed to save thousands of lives thanks to timely evacuations. The third edition of the United Nations Global Assessment Report on Disaster Risk Reduction explores disaster risks as a growing problem for the economic and business communities. The report seeks to engage businesses in a dialogue on disaster risk management beyond the emphasis on response and preparedness and suggests opportunities for the creation of shared value for business and society. It offers businesses as well as investors a review of practices that can assist them in reducing their risk of disaster loss (UNISDR 2013).

2.2 Port and Maritime Cities: Unique Interfaces Between the Land and the Seas

Maritimisation was the first form of globalisation. The ancient myths of the Golden Fleece and the Trojan War testify to the importance of maritime routes in Antiquity. The Byzantine Empire gave new importance to the sea, while the Age of Discovery celebrated the first Portuguese breakthroughs in the Atlantic, the discovery of America in 1492, the ocean route to the East in 1498, and various European naval expeditions across the Atlantic and later the Pacific, which continued until the eighteenth century.

Global maritime powers can become pioneers in moving towards sustainable development. Prominent cities in shipping, maritime finance, law and insurance, may bring together leading actors of the blue economy. Value chains often develop in networks with headquarters in financial world centres, operating units close to the markets and scientific research and development centres located in knowledge hubs (Menon 2012).

Thriving maritime cities are economic hubs, bringing together very diverse ecosystems. Globalisation brought market integration, strong growth in international trade, direct foreign investments, the emergence of transnational companies and a dramatic increase in the interdependence between world citizens. Multinational companies operate across the entire world, taking advantage of economic differences by locating their business activities in the most attractive locations.

The taxonomy of world ports is also very dynamic given the leaps in containerisation and the rise of Chinese ports. During the years 1990–2010, maritime transport by intermodal container increased dramatically. Singapore and Shanghai succeeded Rotterdam as the primary ports on the planet. In Europe, Rotterdam is the largest seaport and a huge industrial complex. London is a strong maritime hub with robust legal, insurance and financial services, while Oslo has developed a solid foundation for offshore industry development. In the Americas, Rio de Janeiro is expected to show the sharpest growth in the future (Menon 2012).

Maritimisation is an essential dimension of Singapore, a comprehensive maritime hub, reputed for its quality ship registry, which has increased steadily and is ranked among the world's top ten largest ship registries. A government strategy promoted shipping through the integration of the maritime and port authority with the tax system, investment, and business creation. As an integrated hub involving shipyards, maritime industries, finance, legal and insurance services, and commercial operations, next to the rich range of urban services and activities, the city is a global innovator offering multiple coastal and offshore marine services.

Hong Kong is a leading maritime cluster and the gateway to the world's manufacturing heart, the Pearl River Delta, which accounted for more than one fourth of Chinese exports in 2011. The city is trying to keep its current position in line with the momentum in Asia's economic development and seaborne trade. Ranked as the third-largest container port in the world in 2011, Hong Kong is in close proximation to the fourth and seventh container ports, Shenzhen and Guangzhou, respectively, its main competitors. Driving factors of success for the Hong Kong cluster include regulation and taxation favourable to global business, an active maritime community and incentives to attract and train a skilled labour force. A major challenge to the port relates to hinterland accessibility and related costs. Quality of life and the environment is an important element.

Given its restricted land resources, Hong Kong has become one of the ports with the highest land productivity in the world. It is also a regional leader in green port policies. A voluntary fuel switch programme, the Fair Winds Charter, was initiated by the maritime industry in 2010. The Hong Kong ship owners association and the Hong Kong liner shipping association had for many years been involved in the reduction of emissions from shipping, both in global negotiations and in local voluntary efforts. The Fair Winds Charter was developed as the world's first genuine voluntary scheme to reduce shipping emissions at berth and at anchor. In 2012, the charter was partially supported by the government with a 3-year incentive scheme. The goal is to create an Emissions Control Area for the Pearl River Delta. The scheme introduced environmentally differentiated port dues providing advantages for ships switching to low-sulphur fuel, prior to the introduction of legislation for all ocean-going vessels berthing in Hong Kong to switch to low-sulphur fuel (OECD 2013).

Long-term integrated models are necessary for sustainable port policy options. These not only have to be economically and environmentally beneficial, but also must respond to public opinions and societal preferences. Creating processes that are inclusive of the diversity and aspirations of coastal cities is arguably the most challenging issue for a sustainable maritime city. With a shared vision and conscious navigation, cities can face diversity and complexity, and ultimately move towards more sustainable marine and maritime futures.

Maritime cities are strong links in the world dynamic chains. Global forces such as containerisation have an impact as important as local forces, such as the evolving relationships between urban policy and port growth and also the symbiosis of the port with the regional and national economy. The business environment in which a port carries out its operation is increasingly reflected in port competition and cooperation, at regional and global scales.

Some ports tell a story representative of competitive struggles for first place. In Europe, Hamburg's port performance has been successful over recent decades, but the global crisis dealt a major blow. Since the 1980s, its market share among north-western European ports has increased significantly, particularly in container traffic which doubled from 13 % in 1980 to 26 % in 2005, making Hamburg the second busiest container port in Europe. Following the crisis, market shares for Hamburg suffered a more significant drop than those of other European ports and the average annual growth over 2001–2010 was 2.9 %, falling behind its main competitors Rotterdam (3.7 %) and Antwerp (4.4 %).

Hamburg has important positive economic impacts. The port-related activities represent a considerable share of metropolitan employment and value added. Moreover, the indirect economic effects are significant. It has been estimated that each additional euro of demand in the port of Hamburg leads to 0.71 € of additional spending in other sectors, a multiplier which is higher than the multipliers estimated for Rotterdam and Antwerp. These indirect effects are particularly high in the transport equipment sector, and the food and petro-chemical industries (OECD 2013).

Most of these positive impacts are felt on a broader regional scale. Overall, only 13 % of the multiplier effects have an impact on Hamburg and its neighbouring states, but almost a third of the multiplier effects spill over to the two large southern states of Germany and more than half to the rest of Germany. Hamburg also has a key role for the import and export of goods for large parts of Germany and for central Europe, especially Hungary and the Czech Republic. In addition, Hamburg is the prime port for transhipment and short sea shipping in the Baltic Sea.

In the Mediterranean Sea, Marseille was the second largest European port in the 1970s before going through a declining phase of stagnating port volumes to become fifth largest port. A port reform was enacted in 2011 to attract new market shares. As a regional maritime services centre, Marseille has, however, significant economic influence, and generates around 40–45,000 port-related jobs. Long shielded from competition due to its quasi-monopolistic position, the port city is in fierce competition with Le Havre and Antwerp (OECD 2013).

The port authority has formulated an ambitious target of handling five million containers in 2030. In order to reach this objective, several of the challenges related to hinterland, maritime connections and port efficiency need to be successfully addressed, and environmental impacts have to be mitigated and prevented to ensure better working and living conditions. The maritime heritage of Marseille could be an asset in the city's quest to develop into a European maritime services centre, improving synergies with other port cities.

New offshore activities are being promoted in the region, such as the experimental project Mistral, expected to host two floating wind turbines. The vertical-axis floating wind turbines promise a reduced cost and visual impact. The Nenuphar project in France develops prototypes for 2 MW floating vertical-axis turbines. Prospects will expand as wind energy gains ground and offshore wind moves into the mainstream. The Mistral project is, however, at the crossroads of marine protected areas, and a moderate environmental impact has been identified, especially for birds risking collision when attracted by the lights of the offshore sites.

2.3 The City and the Sea at the Twenty-First Century: A Symbiosis at Risk?

The future of the world's coastal cities and the threats originating from the sea are of particular concern to human civilisation. Advancing climate change holds more threats for coastal cities, as predictions suggest a global rise of 52–98 cm by the year 2100 in the case of high emissions, which would threaten the survival of coastal cities and entire island nations. But even with aggressive emission reductions, sea rise could reach 28–61 cm. Under this highly optimistic scenario, impacts on many coastal areas could still be serious, especially in relation to coastal erosion and a greatly increased risk of flooding. By the end of the century, densely populated coastal areas around the world could be underwater. The risk of increasing floods creates a major political and institutional challenge for coastal cities and can threaten their symbiosis with the sea. Ambitious and proactive actions at the local level are needed to help prevent large-scale flood disasters (IPCC 2014a).

Coastal areas experience land loss from erosion and sea level rise. The International Panel of Climate Change suggests that, over the period 1901–2010, global mean sea level rose by 0.19 % [0.17–0.21 %] m. The rate of sea-level rise since the mid-nineteenth century has been larger than the mean rate during the previous two millennia. Ocean warming dominates the increase in energy stored in the climate system, accounting for more than 90 % of the energy accumulated between 1971 and 2010 (high confidence), with only about 1 % stored in the atmosphere. On a global scale, ocean warming is largest near the surface, and the upper 75 m warmed by 0.11 % [0.09–0.13 %] °C per decade over the period 1971–2010. It is virtually certain that the upper ocean (0–700 m) warmed from 1971 to 2010, and it likely warmed during the period 1870–1971. Effects could alter landscapes, reduce public access and recreational opportunities, diminish natural habitats, and harm species that depend on these habitats for shelter and food. Mitigating shore erosion requires coastal planning and control that help preserve the natural features of coastal shorelines (IPCC 2014b).

Local subsidence amplifies the effect of global sea level rise on local water levels. Subsidence is accelerated by human activities, like water extraction, and aggravates intrinsic vulnerability. Vulnerable cities need a full protection system, which represents a major investment. In times of scarce resources, the mobilisation of such resources could be a real political and institutional challenge for local authorities, especially in the developing world.

Already in 2007, an OECD study on urban vulnerability revealed that flood risks could significantly influence the future prospects of coastal cities. Present urbanisation and flood defence patterns have been designed for past conditions, and even a moderate change in sea level is sufficient to make them inadequate, thus magnifying flood losses to catastrophic levels (OECD 2007).

Climate change, subsidence and excessive population growth and urbanisation are the main threats to coastal cities. The OECD screening was intended to produce an initial estimate of the exposure of the world's large port cities to coastal flooding due to sea rise, storm surge, and high winds. The study also investigates the impact of climate change on the exposure to coastal flooding by the 2070s, alongside subsidence and urbanisation, focussing on the 136 port cities around the world that had more than one million inhabitants. Most of these largest port cities are found in Asia (38 %), and many of them (27 %) are located in deltaic locations. The latter tend to represent a higher coastal flood risk as a result of their lower locations and experience significant subsidence.

Across all cities, about 40 million urban residents, almost 0.6 % of the global population, were exposed to a one in 100-year coastal flood event. The exposure of population and assets to such events depends greatly on their particular topographical and morphological conditions. The exposure is concentrated in a small number of cities. The ten cities with highest population exposure account for almost half the total exposure and the top 30 cities for about 80 % of the global exposure. Of these 30 cities, 19 are located in deltas.

In terms of exposed population, the list of the proposed most vulnerable cities included Mumbai, Guangzhou, Shanghai, Miami, Ho Chi Minh City, Kolkata, Greater New York, Osaka-Kobe, Alexandria and New Orleans, on the basis of 2005 data. The ten cities with highest population exposure are almost equally split between developed and emerging countries. Alexandria is the only city on the most vulnerable list outside Asia and North America. This taxonomy is, however, highly variable depending on the size of cities, the magnitude and nature of the extreme phenomena and the adopted policies (OECD 2007).

In terms of exposed assets, the list of the most vulnerable cities includes, at the top, Miami, Greater New York and New Orleans, followed by Osaka-Kobe, Tokyo, Amsterdam, Rotterdam, Nagoya, Tampa-St Petersburg and Virginia Beach. These cities contain 60 % of the total exposed assets, but belong to only three countries, the USA, Japan and the Netherlands. The total value of assets exposed in 2005 was estimated to equal around 5 % of the global GDP. The situation is very dynamic and one can expect changes to have already occurred, especially with the rise of many ports on the Chinese shores.

By the 2070s, the total exposed population could grow more than threefold to around 150 million people due to the combined effects of sea-level rise and increased storm surge, subsidence, population growth, urbanisation and socio-economic change. The total asset exposure could grow even more dramatically, more than ten times the current levels, and rise to roughly 9 % of projected annual GDP by 2070.

Symbiosis with the sea can be challenging and has to be carefully planned and monitored. Risk management policies, including upgraded defence engineering, could mitigate the impacts of rising sea levels. Coastal defensive structures should make the best possible use of the rich protection already generously offered by nature. Marshes, sea grass beds, coastal forests, and coral reefs provide natural buffers that absorb the energy from waves and storms, making it easier to protect cities. In addition, artificial constructions could also be required to provide full protection, such as barriers to manage safe access to ports.

Cities can keep their flood risk under control by better managing urban development. Smart coastal protection infrastructure has to be integrated into land-use planning and policy. Emerging countries have a real window of opportunity, as urbanisation is ongoing at a very fast pace and infrastructure has not yet trapped the future into unsustainable schemes. Smart urban policies, based on disaster risk management and governance, could drive new developments toward safer places and prevent the urbanisation of the most vulnerable and costliest to protect areas. Weak institutions, lack of enforcement of building norms, and extension of informal settlements are strong obstacles.

Public awareness is extremely important, as vulnerable cities have to be prepared for a broad range of measures, from individual protection to large-scale evacuation, especially in the most dangerous places. They also need concerted contingency plans to manage crisis and emergency services and recovery and reconstruction schemes, well-regulated insurance systems, and post-disaster measures. Forging awareness, well before extreme events happen, is primordial and the participation of all stakeholders is of the highest importance.

Preventing urban disasters from the sea and taking appropriate measures is vital. In sheer economic terms, the stakes of sea-level rise in urban areas are particularly high. An added 0.5 m of ocean water by the year 2050 would put $8 trillion in assets at risk in the world's 136 major port cities. On the Northeastern coast of the United States, the expected maximum rise in sea level would threaten five cities, including Baltimore, Boston, New York, Philadelphia, and Providence. Without immediate climate action, sea level rise on the East Coast, the shift to an arid climate in California or the dieback of the Amazon rainforest due to increasing drought, are likely to affect hundreds millions of citizens and cost hundreds of billions of dollars (WWF and Allianz 2009).

Cities and their stakeholders, investors and citizens must be prepared for climate change as long as there still is leeway for action. This allows them to work together to develop specific concepts, whether for existing assets or for future projects like renewable energy and water supply concepts. Better models could help citizens understand the frequency and strength of natural disasters. The human component is playing an ever-increasing role in reducing the risk from natural disasters, in terms of both risk management and addressing the anthropogenic causes of climate change.

Coastal and sea defence planning and flood risk governance are the main domains of decisive action. In the past, engineering provided the preferred protection solutions, including dams, dikes, and seawalls, to the prospect of encroaching seas. The Thames Barrier, a 570-yard-wide floating span of rotating gates, is a good example. It permits navigation on the river while protecting the city against storm surges and high tides. But responses to sea-level rise increasingly encompass soft engineering solutions such as green infrastructure and wetland restoration. They could even include measured human retreat from coastlines, allowing rising seas to follow their way (Arup 2014; Wilson 2013).

The Netherlands offer the world's gold standard out of their experience in fighting with sea-level rise. Dutch cities long had to address the challenge of being below sea level. Between 1950 and 1997, the low-lying country constructed vast delta works, including 250 miles of dams and other barriers. Over the years, cities developed integrated, flexible strategies to manage permanent symbiosis with the sea. The country rediscovered the compatibility of hard and soft solutions. In 2007, a new Delta Committee, the second in five decades, recommended raising the flood protection levels of coastal dikes by a factor of 10. The committee advocated two new strategies that rely on sustainability principles instead of engineering. The first, "Building with nature", involves beach nourishment as well as restored natural estuaries and tides to prevent flooding. The second, "A room for the river", asks for strategies requiring adequate administrative, legal and financial frameworks.

The city of Rotterdam, largely below sea level, has built up centuries of expertise in all aspects of the port and its related environment issues. An historic review of the port of Rotterdam encapsulates much of the essence, strengths and weaknesses of major ports. Strategically located at the Rhine-Meuse-Scheldt delta on the North Sea and at the heart of a dense rail, air, road and inland waterways network, Rotterdam is a true gateway to Europe and was the largest port of the world from 1962 to 2004. Its vision for 2030 is exemplary (Port of Rotterdam 2012).

The port of Rotterdam is located near the New Meuse, a channel in the delta formed by the Rhine and Meuse on the North Sea. It stretches over a distance of 40 km and includes the historic city harbour dating from the Middle Age. As the city of Rotterdam developed from a small town into a major harbour city in the fourteenth century, the port continuously expanded. Since the nineteenth century, links between Rotterdam and the North Sea were poor, with a large estuary/delta area and many small waterways. To improve the connection to the North Sea, the "New Waterway" was constructed to connect the Rhine and Meuse rivers to the sea in 1872. Over the years, the port was further developed seaward through docks and harbour-basins. In the first half of the twentieth century, the port activities moved towards the North Sea.

The port of Rotterdam and its surrounding area may have to face a storm surge from the North Sea, a risk largely underestimated until the flood of 1953, which killed almost 2000 citizens and caused widespread damage. The flood acted as a clarion call and a Delta committee was assigned to draw up a plan to protect the areas from future flooding and provide clean drinking water. In 1959, the Delta Law was approved for the implementation of the Plan and the construction of the Delta Works, a chain of flood protection structures within and around the Rhine-Meuse-Scheldt delta from North Sea floods.

The aim was to shorten the coastline and convert it into fresh water lakes. The project helped in draining the low areas which flood frequently and regulated the flow of saltwater from the sea, providing both drinkable water and fresh water for irrigation. The infrastructure therefore provides multiple benefits, including flood protection, drinking water and irrigation. The risk of flooding was reduced to one in 4000 years. The project also improved recreational facilities and generated employment.

The Delta Works flood protection plan considered various options before a unique design was built, the Maeslantkering. This flood barrier consists of two huge doors that normally rest in a dry dock beside the New Waterway. When a flood 3 m above sea level is predicted, the gates are floated into position and sunk. When the water level recedes enough to open the gates, the latter are floated back into their docks.

The Rotterdam harbour functions as an important transit point for transport of bulk and other goods between the European continent and other parts of the world. Large oil refineries are located west of the city. The harbour territory has been enlarged by the construction of the Europoort, its Gate to Europe complex along the mouth of the New Waterway, including the industrial park Maasvlakte. During recent years, the harbour skyline has been enriched with wind turbines, taking advantage of the exposed coastal conditions.

The extension of the port of Rotterdam to the west continued, after years of discussions, protests and planning. The infrastructure has been laid on reclaimed land, and industry, shipping, sea defences and transport places have been created as part of the Main Port Development Rotterdam project. The project also includes some ecological and recreational areas to upgrade living conditions in the region in harmony with the port.

On the other side of the Atlantic, the city of Boston, a great part of which is built on landfill, is confronting a range of policies, from hard engineering to design adaptation and green infrastructure. The city has a long history of design and engineering flood defence projects, which bears witness to the city's vulnerability. In 1978,

23 years after the wettest tropical cyclone in the history of the Northeast, hurricane Diane, flooded New England in 1955, Boston's then-new Charles River dam was barely in place when the city was hit by a 3-day blizzard. The resulting rise in tide came within 0.4 m of overtopping the dam, which was designed to protect against a storm surge of 3.8 m above average sea level. In the absence of the Charles River dam, the dramatic floods of 2010, which inundated parts of Rhode Island, could have been disastrous for Boston.

There is pressure to act, since any solution to sea-level rise could take a long time to plan and build, and urgency has intensified. A forum sponsored by the Boston Harbour Association considered various future scenarios in which Boston is hit by exceptionally strong storm surges. The city can expect the equivalent of previously once-a-lifetime storms every 2–4 years by 2050. The forum considered a range of responses, from flood-proofed buildings and retention ponds to a massive sea barrier that would protect the inner harbour.

Flood scenarios suggest that a sea-level rise of 2 m in combination with a 100-year storm that produces a surge of 5.5 m above mean sea level is possible. Even a modest rise in sea level by mid-century, combined, for instance, with a powerful storm, a high tide, and strong winds, would partially submerge Boston. Flooding from storm surges would be accelerated by modern infrastructure like sewers and subterranean passageways, which channel water into the web of the city. Boston is investing in two steel triangles as high as skyscrapers that would pivot as they swing out into the channel between Castle Island in South Boston and Logan Airport in East Boston. Once pontoons are filling with water, the barrier would sink and then lock into a submerged concrete structure.

A balanced blend of hard and soft engineering responses would be most appropriate for many cities like Boston, which has to reinforce the ageing protections for the Charles River basin. The engineering culture that shaped and equipped Boston is expanding to green infrastructure, which could help build resilience to rising sea-levels. A re-invented Boston Harbour would have improved seawalls, climate parks, retention areas, and green areas that recapture the nineteenth century vision of an emerald necklace surrounding the city. Suggested works include the renewal of the last half-mile of the Charles River through a wetland that extends the interface between saltwater and freshwater and the reconception of Paul Revere Park, an open space at the crossing of the Charles River with the inner harbour (Wilson 2013).

Nature can be a great help in reconciling cities with the sea. It offers protection which is usually less costly, intrusive and perturbing than engineering works and can effectively and efficiently shield cities and the surrounding ecosystems. The use of green infrastructure and the enhancement of natural processes should become an essential systematic part of flood planning. Many cities could benefit from creating urban retention areas for flood waters and transforming unused assets into protective systems. Transport infrastructure, abandoned, or underutilised, can also be enhanced and complement green infrastructure for flood control (Arup 2014).

Emerging megacities face far greater challenges. Rising seas might trigger major disasters that go far beyond ecology and engineering, especially on intrinsically vulnerable short-term equipped urbanised coastlines that are inadequately prepared for widespread inundations. Vulnerability to climate change depends on natural and anthropogenic factors and the policies adopted to protect the concentration of people and assets. Dhaka and Phnom Penh top the rankings of those cities having the lowest adaptive capacity to climate change. Both are capital cities of Least Developed Countries and have grown very rapidly. In Bangladesh, Dhaka, with an estimated population of more than 15 million, setting at sea level in the broad and flat heart of the Bengal delta, represents a case of extreme vulnerability. Known as the City of Mosques, Dhaka is one of the world's most densely populated cities, its population having doubled in just over a decade. Both physical morphology and lack of adequate infrastructure and policies have made the megacity very weak in regards to extreme phenomena. Dhaka is sinking due to ground subsidence and sea-level rise (WWF 2009).

The IPCC projects that salt-water intrusion could reach up to 100 km from the coast and this could influence the availability and quality of freshwater and rice production and affect the natural ecosystems, such as mangroves, that currently safeguard the coast from storm surges. The estimated 10,000 km^2 of mangroves within the Ganges-Brahmaputra delta currently limit the effects of sea-level rise and storm surges, but massive storm surges have already occurred. In the past, tropical cyclone-induced 6 m storm surges have led to 500,000 deaths in 1970 and 150,000 deaths in 1991. Future losses could be even more serious without the current mangrove forests to buffer the effects (WWF 2009).

Dhaka, as with all cities which are low-lying and subject to violent storms, needs to stop building on its waterbodies and wetlands and take better care of its water resources. For Kolkata too, better water management is crucial for minimising the effects of salinity and over-extraction of groundwater. The city needs increased protection, enforcement and restoration of mangroves and wetlands to limit the impact of saltwater intrusion and storm surges. Cities may also have to consider investing in protection versus relocation. The retreat from coastal cities, as happened in New Orleans after Katrina, is an extreme case and policy option, and a powerful inducement for cities to study, imagine and plan alternative options that can be discussed with stakeholders.

Kolkata has a high overall vulnerability. Being situated on the banks of the Hooghly River, and within the Ganges delta, Kolkata is only metres above current sea levels. As it expanded, it reclaimed significant amounts of surrounding wetland. The city is also within a considerable seismic zone and thus prone to earthquakes. Sea level rises, accompanied by ground subsidence, seem to be the most severe threats. Due to the combined effects of sea level rise and subsidence, the saltwater could intrude upon the Ganges delta 100 km from the coast, seriously impacting groundwater supplies (WWF 2009).

Resilience is a key issue for coastal cities that wish to run the race of sustainability and manage their symbiosis with the sea responsibly. Government, public authorities, regulators, investors, businesses and civil society should cooperate at the earliest possible stage to collectively address a maelstrom of legal, social, economic, environmental and other issues related to rising sea levels. Tokyo has much to teach. The Tokyo Metropolitan Area, composed of the capital city and the surrounding three prefectures, enjoys one of the highest levels of well-being. Transport and telecommunications infrastructures, notably a most modern teleport, made this global metropolis highly liveable for a huge number of residents and businesses. Tokyo had to work very hard to strengthen its resilience in the face of severe disasters like the Great Kanto or the Kobe earthquakes and more frequent typhoons and floods.

The Tokyo Metropolitan Government bears witness to storm surge and flood control measures. As complete expansion is difficult for rivers flowing through highly urbanised areas, special situational measures were often taken to quickly ensure safety against flooding. Such measures include regulating reservoirs to store part of the flood water or diversion channels to funnel excess water. The Kanda River provides a good example with its Loop Road No. 7 Underground Regulating Reservoir, which prevented important flood damage when Typhoon Melor struck Tokyo in October 2009, using 94 % of its storage capacity.

The eastern lowlands of the Tokyo Metropolitan Area are composed of weak ground, and most regions are below sea level, due to land subsidence caused by post-Meiji period groundwater pumping. These areas are especially susceptible to damage caused by natural disasters such as storm surges and earthquakes. These lowlands are being bolstered against flood damage through river improvement projects, the representative example being the Super Levee Project. A super levee is an especially wide embankment built in cooperation with riverside urban redevelopment projects. It assists in effective land use, the strengthening of earthquake countermeasures, the development and maintenance of a healthy river and the improvement of the urban environment.

Extreme phenomena like earthquakes and tsunami in Japan illustrate the vulnerability of urban coastal regions, ports and nuclear structures to coastal floods. The disaster of 11 March 2011 caused by the earthquake in East Japan, and the subsequent tsunami and damage to the Fukushima nuclear plant, also had an effect on Tokyo, 400 km away from the epicentre. Risk-management policies, refined over the past two decades, and the improvements brought to the seismic performance of buildings resulted in very few fatalities and limited damage. The city prevented fires, thanks to the "my-com-meter" system, introduced post-1995, shutting down the residential gas supply at the first sign of an earthquake. Transport infrastructures also resisted well after considerable reinforcement in previous years. A European Research Council project is expected to build a unique tsunami generator for testing and quantifying the effects of tsunami waves on built environments and the performance of coastal defences.

The XXIV Congress of the International Union of Architects (IUA) held in Tokyo in September 2011, six and a half months after the historical disasters of March 2011, brought together experts, professionals and students in architectural and urban fields, from more than 100 nations and territories around the world. Deeply affected by the devastating events, participants debated the forward and back casting visions of architectural and urban design for 2050. The IUA Tokyo declaration expressed a commitment to learn from overcoming disasters around the globe to address the challenges and progress towards economic, social and environmental sustainability. The declaration is also an invitation to exchange the many diverse global and local initiatives for equity and solidarity, and promote responsibility beyond the limits of countries and cultures (IUA 2011).

2.4 Coastal Cities, Hives of Possibility for Sustainable Development

Coastal cities have been shaped by human genius enhancing terrestrial and marine assets. The proximity of the sea enriches the concentration and density of diverse people and resources. Each coastal city is a public good, both as an organism living by the sea and a technical construction, including a maritime space. Coastal urban population growth and excessive consumption exert huge pressure on global assets such as soil, water, food, natural resources and raw materials. Furthermore, most cities are located in regions exposed to at least one major risk of natural disaster, especially in Asia, Latin America, the Caribbean, and North America. Each coastal city is unique and has its own developmental trajectory, but they all seem to be increasingly interconnected and interdependent by multiple links and networks, including maritime routes. Citizens throughout the emerging world increasingly ask to be part of the interconnected world and this may be decisive for sustainable development.

The coastal and maritime cities tell multiple inspiring tales of effort and grandeur, neglect and decline. In the Greek mainland, the history of Piraeus, the port of Athens since Archaic times, is a succession of glorious moments and some intervals of waning. Themistocles proposed the fortifications of Piraeus and encouraged the Athenians to take advantage of the potential of the strategic harbour. In 483 BC, the Athenian fleet left the older harbour and was transferred to Piraeus, distinguishing itself at the battle of Salamis between the Greek city states and the Persians in 480 BC. Themistocles initiated the construction of the new port, turning Piraeus into a great military and commercial harbour, which served as the permanent navy base for the mighty Athenian fleet. The port was only occasionally used by the Byzantine fleet and the city was mostly deserted throughout the Ottoman occupation of Greece. Now, in the third millennium, Piraeus is the largest Greek seaport, the largest passenger port in Europe and one of the largest passenger ports in the world. The port is also a major employer and hosts many maritime service clusters. Athens is trying to improve its organic link to the sea, especially in redesigning the connecting artery of avenue Sygrou.

If Piraeus flourished as Athens's gateway to the sea, Tianjin, which has just entered the league of megacities as the shipping gateway to Beijing, is a huge port and manufacturing centre on the Bohai Gulf, strategically located on the Grand Canal linking the Yangtze and Yellow Rivers. Southeast of central Tianjin, the Tanggu New Area Master Plan redevelops an industrial zone, once the old port of Beijing, into a new commercial hub. It features a mixed-use district of high-tech buildings, historic neighbourhoods, an integrated road and rail system and open green and water spaces. A high-speed train linking Northeastern China's major cities is creating an accessible network from the capital to the sea.

A panoramic journey to the seas of the European Union can highlight, in a nutshell, some critical challenges for coastal urban sustainability. One should start from the Arctic, a unique commons case, yet emblematic of regions in rapid transformational phases, at the centre of a complex set of political, socio-economic and environmental dynamics linking actors within and outside the region. The USA (Alaska), Canada, Iceland, Denmark, Norway, Sweden, Finland and Russia all encircle the Arctic. The region is not conflict-free. The Northwest Passage is an international strait open for navigation, however Canada considers it to be within its jurisdiction. There are various local, national, regional and international claims regarding identity, stewardship and sovereignty in respect to the territories and resources of the region. If the aim for the Arctic is to be stable and peaceful over the long term, shaping its political and security arrangements will be critical and committed world cities could use their power to achieve the best possible future for the region (EMB 2014).

In the past 100 years, average Arctic temperatures have increased at almost twice the average global rate. According to the recent IPCC reports, the ice level of the Arctic attained its lowest levels since the first satellite observation in 1978. This is an unprecedented trend, and unless CO_2 emissions are reduced, navigation through the Arctic Ocean will be possible every year in summer before the middle of the century. Since 2005, the number of fishing boats in the Arctic has multiplied by 7. The Arctic, accessible to cargos at each end of the summer since 2007, saw the number of cargos passing by growing more than exponentially, from 4 in 2010 to 71 in 2013. The Arctic accounts for one fourth of the world's reserve of hydrocarbons, and 80 % of its reserves are offshore. The very low rate at which bacteria can break down spilled oil in polar conditions and the general low recovery rate of polar ecosystems means that damage from such pollution would be very serious.

With the Arctic being perceived as a potential new market by the shipping industry, associated activities like maritime trade, tourism and transport are likely to emerge faster than the necessary infrastructures for safe, secure and reliable shipping in the Arctic Ocean. Multi-level interactions of regional actors, states and multilateral organisations shaping political and security developments are crucial. The interests of Arctic and non-Arctic communities and states, the so-called Arctic "insiders" and "outsiders", are difficult to be integrated and have major implications for existing regional governance and security frameworks, notably those in Europe. The European Marine Board highlighted the need for a strategic plan for the Arctic Ocean, including data collection to support ecosystem-based management (EMB 2014).

The North Sea is a semi-enclosed sea and an increasingly fragile ecosystem, mainly due to the intensity of human activities. Since the 1970s, the North Sea has been Europe's main offshore oil extraction site. Its coasts are among the most heavily populated areas and host some of the continent's principal agglomerations, including the three largest ports, Rotterdam, Antwerp and Hamburg.

The North Sea is one of the most important European maritime services areas. The decision-making centres are mainly located in London, Hamburg, Oslo and Rotterdam. The region has numerous shipbuilding yards, mainly in Germany, the European leader, and the Netherlands. The North Sea also hosts an active fishery and aquaculture industry and many maritime service providers.

A heavily exploited marine space, the North Sea faces numerous environmental problems. The main issue is the eutrophication of coastal waters, caused by nutrients carried by the many rivers that flow into it. Other problems relate to chemical pollutants, the illegal emptying of oil tankers' fuel tanks, and overfishing. Climate change has an impact on coastal erosion and requires investment in infrastructure. There are several protected marine areas on the British coastline and along the Dutch, German and Danish coasts.

Another fragile sea is the Baltic Sea, one of the world's largest brackish water bodies, a unique inland shallow sea, partially covered by winter ice and surrounded by eight EU Member States and Russia. Its waters take a long time to be renewed. As a result, the marine environment is very vulnerable, particularly to eutrophication provoked by nutrients in urban waste water, coastal agriculture, and industrial pollution. The Baltic environment has been subject to many pressures, both natural and anthropogenic, and its exceptional biodiversity has been further affected by climate change, non-endemic alien organisms, fishing, maritime traffic, offshore activities, and increasingly expansive coastal populations and activities, including intense agricultural practices. All these pressures seriously compromise the capacity of the Baltic Sea to provide the sustainable goods and ecosystem services upon which the region directly depends.

The Baltic shores host five capitals of Member States of the European Union and various active commercial ports. The Sea is a supply route for oil, coal and natural gas for the European Union, delivered from the Russian ports of Primorsk and Ust-Luga. The Baltic has few large ports, apart from Primorsk and Saint Petersburg, but a number of medium-sized ports, including Stockholm, Copenhagen, Helsinki, Gdansk-Sopot-Gdynia, Riga and Tallinn. Half of the inter-Baltic trade takes place by sea, while external maritime traffic moves through the Danish straits and the Kiel Canal, the world's busiest waterway in terms of small and medium-sized container traffic.

On the Atlantic side, the EU includes the Bay of Biscay and the Iberian Coast, stretching from southern Brittany to the Strait of Gibraltar. The region is the cradle of Europe's maritime power, since it is from the Portuguese and Spanish coasts that intrepid explorers commenced their great expeditions. Major ports include Bilbao, Porto and Lisbon, the city which saw bold ships set sail in the golden age of discovery, at the crossroads of the seas that influenced European civilisation. A number of car ferry routes connect the Spanish coast with British and French ports. The Bilbao-Portsmouth and Santander-Plymouth routes transport between 140,000 and 150,000 passengers a year each. Cruises represent an important activity in Lisbon. Maritime traffic is very heavy, particularly on the route that runs from the English Channel to Gibraltar.

The fishing industry continues its activities in Galicia, South Brittany, the Basque country and the Lisbon Region. Aquaculture of fish and shellfish also takes place in the region, and there are landings for deep sea fishing, especially in Vigo, as well as in Lorient, Lisbon, and the tuna port of Concarneau. Marine aquaculture has been practiced for many years, with major production centres for mussels and turbot in Galicia, oysters in Poitou-Charentes and sea bass and sea bream in Cantabria.

Tourism is an important activity throughout the region, from South Brittany to Andalucía. It is dominant on the French coast and takes a variety of forms, from sports tourism to gastronomy, thalassotherapy and coastal and cruise tourism.

Energy products account for the majority of maritime traffic and half of the wealth of the marine economy. Accidental pollution has had crucial impacts for the marine and coastal environments. Galicia and South Brittany have been the scene of several oil spills and environmental disasters. Numerous protected marine areas already exist or are being planned in the region's estuaries and bays.

Each port city is a unique socio-economic, cultural environmental and morphological case. At one of the corners of Europe, on an amazing topography, the rock of Gibraltar, a British Overseas Territory located on the Southern end of the Iberian Peninsula, is a mythical landmark at the entrance to the Mediterranean Sea. British military naval activity traditionally dominated Gibraltar's economy. This has declined dramatically over recent decades, and is estimated to account for only 7 % of the local economy, compared to over 60 % 30 years ago. Shipping is still one of the four main sectors dominating Gibraltar's economy, together with financial services, Internet gaming, and tourism. The Rock is a popular port for cruise ships and attracts day visitors from holiday resorts in Spain.

The Mediterranean and Black Seas border the European Union from the Southern and Eastern sides. The Mediterranean Sea, the only three-continental space on the planet, is a biodiversity and climate change hotspot, and the epicentre of major civilisations. The Mediterranean Sea is a unique semi-enclosed basin characterised by a combination of coastal and open sea dynamics and often referred to as a miniature ocean and a physical laboratory for marine environmental research. Being a shared sea between European, North African and Middle Eastern countries, the Mediterranean is a strategic area of intense and often contrasting economic, societal and cultural characteristics. Tourism is the most developed activity of the region, which attracts approximately one third of all the world's tourists every year. In association with cultural heritage sites, tourism also generates spill-over effects into other economic sectors and stimulates various activities, such as shipyards for luxury yachts and car ferries. The strong Mediterranean heritage acts as a catalyst and provides many cultural opportunities (CEPS 2013).

The Mediterranean region is an exceptional ecosystem, enjoying a rich biodiversity. But it is a closed sea, bordered by heavily populated and industrialised coastlines that are visited by millions of tourists, crossed by intense maritime traffic and fed by urbanised river basins and industrialised and agricultural lands. Major threats include pollution related to urbanisation and industrial activities, overexploitation of fishery resources, loss of biodiversity and benthic habitat degradation, and invasion of alien species.

One of the major direct anthropogenic pressures on the Mediterranean Sea is from the use of the water resource which is changing the fresh water of the basin and the nutrient loads. Advancing climate change, expanding economic activities and accelerating coastal population growth are expected to amplify the impacts and affect the ability of the Mediterranean Sea to provide ecosystem goods and services. Policy making should take into account a complete range of scenarios for the future state of the Mediterranean Sea based on sound knowledge, observation-based evidence and reliable foresight scenarios for this complex and dynamic sea basin.

The Mediterranean shores host 150,000 million inhabitants. Many cities count over a million citizens, including Alexandria, Athens, Rome, Algiers, Barcelona, Marseille, Beirut, Tunis, and Tel-Aviv. Piraeus, the largest port, welcomes 11.5 million passengers embarking each year for the Greek islands, followed by the port of Reggio de Calabria, which transports 10.5 million passengers to Sicily. The main maritime routes linking the North and South shores of the Mediterranean Sea connect Spain with Morocco, France with Algeria and Italy with Tunisia.

Alexandria extends about 32 km along the coast of the Mediterranean Sea and 5 m above mean sea level. It is the second largest metropolitan area in Egypt by size and population after Greater Cairo. Economic geography contributed much to making Alexandria one of the most important trading centres in the world. It is Egypt's largest seaport, a centre for culture and tourism, and an important industrial centre as well, due to the natural gas and oil pipelines from Suez. The city benefited from the easy overland connection between the Mediterranean Sea and the Red Sea, and the lucrative trade in Egyptian cotton. Alexandria has always been a multicultural city, with its three largest ethnicities being Greek, Jewish, and Egyptian. Already eroded by the waves, the city is facing the worst climate threats on the Mediterranean shores.

The Black Sea is a semi-enclosed sea, only communicating with the oceans through the Bosporus and the Dardanelles Strait. Its ecosystem has been greatly affected by the urban and industrial waste carried by the major rivers that flow into it. A most popular tourist destination, the Black Sea is an important traffic route both for people and goods, including gas and oil from the Caspian Sea. Seaside resorts in Bulgaria, Romania, Ukraine and Russia are still very active, while in Turkey, tourism favours the Mediterranean coast. Shipyards in the Marmara Sea are expanding rapidly and have elevated Turkey to be among the world leaders.

Istanbul is the largest urban agglomeration in Europe and the only transcontinental world city, straddling the Bosporus between the Sea of Marmara and the Black Sea. Founded around 660 BC as Byzantium, the city developed to become one of the most significant cities in history. For nearly 16 centuries, following its reestablishment as Constantinople in 330 AD, it served as the capital of four empires, including the Roman Empire (330–395), the Byzantine Empire (395–1204 and 1261–1453), the Latin Empire (1204–1261), and the Ottoman Empire (1453–1922). The city played a major role in the advancement of Christianity, before the Ottomans conquered it in 1453 AD and transformed it into an Islamic stronghold.

Istanbul enjoys a strategic position along the historic Silk Road, rail networks to Europe and the Middle East, and the only maritime route between the Black Sea and the Mediterranean. Overlooked for the position of the new capital of Turkey during the interwar period, the city has since regained much of its prominence. The population of the city has increased tenfold since the 1950s and city limits have expanded together with infrastructure improvements and the transport network.

Considered a global city, Istanbul is one of the fastest-growing metropolitan economies in the world. It hosts the headquarters of many Turkish companies and NGOs and accounts for more than a quarter of the country's gross domestic product. Art festivals were established at the end of the twentieth century and, in 2010, the metropolis was crowned the non-EU European Capital of Culture. Approximately 11.6 million foreign visitors arrived in Istanbul in 2012, making the city the world's fifth most popular tourist destination. The historic centre, partially listed as a UNESCO World Heritage Site, is a strong magnet, as is the city's natural harbour, the Golden Horn.

Distinctiveness of cities is a great asset on the world chessboard. Venice is probably the most emblematic of them all. Some argue that Venice is a static entity, but it can also be seen as a city in perpetual transformation, with its centuries of splendour and its times of decline. The city, listed in its entirety as a World Heritage Site, along with its lagoon, enjoys an exceptional natural and cultural capital. Strategically located in the marshy Venetian lagoon, stretching along the shoreline, between the mouths of the Po and the Piave Rivers, the city is celebrated for its architecture and art. In 828 AD, the city's prestige was raised by the acquisition of the claimed relics of St. Mark the Evangelist from Alexandria. As Byzantine power waned, the city gained in autonomy and, from the ninth to the twelfth century, it developed into a city state. It became a major maritime power during the Middle Ages and Renaissance, and a springboard for the Crusades, and it evolved as a foremost trade and art centre from the thirteenth century to the end of the seventeenth century. Venice became an imperial power following the Fourth Crusade, which culminated in 1204 with the fall of Constantinople. By the late thirteenth century, Venice was the most prosperous city in Europe. At the zenith of its power and wealth, it dominated the Mediterranean world.

After the invention of the printing press in the fifteenth century, Venice was quick to grasp the opportunity and became the printing capital of the world. However, after C. Columbus discovered the New World and Portuguese explorers found a sea route to India, Venice went into centuries of decline. Much of the Venetian wealth was invested in the grandest palaces and in supporting mighty artists. This reinforced the urban cultural capital which again became a magnet for the world. Since the eighteenth century, tourism has been a major Venetian cultural industry. The city receives more than 50,000 visitors per day, almost as many as the resident population in the historic centre. The revival of the Carnival of Venice in the 1980s, the two international Art and Architecture Biannual exhibitions and the city's Film festival attract a huge number of cultural visitors and artists. And no other city evokes such an intense feeling of wonderment as a three-dimensional water-bound masterpiece.

References

Arup (2014) Cities alive. Rethinking green infrastructure. Arup, London

Centre for European Policy Studies (CEPS) (2013) Prospective analysis for the Mediterranean region (MEDPRO) EC-supported project funded under FP7. CEPS, Brussels

European Marine Board (EMB) (2014) ARCTIC 2050: towards ecosystem-based management in a changing Arctic Ocean. EMB, Strasbourg

(The) Greater New Orleans (2013) Urban Water Plan (Vision, Urban Design and Implementation). New Orleans; references IPCCa, IPCCb

Goddio F (2013) Heraklion. Sunken civilisation
International Panel of Climate Change (IPCC) (2014a) Climate change 2014: impacts, adaptation and vulnerability. IPCC, Bonn
IPPC (2014b) Climate mitigation. Bonn
IPCC (2014c) Climate change fifth assessment report. IPCC, Bonn
International Union of Architects (IUA) (2011) Tokyo declaration. Beyond disasters. IUA, Tokyo
Menon (2012) The leading maritime capitals of the world. Menon, Oslo
Organisation for Economic Co-operation and Development (OECD) (2007) Ranking of the world's cities most exposed to coastal flooding today and in the future. OECD, Paris
OECD (2013) The competitiveness of global port cities. Synthesis report. OECD, Paris
Port of Rotterdam (2012) Port vision 2030. Port of Rotterdam, Rotterdam
UN Convention on Biological Diversity, Stockholm Resilience Institute and ICLEI (2012) Cities and Biodiversity Outlook. Action and Policy. UNEP, Paris
(The) UN Office for Disaster Risk Reduction (UNISDR) (2013) Global assessment report on disaster risk reduction. UNISDR, New York
Wilson M (2013) Shifting ground: redefining Boston's landscape Infrastructure for a changing climate. Thesis at the Harvard Graduate School of Design. Cambridge, MA
World Monuments Fund (WMF) (2006) Watch 2006
World Resources Institute (WRI) (2012) Impacts of Hurricane Sandy and the climate change connection. WRI, Washington, DC
World Wide Fund (WWF) (2009) Mega-stress for mega-cities: a climate vulnerability ranking of major coastal cities in Asia
WWF and Allianz (2009) Major tipping points in the Earth's climate system and consequences for the insurance sector. World Wild Fund and Allianz, Berlin
Zangger E (1993) The flood from heaven: deciphering the Atlantis legend. William Morrow, New York

Selected Internet Links

www.oceansatlas.org
www.cluster-maritime.fr
www.oecd.org/regional/portcities
www.eurocean.org
www.ifremer.fr
www.isemar.fr
www.hcmr.gr
www.imo.org
www.oml.fr
www.mar-economists.org
www.marine.ie
www.kg.eurocean.org
www.worldoceanreview.org
www.coastalchallenges.org
www.stateofthecoast.noaa.gov/population/
http://www.preventionweb.net/english/hyogo/gar/2013/en/home/
http://ec.europa.eu/maritimeaffairs/
www.seadatanet.org
www.geoportal.org
www.emodnet.eu/

Watercolour 3
Alexandria, a Deltaic City next to Sunken Cities

Chapter 3
Resourceful and Resilient Cities and Marine Ecosystem Services

Abstract The great wealth of coastal areas, whether in terms of access to the sea, maritime trade, tourism, fishing, or natural resources, increasingly attracts urban populations. However if this precious coastal capital is unsustainably managed, the potential of its benefits can be easily undermined. Coastal urbanisation affects land use and cover of the shore, biodiversity, soil, water and air quality and global climate. Responsible coastal cities try to reduce their ecological debts on land and the sea. Urban coastal organs and functions have to boost the resilience of places and assist in transition to a civilisation of sustainability. Sustainable cities and oceans, among the priority areas of the Rio+20 conference and the dialogue for the post-2015 sustainable development goals, need ecological and environmentally friendly cells and neurons to thrive in harmony with the planet and the seas. Together with their citizens and stakeholders, they can play a cardinal role for exploration and protection of our precious marine resources.

This chapter focuses on coastal cities as vital and interdependent ecosystems able to manage crucial amounts of scarce resources and materials, ensure food security, offer sustainable goods and services and reduce greenhouse gas emissions and waste, especially preventing marine debris. Many coastal cities provide outstanding models of emission-neutral, waste-free eco-neighbourhoods and intelligently designed waterfronts to celebrate the sea which links them to the conscience of the world. Furthermore, they can send an impulse for maritime routes to become vectors of responsibility for sustainable development.

3.1 Coastal Biodiverse Cities and the Well-Being of the Ecosystems

Coastal cities are required to offer citizens the possibility of leading a happy and fulfilling life in balance with nature and the sea. They have to provide fresh air and water, healthy food, shelter, energy and transport, materials, services and opportunities for education, employment and leisure. For this, cities interact with the surrounding land and sea and impact a broader terrestrial and marine environment. Their

V.P. Mega, *Conscious Coastal Cities*, DOI 10.1007/978-3-319-20218-1_3

ecological footprints, which depend on the nature and nurture of cities, reflect their resource consumption over multiple terrestrial and sea value chains. Cities can collectively influence over 70 % of the global ecological footprint.

Marine ecosystems offer ample possibilities for food, recreation, health and wellbeing. The global ocean produces almost half of the oxygen supporting life on Earth and absorbs more than a quarter of the carbon dioxide emitted into the atmosphere. It has been described as the kidney of the planet or as a huge biological pump at the heart of global atmospheric and thermal regulation and water and nutrient cycles. But the ability of the ocean to continue providing these essential ecosystem services is being compromised due to marine pollution, environmental degradation, overfishing, and rising temperatures and absorbed emissions (GOC 2014).

Greenhouse gas emissions are changing the chemistry of the planet, raising ocean temperatures and acidity levels, which in turn are endangering coral reefs and marine life. In the future, climate change is expected to have the greatest impact on the ocean capacity to provide its ecosystem services. Humanity is rapidly approaching unprecedented tipping points that, if unheeded, may inexorably lead to systemic collapse. Since 2008, the World Ocean Day, celebrated each year on 8th June, emphasises various aspects of individual and collective responsibilities in sustainably managing ocean resources.

Food from the ocean first drew humans to urban settlements by the sea. Coastal cities are the places where human, atmospheric, terrestrial and water ecosystems interact and influence each other. Their natural capital is very diverse and most promising in an era in which depletion of the global natural capital becomes less and less reversible. The sustainable use of resources, respecting the carrying capacity of the biosphere and supporting the provision of ecosystem services, is a major challenge for the world's cities and citizens.

Humanity consumes far more than the ecological resources and services that the planet can sustainably regenerate every year, and, if current trends continue uncurbed, the biocapacity of two planets will be needed for its survival before 2050. Earlier and earlier each year, the Global Footprint Network informs us about the overshot day, a recently coined term, when global demand for natural resources and ecological services starts surpassing annual supply. In 2014, the Earth Overshoot Day was on August 19th. In less than 8 months, humankind had already exhausted the equivalent amount of natural resources that the planet generates during the entire year. The ecological deficit, signalling the gap between supply and demand of natural resources, is seriously eroding the planet's ecological capital assets (Ecological Footprint Network et al. 2012).

Humanity is jeopardising its very future, on land and sea. A critical threshold was already crossed in the 1970s when global demand began outstripping our planet's biocapacity. Since then, ecological overspending has entered a vicious cycle. As resource deficits become larger and resource prices remain high, pressures become unbearable and accumulation of the debt weakens the ability of communities to initiate virtual global cycles. Years of unbridled consumption have led to the depletion of many finite resources, on land and sea, and the consumption of renewable resources at paces have become more rapid than their regeneration rates (WWF 2014).

To curb these trends, responsible cities started introducing ecosystem-based management involving local approaches to conscious resource management, recognising that humans are an integral part of their ecosystems, which are vital in supporting human life. The process emphasises the necessity of sound scientific understanding and strong citizen participation in addressing complex and often contentious issues. The future role of ecosystems for human well-being depends increasingly on their health and self-regenerating capacity (UNEP 2011).

Marine resources are often perceived as "less finite" than terrestrial ones. Deterioration of marine ecosystems has long been hidden by the apparent immensity and relative, until recently, inaccessibility of large parts of the global ocean. The need to reduce greenhouse gas emissions has driven both the deployment of offshore renewable energy, but has also provided a further impetus to seaborne transport over land transport. Improvement of the energy efficiency of ships and the use of cleaner fuels hold a significant potential to further reduce greenhouse gas emissions, and increase the sustainability advantage of maritime transport.

Coastal cities are among the ecosystems of highest concern. This has, with participative foresight, been highlighted by international organisations such as the 2010 UNEP. The UNEP foresight panel, convened to identify crucial emerging environmental issues, has succeeded in shedding light on many cities and land-sea interactions. Cities and the seas were recognised as critical emerging issues for the global environment, of almost universal spatial scale, and subjected to new developments, scientific knowledge, scales and impact. A number of other issues were declared by the scientific community to be "emerging", but which had not yet received strategic attention from the research and policy community. The process also suggested that the linkage between the policy and science communities is inadequate or even deteriorating, and this broken bridge is detrimental to positive global environmental change (UNEP 2012a).

The UN Conference on Environment and Development (Rio de Janeiro 1992) and Agenda 21 had already underlined the role of cities and local governments for sustainable development. Irresponsible land use planning, rigid zoning and urban sprawl, and low-density urban development into green areas, resulted in increased consumption of land, energy and resources, air, soil, water and noise pollution and greenhouse gas emissions. The impacts of climate change on oceans and the cities, the role of seas and green areas as carbon sinks and the effects of compact cities versus urban sprawl are subjects of ongoing research. The growth rate of urban land cover has been double that of the growth rate of the urban population between 1990 and 2000 (Angel et al. 2011).

In 2012, cities and the oceans were two among the seven critical issues of the Rio+20 conference that intended to give a new impetus and turn sustainable development from aspiration and irregular implementation into a genuine systematic path to responsible prosperity for the present and the next generations. Crucial links among cities and the oceans were also reflected on other critical issues, including food, water, jobs, disasters and energy. Governments and businesses underlined issues of concern, including threats from climate change and ocean acidification, overfishing, illegal fishing, and subsidies that drive unsustainable consumption. They also debated the need to conserve and protect marine ecosystems to both restock the ocean and build its resilience to change.

The 2012 Rio+20 declaration "The Future We Want" encouraged nations to introduce green growth policies and strengthen efforts to eliminate poverty and inequality. By the end of the conference, hundreds of voluntary commitments for sustainable development had been registered by governments, businesses, civil society organisations and universities. More than 80 companies and 50 countries also committed to boost natural capital accounting, after a proposal by the World Bank to take into account the value of assets, such as clean water and marine ecosystems, into national accounting systems and government and business decisions.

However, in spite of hundreds of internationally agreed goals and voluntary commitments, the world continues on an unsustainable path. Inaction is considered to be a major bottleneck. UNEP's 5th Global Environmental Outlook (GEO 5) stressed that drastic large-scale actions are needed to reverse the trends of resource overconsumption. Action has to follow commitment. Some goals that were crowned with success include elimination of substances that deplete the ozone layer, removal of lead from fuel, access to improved water supplies and research to reduce pollution of the marine environment. The assessment emphasised that sustainability targets can still be met if current policies are radically changed and best practices are scaled up to optimise their overall effect (UNEP 2012b).

"Business as usual" approaches can have an unbearable weight on the future, especially in times of a worrying deficit in global governance (Gore 2013). The OECD Environmental Outlook to 2050 suggests that the potential cost of inaction on climate change could be as high as 14 % of average world consumption per capita in 2050. Continued degradation and erosion of the natural environmental capital, together with irreversible effects, could endanger two centuries of continuous upgrading of living conditions. Multiple-benefit best cases confirm that green growth is not prohibitively expensive and can help humanity to strike a renewed green deal (OECD 2012).

Dynamic world cities emerge as both the main drivers and beneficiaries of a paradigm shift towards green growth, proposed as a shortcut to sustainable development. Inclusive green growth brings opportunities for new environment-friendly businesses and jobs, while managing the necessary structural changes for transition to a greener society. Greening growth is necessary and can be efficient and affordable (World Bank 2012). A concert of international reports on green growth suggests that the way forward requires integrated innovative solutions to tackle political economy constraints, change deeply entrenched behaviours and develop the methodological instruments to monitor progress (UNEP/ILO 2012).

Sustainable management of natural resources, including marine resources and ecosystem services in an increasingly urbanised world, is a major challenge for cities, long typified as concrete jungles inhospitable to flora and fauna, ecosystems and habitats. Depending on their values, performance and governance, cities can be an opportunity or a threat for biodiversity. Seizing the opportunity asks for a mix of high quality green and blue public spaces in dialectic symbiosis with dense and compact built-up areas (EEA 2010).

The sustainable development of terrestrial and marine resources is cardinal for coastal cities in search of new models to impregnate and transform local production and consumption patterns, social values and lifestyles, socio-economic and environmental policies and citizen awareness and participation towards sustainable development. Many cities have conducted assessments of their ecological footprints on land and the sea and identified ways to influence and reduce them.

Furthermore, many cities together with their main stakeholders have declared their will and readiness to act and correct their enormous ecological footprints on the indebted planet. Urban design and infrastructures have crucial links to footprint trends. Population density and public transport significantly reduce the per capita ecological footprint, while the affluence of citizens often leads to higher final consumption. This was the case of San Francisco, the city by the Bay, which tried to evaluate the ecological footprint of the metropolitan area. The assessment concluded that the overall impact of average San Franciscans had absolutely to be reduced and provided options for more environmental-friendly lifestyles (Ecological Footprint Network 2011).

Vancouver is reputed to have the lowest carbon footprint among major North American cities through high density built environment, limited urban sprawl, sustainable public transport, and the provision of many green and open spaces. The Greenest City Initiative, involving all levels of city government, citizens and business in a range of projects, aims at Vancouver becoming the greenest city in the world by 2020. Long-term planning, a 15-year long investment in multimodal transport, and an open attitude towards migrants, made the city environmentally friendly and culturally diverse (Government Office for Science 2014a, b).

The Cities and Biodiversity Outlook, the world's first global analysis of the links and opportunities between urbanisation and biodiversity, highlights a wide range of worthy initiatives by cities, local authorities and sub-national governments in both developed and emerging countries. The assessment underlined the potential benefits from taking into full account terrestrial, freshwater and marine ecosystem dynamics in urban planning and policy. This should be seen as a major opportunity to make giant leaps for the reduction of biodiversity loss and the improvement of quality of life. The 10-year global initiative "Cities in Biodiversity Hotspots", involving more than 250 cities located in and around the 35 biodiversity hotspots of the world, provides a platform for action and mutual learning (UN Convention on Biological Diversity et al. 2012).

Urban regions must take increased responsibility for conceiving, designing and implementing actions that take into account their profound connections with and impacts on the rest of the planet. Urban areas are expected to continue to expand faster than urban populations. The total urban area is likely to triple between 2000 and 2030, while urban populations are likely to nearly double. Urbanisation draws heavily on natural resources on a global scale, with serious effects on biodiversity and ecosystem services. Food, materials and energy are drawn in great quantities from all parts of the world, despite the sustainability calls for giving priority to the local and seasonal food resources.

However, there are untapped opportunities for cities to sustainably manage ecosystem services. For example, a portfolio of measures can be used to decrease waste and reduce consumption, while at the same time invest in protecting biodiversity, water quality, local food production and carbon-sequestering ecosystems. A global system of cities wishing to lead the world towards sustainable futures has to closely cooperate with other actors such as national, regional, and local governments, multinational corporations, and civil society. Each of these actors has important roles to play in sharing and managing planetary resources.

Many cities have begun to take an increasing responsibility in the management of terrestrial and marine resources and impacts at regional or even global scale. Actions by a consortium of municipalities or state governments operating at larger scales are likely to accomplish even more in addressing challenges of ocean sustainability. Partnerships across urban and rural communities are also expanding to address multiple global sustainability issues, and the inter-connections and impacts on broader variable geometry territorial scales.

In 2002, the signatory countries of the Convention on Biological Diversity (CBD) committed to achieve, by 2010, a significant reduction in the rate of biodiversity loss. This year of high and noble intent was defined as the International Year of Biodiversity and expanded later to the Decade of Biodiversity to frame efforts with a longer perspective (2011–2020).

Effective stewardship of ecosystem services must consider the interconnectedness of resources that link cities to ecosystems outside of their boundaries, and the multitude of actors and factors that shape and sustain the resource flows. This responsibility includes implementing the ecosystem approach of the Convention on Biological Diversity and supporting local governments in addressing the Aichi Biodiversity Targets, the 20 ambitious goals of the CBD's Strategic Plan for Biodiversity 2011–2020. The targets provide a framework of action for all stakeholders striving to preserve biodiversity and enhance its benefits for citizens. The first target suggests that by 2020, at the latest, people will be aware of the values of biodiversity and the steps they can take to conserve and manage it sustainably.

Some of the Aichi Biodiversity Targets concern directly the marine environment and are crucial for coastal cities. Target 6 asks for all fish and invertebrate stocks and aquatic plants to be managed and harvested sustainably by 2020 and the impacts of fisheries on stocks, species and ecosystems to be within safe ecological limits. Cities are also concerned by the development of marine protected areas and should deploy efforts to achieve the Aichi Target 11 asking for at least 10 % of coastal and marine areas and 17 % of land, especially areas of particular importance for biodiversity and ecosystem services, to be preserved by 2020, through effectively and equitably managed, ecologically representative and well-connected systems of protected areas and other area-based conservation measures (CBD 2010).

Coastal cities, bridging human, land and sea biodiversity, are pioneers in bringing forward new initiatives. Connecting fragmented ecosystems on land and the sea is particularly useful to increase ecological function. This can be achieved through planting trees by the waterfront to expand the tree canopy, establishing corridors of multilayer plantings along the seaside, as well as creating ocean gardens or water parks. Cities can also focus on improving and restoring their blue and green spaces to draw increased attention to local wetlands and hybrid spaces. Sustainable planning, design and management of urban areas are crucial factors for the preservation of biodiversity and healthy ecosystem services.

Monitoring and performance evaluation are fundamental to understanding the impacts of biodiversity management. The Cities and Biodiversity Index (CBI) is proposed as a self-assessment inventory that enables cities to track their progress in conserving and enhancing biodiversity. Other initiatives to benchmark environmental stewardship include notably the Environmental Sustainability Index. After a preliminary assessment of data availability and feedback by several world cities, indicators and scoring systems were developed, tested, revised and improved upon and a user's manual was prepared. Efforts continued towards a single index aggregating the contributing indicators.

Every coastal city and region should have a marine biodiversity strategy with shared objectives and a policy portfolio for achieving them. Lisbon established a Cooperation Protocol with the pioneering goal of enhancing its biodiversity capital by 20 % until 2020. In this Protocol, the cooperating parties also committed to prepare an Urban Biodiversity Matrix of Indicators.

The "Fibre City: Tokyo 2050" is a vision for the future of the first world metropolis aiming at a radically new balance between nature and the built environment. The vision responds to four urban challenges, including a shrinking population, an ageing society, environmental crisis and the risk of earthquakes. While unprecedented population declines are expected in Japan due to falling birth rates and continued resistance to immigration, the metropolis could benefit from preserving the freed up land as a precious resource. The concept of the Fibre City 2050 focuses on the ways in which urban fibres can be used to construct an alternative image of a metropolis. As a visionary view of Tokyo, from macro to micro, Fibre City intends to inspire global cities retrofitting for greater harmony with nature, better access to public transport, and improved liveability with rediscovery of historic features like Edo canals and bridges that have been covered by elevated expressways.

New York took a comprehensive and integrated approach to its ocean and coastal resources, leading to better management decisions and healthier communities and ecosystems. The New York Ocean Action Plan (OAP) is a coordinated and inclusive effort for improving the health of ocean ecosystems and their capacity to provide sustainable benefits to citizens. Through concerted action, the goal of the OAP is to achieve sound ocean ecosystems for the benefit of citizens, communities and the world. Grounded in stakeholder participation and short-term actions to reach long-term goals, the Plan guides State government funding, research, management, outreach, and education.

The necessity of the Ocean Action Plan was highlighted already in 2009 by the New York Ocean and Great Lakes Ecosystem Conservation Council, which aimed at promoting healthier ecosystems in New York through the use of ecosystem-based management and greater coordination of stakeholder activities. The geographic action area encompasses the State's ocean waters stretching from New York City to the end of Long Island, including ecological connections to offshore waters out to the edge of the continental shelf. The New York Department of State, working with all concerned stakeholders, drew a map of offshore uses and compiled resource data to identify offshore habitat areas and locations that may be best suited for offshore wind energy development. Additionally, it explored the ecological relationship with several estuaries and their habitats, given the interconnectedness of ocean waters with waters near shore.

As a contribution to the UN Decade on Biodiversity, the City of Brest hosted, in 2012, an international meeting on Marine and Coastal Biodiversity. It provided a multi-disciplinary platform for scientists, international organisations, and cities to discuss interactions between the development of marine territories and biodiversity management. UNESCO's Intergovernmental Oceanographic Commission presented projects for managing and protecting marine and coastal biodiversity, including the Ocean Biogeographic Information System.

Many cities strive to provide citizens with easy access to high-quality wetlands and local nature reserves. Singapore and Perth have built elaborate urban forest canopy walks. More cities designed imaginative parks in aquatic and marine settings. Central Park in New York City, Park Guell in Barcelona, Pharo gardens in Marseille, Griffith Park in Los Angeles and Stanley Park in Vancouver are prime examples of accessible green urban oases. Stanley Park is a magnificent green oasis in the heart of the built urban landscape of Vancouver, which has already celebrated its 125 years. It includes 400-ha natural West Coast rainforest and enjoys scenic views of water, mountains, sky, and trees along the famous Seawall. Since its opening, in 1888, the City and the Park Board have worked together to preserve and protect the ecology of the park and ensure its sustainability, while enhancing Vancouver's liveability through the enjoyment of Stanley Park.

Sydney intends that each and every citizen will live less than 250 m distance from a green area. According to the concept of plaNYC, the city's ambitious plan for 2030, all New Yorkers should live within a 10-min walk of a park, usually with a water pond. A Greener, Greater New York has been the consensual vision of many organisations and public agencies, which committed to prepare the city for one million more residents, strengthen the economy, combat climate change, and enhance the quality of life for all New Yorkers. Public spaces and parks are among the plaNYC areas of highest interest (New York City Mayor's Office of Long-Term Planning and Sustainability 2011).

In Porto, the waterfront, both along the river and the ocean, is teeming with happy pedestrians and visitors, enjoying all the way from the city centre to the ocean. The Passeio Alegre echoes well its name. The City Park is a most spectacular landscape expanding 83 ha near the sea. Opened in 1990 after a long gestation, the park is a meandering labyrinth of winding paths that extends over 10 km, covered by trees, shrubs, and aquatic plant species. From physical exercise to kite flying, the park supports a very diverse array of activities bridging the city with the Atlantic coast. The city also has several themed gardens including the garden of feelings, the garden of aromatic plants, the garden of roses and the apple tree estate for citizens to rediscover their environment.

Many international initiatives try to identify and celebrate the brightest best cases. The Local Action for Biodiversity Initiative introduced by the International Council for Local Environmental Initiatives (ICLEI) highlighted the achievements of pioneering coastal cities. The City of Amsterdam is a biodiversity world leader, home to a vast network of rivers and lakes, as well as urban beaches and marine ecosystems with hundreds of species. From a small fishing village in the twelfth century to a world class city in the twenty-first century, Amsterdam had made sustainable development its utmost goal. Although highly urbanised, the city has a wide diversity of flora and fauna and a unique network of urban parks which provide recreational ecosystem services. The city's environmental policy plan, reviewed every 4 years, ensures that the environment is given dynamic priority and is well integrated with its economic and social development.

Just 6 km away from Amsterdam's central station, the Ilperveld fen is part of the Natura 2000 protected sites in Europe, the larger world coordinated network of protected areas, established according to the Birds and the Habitats Directives that collectively form the "nature legislation" of the European Union. The area offers uninterrupted views of large flat fens, criss-crossed by a myriad of ditches and canals. Originally used for hay-growing and pasture, the area was degraded as a municipal dump in the 1980s and was abandoned by thousands of meadow birds. The site has since been protected within the framework of Natura 2000, and a large restoration has been undertaken. Farming and cattle grazing has also been re-introduced to maintain the wet meadows in good condition for the benefit of the birds which have since returned. The area is also very popular with bird watchers, nature lovers and school children (EC 2012).

Water parks are amongst the most vital urban commons and they can beautifully integrate with green parks. Biodiversity and ecosystem services are global public goods, but local and regional authorities have the legal power to designate conservation areas and the ethical obligation to protect them through incorporation of biodiversity concerns into their spatial planning processes. Public commitment is essential for sustainable communities which identify biodiversity as a precondition for resilient cities. The designation of marine protected areas and the development of green infrastructures offer cities an opportunity to integrate biodiversity and resilience into local and regional plans taking into account the land and the sea dimensions (EEA 2010).

Protected marine parks can be of special interest for cities and the recreation and well-being of citizens. The Arcachon Park, opened in 2014, is the sixth French marine park. Governed by local actors, the park enables different activities such as tourism, oyster farming, fishing or scientific observation to sustainably coexist in a natural environment. Marine protected areas are part of the French strategy for the sea that was launched following the Grenelle of the Sea in 2009 with the ambitious goal of protecting 20 % of the waters under French jurisdiction by 2020. All French protected areas accounted in 2014 for about 4 % of these waters. Their geographical distribution is quite uneven, with most of them located overseas. The Basin of Arcachon is a place of exchange for many species and has the particularity of being empty of two thirds of its capacity at low tide, revealing sandy islets and tidal marshes. As the second global maritime domain, France tries to lead by example by establishing effective means to protect and enhance its marine areas.

Attractive green areas and blue waterfronts are essential for urban populations and wildlife to thrive. Copenhagen provides a prime example. An ancient Viking fishing village, strategically located at the entrance to the Baltic Sea, the Danish capital/European Green Capital in 2014 made the Blue and Green City one of its priorities. Swimming races and cliff-diving contests with athletes jumping into the harbour from the roof of the opera pay tribute to the city's long symbiosis with the sea. Copenhagen's ambitious goal of being carbon neutral by 2025 and the eco-metropolis pathway are among the world best practices.

Many more cities work hard to transform urban grey by the sea to urban blue green spaces. All roofs could become candidates for micro-farms and all walls could support vertical gardens and micro-farms. Inventors already proposed self-regulated urban farms to the size of a parking place. Some cities are equally interested in ecosystem services delivered by biodiversity in green areas in the hinterland and in faraway lands and oceans. They can cooperate with the farming sector, responsible for 80–86 % of all food-related global GHG emissions and 14–24 % of total global emissions, to improve its performance. It has been estimated that as much as 30 % of all food grown worldwide is wasted before reaching the consumer. In industrialised countries, more than 40 % of food losses occur at retail and consumer stages.

Urban agriculture has the potential to become pervasive within cities and could even provide citizens with a considerable part of the needed food. Vertical farming in tall buildings could replace less productive single-story greenhouses as the source of all city-grown produce. Some forms of vertical farming already exist and Chicago claims the largest multi-storey greenhouse of special interest to densely constructed cities. Urban aquaponics, combining plants and vegetable farming with fish farming in a clean water environment, promises to further expand the limits of the possible (van Veenhuizen 2006).

Cities and agriculture are responsible for nutrients and waste entering the marine environment. Furthermore, many hundreds of fishing nets are lost or abandoned every year, often due to conflicts between trawlers. The problem of lost or abandoned fishing nets is severe due to the non-degradable netting materials which can maintain 10 % of their catching capacity once lost. Much waste comes from food packaging, which does not always meet the performance standards required at equivalent cost, and the industry remains reliant on fossil-based coatings and films. Various research and innovation projects try to develop sustainable, eco-friendly food packaging coating products made from seaweed extracts and starch and applied to paper and cardboard in the form of a spray. This is a prime field for cities and businesses to join efforts.

Coastal litter, a prime symbol of resource inefficiency and inaction on land and the sea, is a huge problem, as it can be ingested by and entangling wildlife. Marine debris comes mainly from land activities and only one fifth is due to shipping. Only 15 % of the debris remains on the surface, while 70 % goes to the seafloor. It pollutes beaches, causes harm to ecosystems, undermines the recreational value of the places and prevents the optimal enhancing of marine resources and the development of the blue economy (GOC 2014). If no urgent action is taken, marine litter is projected to increase by 44 % by 2030.

Plastics are by far the most prevalent debris contributing an estimated 60–80 % of the total marine debris. It seems that eight million tonnes of plastic waste are entering the world's oceans every year. The worst offenders are in Asia, with China responsible for 28 % of all plastic entering the ocean, followed by Indonesia with 10 % of it. The global plastics sector, of which bio-plastics make up only 1 % of the market, is a large sector leading to large amounts of non-biodegradable plastic waste. Plastic marine debris contains highly toxic substances in concentrations that are more than 100 times higher than those normally found in ocean waters, affecting marine life and ecosystems.

Synthetic polymers in the ocean should be regarded as hazardous waste (Eriksen et al. 2014). Marine debris affects all marine habitats, from densely to remote populated regions. It has been estimated that the average density of marine debris varies between 13,000 and 18,000 pieces per square kilometre, with far higher concentrations, more than 200,000 pieces per square kilometre, in the convergence zones between two or more ocean currents. Computer model simulations, based on data from about 12,000 satellite-tracked floats deployed since the early 1990s as part of the Global Ocean Drifter Program, confirm that debris is transported by ocean currents and tends to accumulate in a limited number of sub-tropical convergence zones or gyres.

The accumulation of floating garbage in the oceans led to potentially five garbage patches scattered globally, located in the North and South Pacific Ocean, North and South Atlantic Ocean and Indian Ocean. The total amount of plastic garbage, estimated around 36,000 tonnes, is unequally dispersed, and twice as much in the North Pacific as in the North Atlantic. The ecological impacts range from over a million of seabirds and one hundred of thousand marine mammals killed by ingestions of plastics or entanglement. Economic impacts, estimated after beach clean-up costs, tourism losses and damages to fishing and aquaculture industries, are also very important (Sesini 2011).

In the North Pacific, the immense cauldron known as the Great Pacific Garbage Patch contains an estimated 100 million tonnes of plastic waste and chemical sludge, much of which is produced by cities. Cities from all sides of the Pacific and beyond can make innovative alliances and engage in research and innovation to jointly address this major challenge. Technologies for recovering ocean garbage and converting it into fuel for energy production would only have multi-beneficial impact.

Demonstration programmes were developed to support large scale removal and exploitation. A NOAA's programme collects derelict fishing equipment and nets and supports the recycling scheme "nets to energy" which generates power from the collected debris. In Hawaii, the nets, once removed by fishermen, communities, and trained divers, are transported to a scrap metal recycler facility and are chopped into pieces suitable for combustion. The nets are then burned, producing steam which drives a turbine that creates electricity, enough to power nearly 350 homes for a year.

In the EU, the 7th EU Environmental Action Programme calls for a quantitative waste reduction headline target, supported by source-based measures. Research projects have been trying to better shed light on not just reducing waste from marine activities but using it for new products, e.g. bio-based polymers from shell waste from the fishing industry. The EU-funded Ocean Sampling Day is another inspiring initiative which took place on mid-summer day (21 June 2014), the longest day of the year. Science teams around the world with the help and involvement of citizen scientists gave oceans a health check. Schools are often associated with such initiatives and coastal cities can organise their regular sea health checks while promoting science for the youth (EC 2013d).

Advanced cities have to show the way forward. San Francisco has set the ambitious goal of zero waste by 2020 and deployed an ambitious solid waste management programme. The city's zero waste goal implies that products are conceived, designed and used according to the principle of highest and best possible use. Zero Waste also means that discarded materials cascade through reduce, reuse, and then recycle or compost. To meet its zero waste goals, San Francisco developed a long-term strategy and a threefold approach to address the legal, administrative, and social dimensions of waste management. The City enacted strong waste reduction policies and partnered with Recology, a materials management company, to innovate and promote a culture of recycling and composting. Policies embraced a wide-range of fields, from production and packaging to consumption, public and private management and government procurement.

A culture of Zero Waste has been created through strong citizen engagement and the participation of many organisations. The municipality provides policy guidance and oversight, sets framework conditions and financial incentives for product and service providers, conducts outreach and enforcement, while partner Recology tests and operates infrastructure to collect and process recyclable, compostable and landfilled waste. The achieved results are significant. From 1990 to 2010, landfill diversion increased from 35 to 80 %. A construction and demolition debris ordinance has recovered tens of thousands of tonnes of material. Mandatory recycling and composting increased organics collection by 50 % to more than 600 tonnes per day.

Pollution from shipping is estimated to about 20 % of the total pollution. Dumping of garbage from ships is a serious element of the problem of marine debris. In 2013 new, more stringent controls under International Convention for the Prevention of Pollution from Ships (MARPOL) came into force. Port waste reception facilities have to play an important role. The European Union Member States have introduced requirements for delivery of waste ashore before a ship leaves port. Sewage pollution from ships is mainly a problem with cruise ships: with up to 7000 passengers and crew, they are the equivalent of a small town, and can contribute to local eutrophication problems. The local conditions around the ship are significant for the impact of any sewage discharges. The increased requirements under MARPOL on discharges near the shore are likely to reduce the problems.

Clean seawater can also be precious as a drinking resource. Coastal aquifers may contain seawater and salty groundwater from ancient times. State-of-the-art desalination technologies can remove salt from both seawater and brackish water, thus providing a new source of freshwater. Techniques used to desalt water are also useful for removing other impurities common to contaminated source waters. The process tends to be very costly and energy intensive. Desalination can impose an environmental toll and has to be managed according to strict sustainability principles. Research suggests that combining alternative energy sources with desalination technology can lead to more economical and environmentally friendly systems. Desalination is high in the agendas of water-poor countries such as Kuwait, Saudi Arabia, and Israel. In Tunisia, four desalination plants produce nearly 4 % of the national water resource. The plants use reverse osmosis, a process that orients water through a membrane with tiny pores that retain most of the dissolved salts as well as most organic compounds and microbiological contaminants.

3.2 Sustainable Aquatic Resources and Urban Food Security

During the industrial revolutions, increases in agricultural productivity have served as a driving force for urbanisation. As farms became more profitable, excess labour from agriculture moved to industrial jobs in cities. However, while past urbanisation promoted growth and prosperity, the current urban explosion in the emerging world is occurring faster than the pace that allows cities to integrate newcomers without major shocks and risks and to provide them opportunities for a better life. Poverty, food insecurity, and malnutrition are shifting rapidly from rural areas to urban centres. Furthermore, political instability in many emerging countries make cities inadequately equipped in infrastructures, and vulnerable to natural disasters, urban poverty, food insecurity, high malnutrition, and the least able to sustainably address these challenges.

Food from the ocean could be decisive for the food security of coastal cities, which often are highly dependent on imported foods. In many cities in Africa, 30–50 % of food staples and vegetable oil found in the markets are imported. Imported staples most often cost more than urban garden produce or street vendor foods. Urban dwellers, and especially the urban poor, are more affected by international food prices than small farmers. Food security in urban areas tends to be tied to consumption patterns and affordability, while food security in rural areas is related to the availability of food.

In many African cities, the urban poor spend up to 90 % of their household income on food. Up to 70 % of the caloric intake of a poor urban household is from street food. Street vendors are largely unregulated, lack access to clean water and refrigeration, and do not respect standards of hygienic food preparation and packaging. Cutting back on the amount and diversity of food is the most common solution for poor households. If such solutions last long, food insecurity increases and malnutrition rates climb quickly, especially among children.

The link between proper nutrition and normal physical and cognitive growth is a hardwired system. Access to sufficient, nutritious and affordable food is not only a basic human right but it provides the foundation for citizens' ability to grow, learn, prosper, and reach their full potential. Emerging cities have to ensure food for all slum populations, especially the youths less than 18 years old facing high malnutrition risk and expected to form 60 % of the urban slum population in 2030. This has huge implications not only for national wealth creation but also for political stability at often regional scale.

Fisheries and aquaculture are a vital source of nutritious food, economic opportunities and jobs (World Bank et al. 2014). They already account for 16.6 % of all animal protein consumed globally and 6.5 % of all protein for human consumption. They are a precious food for the full spectrum of populations, rich and poor, urban and rural. The contribution of fish to food is likely to increase as consumers become more conscious about the nutritional value of seafood. Stronger sustainability performance of the value chains could also have an important impact in terms of jobs and growth, in particular in coastal areas (FAO 2013, 2014).

One out of two citizens of the emerging world takes its animal protein from the sea. Ninety-seven percent of the world's fishermen are from the emerging world. Urban fish and wet coastal markets concentrating on the catch of the day are progressively replaced by supermarket chains with aquaculture products. The global development of aquaculture has been spectacular. Since 2014, aquaculture contributes more than capture fisheries to seafood at the world level. Food security programmes specifically tailored to meet the local conditions, and designed and implemented by the producers and consumers associations, are more likely to have a positive impact (OECD–FAO 2014).

For hundreds of years, fisheries had provided humankind with good quality food; then overfishing became a major global problem. It is estimated that 53 % of the world fisheries are fully exploited and 32 % overexploited, i.e. 85 % of the world's ocean fisheries are fully exploited, over-exploited or depleted. The world's fleet was evaluated to be 2.5 times larger than is necessary to sustainably catch global fish stocks. The combined engine power of the global fleet has grown tenfold since the 1950s and continues to rise. Illegal, Unreported and Unregulated Fishing (IUUF) and the negative financial incentives that maintain a global fishing fleet, with too many boats for an ever- diminishing supply of fish, represent a world plague. The global economy has lost $2.2 trillion in the last 30 years from fisheries mismanagement, reflecting technological advances and increased fishing subsidies (GOI 2013). Enshrined in the Rio+20 Declaration of 2012, the need to eliminate subsidies that contribute to unethical fishing and to overcapacity has to be thoroughly addressed. On the high seas, it is largely only States, mainly from the developed world and China, which can afford to subsidise their fleets with public funds (GOC 2014).

Other major problems for fisheries include eutrophication, as algal blooms can produce toxins which infect seafood. Climate change is also having an impact, since changing ocean temperatures affect the health and distribution of valuable fish stocks, and make efforts to sustainable management of fisheries more difficult. Cities could cooperate with States in eliminating harmful subsidies and close their harbours to illegally harvested fishes. The acceptable balance is a crucial challenge and cities should not undermine the health of global fisheries by ensuring food security through products of subsidised industrial fisheries distorting the seafood markets.

The concept of "maximum sustainable yield", embedded in the United Nations Convention on the Law of the Sea, can be instrumental for the restoration of fisheries stocks. Managing capture fisheries according to maximum sustainable yield has been successfully implemented in many single-species fisheries. However, sustainable management taking account of fishery interactions, the way different fishing gears catch different mixtures of fish at certain place and times, and ecological interactions among different fish stocks predating on each other or competing for food, is still just beginning.

Sustainable fisheries can contribute much to the food security of cities. Food security issues have been among the most forgotten urban issues. Hardly raised during Rio+20, they are of paramount importance for eradicating extreme hunger and poverty, instability and uncertainty. A new framework for addressing food and nutritional security in both rural and urban areas is critical. The Chicago Council on Global Affairs proposed an Urban Food Model designed for policy makers and city leaders as an analytical tool to evaluate a city's current food security situation and to examine policy options (Chicago Council on Global Affairs 2013).

The importance of addressing urban food security is increasingly highlighted by studies indicating policy recommendations for the nutritional security of cities. Rapid demographic growth and urbanisation, dietary consumption changes, and more frequent and extreme weather phenomena linked to climate change bring enormous challenges to cities, many of which already face the overwhelming burden of providing basic services to poor citizens. Urban food security should advance hand in hand with sustainable trade practices. The global community has witnessed a revival of efforts since the 2007–2008 food price spikes. Discontent over poverty and food insecurity can be politically explosive in cities (FAO 2009).

The international community should acknowledge that poverty and food insecurity and malnutrition issues are serious urban problems. The need to address urban poverty, food security, and nutrition is equally critical in fragile states. The private sector should build linkages between a country's domestic food production and its urban markets, while public policy should pay special attention to the employment, health, and nutritional needs of the poor. More research is needed on the food security and nutrition needs of the urban citizens in the emerging world (CSIS 2013).

Fish production could well change due to global climate change. Warming oceans could result in distributional changes, with fish moving polewards or into deeper and warmer waters. Substantial changes in the dominant species are possible in the world's most productive marine ecosystems. In addition, ocean acidification is likely to change the composition of the plankton, with unpredictable effects on the fish populations.

Given the limited potential for further growth in world capture fisheries production, future demand will mainly rely on a substantial increase in aquaculture production. Global aquaculture has grown at an impressive rate over the past decades and could provide two-thirds of world fish production by 2030. Further substantial expansion of this production is possible, and even essential if world supplies of fish products are to keep pace with human population growth. China and other countries are increasing their investments in aquaculture to help meet this mounting demand. European aquaculture is subject to strict environmental and animal health rules and follows very high standards of safety and sustainability (FAO 2013, 2014; OECD–FAO 2014).

Global freshwater aquaculture production is currently around 37 million tonnes annually and is growing at a rate of around 7 % per year. This production is important for local food security, but has probably little scope for expansion as the productivity of these areas is close to fully exploited. Marine aquaculture production is around 18 million tonnes, of which around 14 million tonnes shellfish and 4 million tonnes finfish. Marine finfish farming has a high potential for expansion but presently needs substantial marine protein inputs.

Aquaculture and capture fisheries are co-dependent, as feed for many high-value cultured fish is provided in large part from capture fisheries. They are also competitors for space in coastal areas, markets, and potentially for labour, and governmental support. Marked progress has been made in replacing feed sources from capture fisheries with agricultural production. Feed conversion and substitution of fish-based feed with products and especially waste from land-based agriculture and forestry could be instrumental.

Sustainable aquaculture promotes consumption and production patterns that preserve the natural resources basis. Main challenges include adequate space, feed, and the development of breeding technology. More space for marine aquaculture can be freed up by the development of extensive, offshore production systems that can use remote locations. Progress in feed technology could help marine fish to be produced more efficiently and with a higher than present proportion of feed of plant origin. Improvement of the breeding technology could make new species available for sustainable aquaculture. Marine shellfish farming requires limited inputs and provides important environmental services by removing nutrients from the water. This type of farming is particularly sensitive to water quality and research is focusing on selective breeding of the most resistant shellfish families.

Further development of aquaculture should consider many synergies. The development of offshore wind farms and/or tidal energy harvesting systems could create areas suitable for siting aquaculture installations, so allowing increase productivity. Integrated multi-trophic aquaculture, e.g. the integrated farming of algae or shellfish and finfish, could be further developed to ensure the optimal use of available space. Biofuels from cultivated algae could be developed in integrated multi-trophic aquaculture. The integration of activities such as angling and tourism in extensive aquaculture ponds could add environmental and landscape management services.

Both finfish and shellfish farming need appropriate sites and can be in competition for space with other coastal economic activities. Strategic planning for sustainable urban development tries to develop synergies among activities which are often seen as antagonistic, such as shipping, tourism, recreational fishing and aquaculture. Possible adverse environment impacts of aquaculture have also to be prevented and addressed. They include localised eutrophication events, changes in benthic communities, contamination from antibiotics and introduction of non-indigenous species. The development of genetically-modified marine organisms tailored for high aquaculture productivity may offer benefits but also raise ethical issues for marine environmental protection and for social acceptance.

The development of offshore wind farms could create new areas suitable for aquaculture installations. Integrated multi-trophic aquaculture, i.e. the integrated farming of algae or shellfish and finfish, can ensure the optimal use of available space. Second-generation biofuels from residues of fish and algae could be developed and provide energy for shipping or coastal activities. The integration of activities such as angling and tourism in extensive aquaculture ponds could add environmental and landscape management services.

Coastal cities can also support the development of alternative fish production systems, investing for example in demonstration project of urban aquaponics, combining aquaculture with hydroponics, growing plants and vegetables in mineral nutrient solutions in a symbiotic environment. In an aquaponic system, water from an aquaculture system is fed to a hydroponic system and the by-products are used by the plants as nutrients. The water is then recirculated back to the aquaculture system. Coastal cities can offer the testing grounds for simulated wetlands with fish waste fertilising plants and the plants filtering the water. Aquaponics has been suggested by the STOA service of the European Parliament in 2015 as one among the ten technologies which could change human lives (STOA 2015).

3.3 Exploration and Preservation of Precious Marine Resources

Oceans cover most of the planet's surface at an average depth of 4000 m, forming earth's largest ecosystem and serving as the primary regulator of planetary chemistry. Every second breath that humanity takes comes from the oceans, which also provides invaluable marine resources and ecosystem services. Environmental health, food and energy security, transport, temperature and climate regulation, and carbon sequestration are dependent upon the global ocean.

Urban biodiversity maps of coastal cities traditionally include only what is above the land and omit the rich and diverse world under the waves and the marine wealth of the nearby seas and of the oceans. The 2010 the decade-long Census of Marine Life, a major international effort, brought impressive results and produced a comprehensive inventory of known marine life. The census investigated life in the global ocean from microbes to whales, from top to bottom, and from pole to pole and helped discovering more than 1200 new species, with another 5000 or more awaiting formal description (Snelgrove 2010).

Many of the discoveries necessary for the future of humanity are still immersed in the deep ocean. As little explored as the space that surrounds the Earth, the global ocean is a huge biogeographic puzzle. Some million species have been estimated, of which science has only described some 10 %. The first census established a baseline of marine life abundance, diversity and distribution, to help understand, assess and foresee changes in the global marine environment, as well as to inform the conservation and exploitation of marine resources. New vision-expanding techniques included genetic barcoding for the identification of species, tracking methodologies and acoustic innovations. While advancing technology, scientists also set to work on standardising sampling protocols comparing species across the global ocean, and offering a clearer picture of marine populations (McIntyre 2010).

The census has created major legacies such as the data base OBIS (Ocean Biogeographic Information System), the world's largest open access, online repository of geo-referenced marine life data under the auspices of the UNESCO International Oceanographic Commission's programme and the International Oceanographic Data and Information Exchange. It has also supported frameworks to aggregate information about marine life, making it possible to estimate the number of species newly described by the entire marine taxonomic community.

Coastal cities can direct resources toward ocean exploration and health, leveraging assets at every level, from the municipality board to the chamber of commerce, citizen associations and the public transport and energy enterprises. Schools can be very creative in promoting ocean literacy that depicts the close relationship between cities and oceans. The spectacular global map from the Census of Marine Life that flips the perceptual field to emphasise oceans is a good example. Terrestrial areas are black or empty, while oceans are presented in blue, with underwater topography, habitats and migration patterns. Other maps are being made possible by new technologies that enable researchers to track previously invisible marine animal movements.

Cities across the globe interact with a significant diversity of species, habitats and ecosystems. Urban biodiversity surveys are essential and they have to equally consider the underwater and the terrestrial environment. The Comprehensive Marine Biodiversity Survey of Singapore launched in 2010 was the first concerted effort to comprehensively catalogue its marine biodiversity. Carried out in phases over 5 years, the survey brought together the broader community of experts, citizen organisations, government agencies and volunteers to collect biodiversity information on mudflats, intertidal areas, coral reefs and the seabed of Singapore. A diverse group of citizens, experts and organisations contributed to this large-scale, multi-disciplinary purposeful project.

Singapore hosts more than ten ecosystems within its bounds and the recent surveys have recorded more than 500 species of plants and animals new to the island, of which more than 100 are new to science. Cape Town is host to almost 50 % of South Africa's critically endangered vegetation types and 3000 indigenous plant species. Such urban development trajectories are critical for rich biological resources and their potential ecosystem services.

Ocean literacy can be a noble cause for coastal cities and link them to the conscience of the planet. Conscious coastal cities can raise awareness about the most crucial marine biodiversity threats. Each blue city can adopt a marine habitat or waterfront space or an exceptional underwater heritage site. School children can learn about the protected resource, neighbourhood associations can support cultural exchanges, and corporate donors can fund research and intervention actions (UNESCO World Heritage Centre 2013).

Public institutions can promote a blue ethic that encourages informed personal engagement and enables shifts in public perceptions, policies and incentives. In Lisbon, the Oceanario, 1 of some 140 public aquaria in Europe, offers a promising model. The Oceanario attracts about one million visitors annually with its message of "One Ocean". Exhibits include information about lifestyle changes to reduce impact on the seas. The aquarium works with schools throughout Portugal to help teachers educate children about themes such as overfishing, or ocean acidification. Vasco, an ocean mascot recognised by children across the country, serves as a model of sustainable living.

With advancing climate change, the evolving role of the ocean merits particular attention. Blue carbon sequestrated by the vast sinks of the oceans and coastal ecosystems amounts at 20–35 % of anthropogenic CO_2 emissions, and plays an important role in mitigating the effects of climate change. Blue carbon sinks, including mangrove forests, sea grass beds, and other ocean habitats, can sequester up to five times as much carbon as tropical forest. Mangroves are reputed to be outstanding carbon sinks, able to accumulate, store and sequester carbon, but 20 % of these have been lost between the years 1980–2005. Tidal marshes and sea grasses are also being destroyed at a rapid pace. These ecosystems have stored carbon for centuries and also provide other ecosystem services, including supporting fisheries, contributing to clean coastal waters, and protecting coasts from floods, storms and erosion. Coastal cities can play an important role to enhance coastal blue carbons through the prevention of threats to these ecosystems, including pollution and unsustainable coastal developments and the restoration and sustainable use of coastal and marine ecosystems (Rockefeller Foundation 2014).

Ocean acidity, "the other CO_2 problem", has increased by 26 % globally since the industrial revolution and is projected to further increase unless global carbon emissions are significantly curtailed. It could cause irreversible damage to marine species. Informing society about the growing threats of ocean acidification through education and outreach is essential (IPCC 2014).

Global marine ecosystem services are likely to be impacted by the acidification of the seas. The effect of ocean acidification on marine crustaceans can be very harmful, as decreases in survival, growth rate, and egg production have been reported for some species. Effects on non-calcifying organisms have also been demonstrated, including on the development of larval stages of some fish and on the ability to detect predators. Reduced survival and growth of sea urchins, sea stars, sea cucumbers, and brittle stars may also occur.

Ocean acidification is projected to differ among regions. Tropical waters hosting coral reefs are expected to experience the greatest overall change and polar waters may become corrosive by 2100. Socio-economic modelling suggests that the impacts on marine habitats and marine resource availability can be serious and cause substantial revenue declines, job losses, and economic costs. Effects to human communities would include changes in shellfish harvest, coral and oyster reef ecosystem services and indirect impacts across marine food webs. Coral reefs provide a habitat for an estimated one million species and offer food, income generation for tourism, and coastal protection against storms for about 500 million people globally (Gore 2013; NOAA 2013). The alterations could be irreversible and unbearable.

Some started already grieving for exceptional reefs. The Economist previsions for 2015 alerted that the Great Barrier Reef, the world's most complex expanse of coral reefs extending about 2300 km along Australia's eastern state of Queensland, is in danger. The Great Barrier Reef is a natural monument, hosting 400 types of coral, 1500 fish species, whales, dolphins, seabirds and turtles. But the reef has never been more threatened. Over the last decades, the reef's coral has been disappearing. Marine scientists largely blame rising sea temperatures and acidification, linked to climate change, and nutrients and pesticides entering into the Coral Sea and the Pacific Ocean. Thirty-four years after having listed the reef as a World Heritage Site, UNESCO will decide in June 2015 whether to add it to the World Heritage in Danger list (UNESCO World Heritage Centre 2014).

The federal and Queensland governments produced a draft plan in late 2014 for protecting and managing the reef up to 2050. Its tourism industry, worth about $4.6 billion a year to Queensland alone, is founded on the health and the natural beauty of the Great Barrier Reef. Visitors to the Great Barrier Reef can enjoy many experiences including snorkelling, scuba diving, aircraft or helicopter tours, glass-bottomed boat viewing, semi-submersibles and educational trips, cruise ship tours, whale watching and swimming with dolphins. All these activities are undermined from the risks to the sustainability of the Reef (Economist 2015).

3.4 Building Strong Urban Resilience to Climate Change

Planet Earth is already 0.8 °C warmer in average than in pre-industrial times. As all average values, this conceals an array of unequal realities and impacts. Communities around the world could experience the harsher impacts of a 2 °C warmer world within 20–30 years, and 4 °C is likely to happen by the end of the century in the absence of decisive concerted global action. International scientific cooperation has been intense and global governance was organised around the Kyoto Protocol, which became law for the 141 countries that had ratified it by 16 February 2005. The Lima Accord, reached at the conclusion of COP 20, marks the first time that all nations have agreed to cut carbon emissions. Although non-binding, it is an encouraging development that sets the stage for more action ahead towards the next steps which will define the post-2015 horizon.

Many analyses describe the COP 21 (Paris, December 2015) as the last chance milestone and suggest that the sooner countries prepare for the inevitable march of time towards a low-carbon future, the better and the less costly. The longer countries wait, the more their carbon-intensive assets will lose value in a low-carbon future. Economies that are not planning their transition to a low-carbon future are vulnerable to unpredictable events such as consumers and investors turning against carbon-intensive assets.

Political decisions involve value judgments, but scientific knowledge can play a key role in, for example, analysing the potential effects of climate change, and their likelihood, identifying and evaluating the potential consequences of various policy responses, expanding the portfolio of possible options, and improving the effectiveness of policies. Robust and long-term climate observations are crucial for reliable foresights and evaluations to illuminate and orient political action. A call for a new era of climate change science and policy highlighted the importance of "citizen-inspired" research, which not only improves understanding of the causes and consequences of climate change, but assists decision makers at the local, regional, national, and international levels in mitigation and adaptation actions (NRC 2010).

Cities are both drivers of climate change and privileged places of concentrated climate responses. The decade 2001–2010 was the warmest period ever recorded. The past few decades have been warmer than any other comparable period for at least the last 400 years. Limiting the global average warming to 2 °C above preindustrial levels and ensuring the survival of humanity on Earth is likely to require emission reductions larger than 80 % below peak levels (IPCC 2013; Footprint Network et al. 2012).

Cities represent a leading force for global action on climate change. Economic growth mainly taking place in cities has been the main cause of the continuous increase of CO_2 emissions since 1990. Climate change is partially resulting from millions of decisions made by the world's citizens within their immediate environments. In 2007, urbanising China overtook the US as the biggest emitter of the world. It is not urbanisation per se, but rather higher levels of income and the related consumption patterns that drive the higher GHG emissions. Urgent action is needed in both the developed and the emerging world, in anticipation of a new investment cycle of capital renewal and expansion, mainly for urban infrastructures.

Cities have a responsibility to create solutions to climate change. Acting both locally and in network, they can have a meaningful global impact. Each city is unique in its morphology, culture, infrastructure, municipal services, and potential impacts to and from climate change. However, the underlying drivers of emissions in cities largely reflect the same inefficiencies and include energy losses, traffic congestion, unsustainable resource and waste management, water leaks and biodiversity losses.

The impacts of climate change pose significant social, environmental and economic threats and risks to the urban, national and global communities. Coastal cities are particularly vulnerable. More frequent and extreme weather events, especially rains and floods, heat waves and droughts, increasing temperatures and rising sea levels seriously affect livelihoods, food and energy supply, infrastructure, ecosystems and society and the economy as a whole. Concerted responsible action guided by unshakable commitments and assisted by performance monitoring and reporting, can help build a strong case for local climate action. Most climate-conscious cities involve citizens and local stakeholders in climate commitments to mobilise action and provide guidelines for implementing policies to reduce greenhouse gas emissions.

High urban densities and short distances yield many opportunities for low-carbon lifestyles, such as the use of cycling paths or public transport. In most countries, energy use per capita of urban residents is lower than the national average. In cities in the emerging world, climate impacts may aggravate poverty conditions and social inequalities. Climate change could also exacerbate other existing knotty environmental problems such as low air quality and poor water supply. More and more climate reports warn that poor cities and communities are the most vulnerable to climate change. As the coastal cities of Africa and Asia expand, many of their poorest residents are being pushed to the edges of liveable land and into informal settlements often clustering in low-lying areas with poor public infrastructures and services and inadequate protection from storm surges, and flooding (World Bank 2013).

In the industrialised world, many cities engineer advanced protection measures. Venice plans ingenious gates, the Experimental Electromechanical Module, a flood barrier system, to seal the lagoon and tame the destructive tides. The city is notoriously vulnerable to flooding and “aqua alta”, the Adriatic’s high tide water, which partly caused its sinking by 23 cm during the twentieth century and became more frequent over the last years. The rise in sea level could be very impactful, since it is estimated that, in the absence of drastic measures, a 30 cm rise in sea level would flood St Mark’s Square 360 times a year. The completed project is expected to include 78 mobile barriers blocking the three inlets to the Venice lagoon. The barriers will be housed in enormous tanks anchored to the seafloor and will rise whenever high water threatens to flood the city. Once the threat is over, the released air will allow the barriers to sink down again. The system would only be mobilised to address the most extreme risks.

Climate change happens everywhere but it impacts very differently the cities of the various world regions. Water scarcity in some areas and overabundance of water in others are the hallmarks of climate change in many parts of the globe. Inconsistences in the monsoon season and unusual heat extremes may have extremely serious consequences. Dhaka, Kolkata and Mumbai may be confronted with increased flooding, intense cyclones, sea-level rise and warming temperatures. In South East Asia, coastal cities may have to face intense stress. A sea-level rise of 30 cm, possible by 2040 if inaction continues, would cause massive flooding in cities and inundate low-lying cropland with saltwater corrosive to crops. The Mekong Delta, in Vietnam, a global rice producer that is particularly vulnerable to sea-level rise, could lose much of its crop production (World Bank 2013).

The most affluent global cities have to show the way forward. In Tokyo, the top coastal world city, most major corporations, such as Sony and Shimizu, are proud of green low-emission headquarters. The government suggests that the era of eco-buildings is rising on the megacity. In 2010, Tokyo introduced the first urban cap and trade programme in order to reduce its GHG emissions and reach its target of 25 % less emissions by 2020 comparing to its 2000 emissions. The scheme covers 1400 installations, including 1100 business facilities and 300 factories which are large CO_2 emitters. Although these account for only around 0.2 % of some 700,000 industrial and commercial facilities, their carbon dioxide emissions in 2007 stood at about 20 % of total metropolitan emissions. Carbon revenues have to be truly and transparently invested in sustainability projects. Most successful campaigns insist on the good communication of tax reforms so that the carbon taxes are not perceived as part of an intelligent structural tax reform (Tokyo Metropolitan Government 2010).

The cap for the first compliance period (2010–2014) has been set at a level of 6–8 % below 2000 emissions and for the second compliance period (2015–2019) at a level of 15 %. Monitoring is permanent and compliance assessment follows the completion of each phase. Mandatory reporting of emissions is an important prerequisite and asks for the cooperation of all, in order that consensual action is implemented. Annual reports insist on the importance of a well-designed consultation process and the capacity of the programme to bring together developers, owners and tenants to curb CO_2 emissions in buildings. The second year report presents remarkable results, far beyond obligations. In total, a 23 % reduction was achieved versus the obligation for 6–8 and 93 % of the participating facilities overpassed their obligations (Tokyo Metropolitan Government 2012).

Coastal cities, like Amsterdam and London, acted early and formulated climate objectives and targets, conceived mitigation and adaptation policies and created dedicated agencies to monitor progress in achieving the objectives. Many cities see adaptation as an opportunity for better urban planning and policy to develop the adequate infrastructures, improve quality of life and create new innovation trajectories and employment possibilities.

In order to confront the challenge of climate change as an opportunity rather than a threat, the City of Rotterdam launched the Rotterdam climate-proof programme in 2008 with the aim of making the port-city resilient to climate change by 2025. Permanent protection and accessibility of the Rotterdam region are key elements. Rotterdam has started to adapt in order to strengthen its position on the global chessboard. The central focus of the adaptation programme is to create and grasp further opportunities to make Rotterdam a more attractive city for citizens, visitors and investors. The Rotterdam Climate Initiative aims at improving the climate for the benefit of people, the environment, and the economy. The initiative adopted an integrated approach and ten sustainability tasks (City of Rotterdam 2010). Since the end of 2013, Rotterdam has its Climate Change Adaptation Strategy to prepare for Climate Change. Furthermore, the city, which lies 6 m below sea level, making it the lowest city in Europe, is helping Jakarta in developing a dredging plan. Amongst other things this includes Dutch floating bulldozers being used to deepen rivers, decreasing the impact of flooding during extreme rainstorms.

In 2007, Boston developed a city-wide Climate Action Plan. The plan aims at reducing greenhouse gas emissions 25 % by 2020 and 80 % by 2050, through taking into account climate in all formal planning and project processes, engaging all communities and developing innovative businesses and skills to take advantage of climate action opportunities. The city was the first in the region to adopt Green Building Zoning, and to include climate resilience to the large new construction review process. Additionally, a Green Ribbon Commission brings together businesses, non-profits, and community leaders from a variety of sectors to develop shared strategies for fighting climate change in coordination with the city's Climate Action Plan, which is being updated every 3 years.

Boston has been designated among the cities of the first cohort of the US Climate Action Champions (White House 2014). A Climate of Progress, launched on Earth Day 2011 encompasses the recommendations from the Climate Action Leadership

Committee and the Community Advisory Committee. Boston's update climate action plan, launched in 2013, focused on climate preparedness and community engagement. The city is interested in developing indicators to assess progress in achieving its climate objectives. The City of Boston, the Boston Harbour Association, EcoAdapt, and SeaPlan are working together to develop and vet a climate change preparedness indicators framework to track and evaluate progress (Boston Foundation 2011).

The climate of the New York metropolitan region is changing, with annual temperatures getting hotter, heavy downpours increasingly frequently, and sea levels rising. In New York City, the creation of the Office of Environmental Remediation and the release of the Sustainable Stormwater Management Plan marked more steps forward. The second phase of "Schoolyards to Playgrounds" was launched for municipal buildings as part of the commitment to reduce City government greenhouse gas emissions. The New York City Panel on Climate Change (NPCC), an independent body that advises the City on climate risks and resilience, works in partnership with the City, to assist achieving the most ambitious target of reducing GHG emissions by 80 % by 2050.

The NPCC latest report includes climate projections through 2100, new coastal flood risk maps to the end of the century for the current 100- and 500-year coastal flood events, enhanced dynamic flood inundation modelling of future coastal flooding that includes the effects of sea level rise, a review of key issues related to climate change health risks for citizens and a process for enhancing a New York City climate resiliency indicators and monitoring system. Mean annual precipitation has increased at a rate of approximately 0.8 in. per decade over 1900–2013 in Central Park. Sea level rise in New York City has averaged 1.2 in. per decade since 1900, nearly twice the observed global rate of 0.5–0.7 in. per decade over the same period. These trends are projected to even worsen in the coming decades. The mean annual temperatures are projected to increase by 4.1–5.7 F by the 2050s, and the mean annual precipitation to grow from 4 to 11 % by the 2050s. The frequency of heat waves is projected to increase from two per year to roughly six per year by the 2080s. Projections for sea level rise in New York City, in relation to the 2000–2004 base period, suggest an increase between 11 and 21 in. by the 2050s and between 22 and 50 in. by 2100, with a worse case projection of up to 6 ft by 2100 (NPCC 2015).

Copenhagen's 2009 Climate Plan was a conscious milestone and the first in Scandinavia. Copenhagen pledged to become CO_2 neutral by 2025 and provide the world an international gold standard for sustainable cities. In order to achieve its ambitious goal, the city has established precise targets including for energy efficiency and renewable energy sources, and green building standards. All new buildings have to be carbon neutral by 2020. Cycling rates exceed 40 % of all commutes. The city also developed a smart bike equipped with sensors to provide real-time information not only to riders but also to administrators for open data aggregation on issues of air quality and traffic congestion.

As most coastal cities, Copenhagen has a symbiotic relationship with the sea, strengthened and demonstrated in times of heavy thunderstorms and exceptional rain levels. A network of temporary reservoirs was built, already in the 1990s, to store excess rainfall and wastewater, preventing the overflow of sewage systems and the risk of flooding in the city. The works also improved the surrounding environment. Since then, the water in the harbour is clean enough for swimming. More frequent intense rain could require infrastructure expansion in the future to maintain the achieved high standards, along with other measures to increase the city's resilience to climate change. A torrential rain in 2010, a once in 100 year's event, stressed the vulnerability of the city and put enormous pressure on its drainage infrastructure.

The Climate Plan asked for green roofs to be integrated in city policy instruments. One of the district plans focuses on North Harbour, one of the largest urban developments in Northern Europe. The long-term vision for the area is to turn it into a true and diverse city district, hosting up to 40,000 citizens. The creation of North Harbour made strides towards the combined model of an eco-friendly city, a vibrant city, a city for all and everyone, a city by the water, a dynamic city and a city with sustainable energy and mobility (EC 2013a).

The Copenhagen Green Roofs offered an inspiration for Recife, in Brazil. The Green Roof Law approved by the Municipal Council of Recife in 2014 requires buildings with more than four floors to have their roofs covered with native vegetation. The Torre Charles Darwin, a 35-storey building in the city centre, serves as the first example of a green roof building. The building will have a cover crop of 2.8 million square feet, along with a rain harvesting tank to power the air conditioning system. The city has more to showcase as member of a world league of excellence. As much as 46 % of the city's total area is green, 60 % of which is protected under conservation laws. A plan for enhancing bike lanes and the creation of bus corridors to facilitate commuters and reduce private cars are among the initiatives undertaken by the city government to reduce its impact on the environment and greenhouse gas emissions. The Apibaribe River Navigability Project is focused on ensuring that the 6 rivers and 66 canals of the city are used as alternative sustainable routes.

The national-local dialogue is crucial to reduce GHG emissions. The Local Government Climate Change Leadership Summit, in June 2009, was a prime milestone in the process leading up to the climate change COP 15 in Copenhagen, in December 2009. It advocated for a national-local dialogue and a partnership with cities to reduce GHG emissions. The City Climate Catalogue, an interactive instrument highlighting accomplishments versus objectives, was launched 5 months earlier to provide a substantial contribution to national governments in the international climate negotiations. All cities are invited to contribute their greenhouse gas reduction targets and engage in concerted actions to achieve and overcome them.

The Durban Adaptation Charter, adopted in 2011, recognises that the majority of climate change impacts and the appropriate proactive and reactive responses will occur at the local level. It marks a clear milestone in inviting local governments to adopt a holistic ecosystems-based approach for developing secure city-region food systems, work for urban poverty eradication, protect and enhance local biodiversity and strengthen urban resilience. The charter stressed that the costs of adaptation and loss and damage resulting from climate change should be covered from local, subnational, national and global sources. It called for new modalities of interaction for local and subnational governments in order to fulfil their role as critical stakeholders and decisively contribute to national and global effort for a better future for all.

Alliances among cities working together to decrease emissions are extremely important for mutual awareness and noble emulation. The World Mayors Council on Climate Change is an alliance of committed local leaders concerned about the contribution of their cities to climate change. The Council was founded in December 2005 by Y. Masumoto, Mayor of Kyoto, soon after the Kyoto Protocol entered into force in February 2005. The members of the Council advocate for enhanced engagement of local governments as stakeholders in multilateral efforts addressing climate change and related issues of global sustainability.

The C40 coalition, also created in 2005, forged a partnership in 2006 with the Cities Programme of Clinton's Climate Initiative (CCI) to reduce carbon emissions and increase energy efficiency in large cities across the world. In less than a decade, the coalition positioned itself as a major global actor for implementing meaningful and sustainable local actions to help address climate change. In 2011, the C40 created new partnerships with the World Bank and ICLEI to accelerate climate action in cities through streamlined financing, greenhouse gas accounting and uniform reporting. The release of two reports developed in collaboration with the Carbon Disclosure Project and Arup respectively, emphasised the critical role of measurement and transparency in tackling climate change in megacities.

Both for cities and companies, measuring and disclosing the amount of emissions are the first steps for preventing and mitigating global warming and also creating a new culture for adaptation. The Carbon Disclosure Project (CDP), a non-governmental international endeavour, launched a transformative global system for cities and businesses to measure, disclose, manage and share climate change and water information. The CDP brought together over thousands of organisations across the world's largest economies to report their greenhouse gas emissions and share their assessed climate change risks and opportunities, in order to set reduction targets and improve performance. Some leading companies have moved to become carbon neutral, while other organisations managed to reduce greenhouse gas emissions by adopting ambitious initiatives.

In 2014, 207 cities disclosed their climate mitigation, adaptation and water management data. Cities also reported 757 adaptation activities and 102 urban climate adaptation plans. Cities are reducing the climate risks faced by citizens and businesses through investment in well-performing infrastructure and services and by developing impactful policies and incentives. The benefits that business brings to cities, including jobs, tax revenue and services, are among key drivers for cities to improve their climate resilience. Similarly, businesses are reliant on public infrastructure and policies to support and guide their operations. Both public and private sectors can benefit from a greater understanding of each other's climate change risks, and companies can help reduce city-wide risks by embedding local adaptation needs within their operations (CDP 2014a).

Transparent disclosures, including to investors and citizens, enables local and regional authorities to underline commitment, reduce risks and associated insurance costs, and demonstrate their visions and values in a competitive globalised world. There is significant opportunity for cooperation between governments and business to improve climate resilience. Companies worldwide seem already ahead of their governments in planning for climate change risks, costs and opportunities. They are calling for clear pricing and regulatory certainty in order to better plan their activities and their climate-related investments, and more secure, interconnected carbon markets (CDP 2014b).

Peer pressure and network emulation are crucial for mitigation and adaptation. The CDP Public Procurement Programme channels information through five distinct programmes, one of which focuses on cities. The project is designed to enable local and national governments to create new climate-friendly markets. This is an effective way for local and national governments to ask their suppliers about energy consumption patterns and climate implications. This information allows governments to better understand climate change risks, which in turn can help work towards building a low carbon government supply chain.

Climate change risk assessment is increasingly being adopted by coastal cities. In search of an advanced methodology, the EC-supported RAMSES project presented urban climate impacts and estimates of adaptation costs and benefits using a common currency. The aim is to empower policy makers, businesses and the civil society to make direct comparisons across cities and activities. The project offers a generalised approach on key infrastructures and characteristics and their relation to climate mitigation and adaptation. This is complemented with selected case studies and simulations of the effects of climate change. The analysis of the institutional and political context can help establishing the transition strategies for specific cities including Antwerp, London, New York and Rio de Janeiro. The project is expected to improve understanding of urban systems and provide quantitative evidence of the costs and benefits of mitigation and adaptation in cities.

Awareness about the effects of climate change before extreme events occur is necessary for drastic changes in city and regional management. Innovations bringing multi-dividend responses are the most powerful. Strong ecosystem-based spatial planning especially for risk-prone areas can be an effective and sustainable way to deal with risks. Keeping public space and buildings cool by using for example green roofs, rather than air conditioning, could help saving energy and resources.

Extreme weather events resulting in hazards such as heat waves, floods and droughts are expected to happen more frequently and affect quality of life for citizens and attraction of visitors. Urbanisation, population ageing and other socio-economic trends interact with climate change and can compromise public health, reduce productivity and constrain the functionality of infrastructures. An ageing population in OECD cities increases the share of citizens vulnerable to heat waves and asks for additional urban services.

Urbanisation can increase the vulnerability of places and ecosystems. The replacement of natural vegetation with artificial surfaces and buildings impacts temperature, moisture, wind and rainfall patterns. Excessive amounts of rain water cannot drain into the ground, especially in sealed urban areas, and this can generate or worsen floods. Artificial surfaces store heat and cause higher temperatures in cities compared to the surrounding agricultural or forest land.

Most cities initiated their adaptation journey and adopted plans to reinforce their capacity to withstand threats. Copenhagen's Climate Plan includes an adaptation plan with a range of interlinked environmental initiatives. To reduce the risk of flooding, vegetation throughout the city expanded through green walls and roofs, as well as through the creation of a network of "pocket" or "climate parks", i.e. small green spaces artfully integrated in the urban fabric. Well-designed small green areas dotted throughout the city can slow rainfall run-off and reduce the risk of flooding.

Green and blue areas help the city to cool down during summer months and cope with the expected higher temperatures. Sea and vegetation areas have an immediate refreshing effect as they hold moisture and release it into the air. Conversely, artificial surfaces tend to absorb and retain heat. Furthermore, beaches and parks also increase recreation spaces for residents and visitors and improve conditions for biodiversity and human well-being. Clean urban beaches, often after "dirty waters", as in Boston, provide clean bathing waters.

Melbourne's adaptation action began in 2008 with the publication of "Future Melbourne—City of Melbourne" which acknowledged that climate change presents key strategic risks for the community and must be collectively addressed. By 2030, Melbourne is likely to be significantly affected by warmer temperatures, drought and heat waves, lower rainfall, intense storm events, sea level rise and flash flooding. The plan for the future direction of the city was developed through open collaborative public engagement with key stakeholder groups including professional networks, other city councils, the Victorian and Australian Governments, universities and non-government organisations. Future Melbourne provided the impetus to undertake a comprehensive climate change risk assessment, culminating in an integrated City of Melbourne Climate Change Adaptation Strategy and supporting Action Plan and the ongoing implementation of action plans enhancing the resilience and diversity of urban forests, and design strategies for cool roofs to reduce the Urban Heat Island effect (C40 Cities and Siemens 2014).

In the European Union, the Adaptation Strategy focuses on promoting action by Member States, devising climate-proofing action at EU level ensuring that Europe becomes more resilient and decision-making better informed. The "EU Cities Adapt" project was launched in 2012 together with the European Climate Adaptation Platform (http://climate-adapt.eea.europa.eu), the "one-stop shop" for information and policy advice, to help cities adapt. The Initiative "Mayors adapt—the Covenant of Mayors Initiative on Climate Change Adaptation" was promoted to engage cities in taking decisive action to adapt to climate change. Cities signing up to the initiative commit to contributing to the EU Adaptation Strategy, through a comprehensive local adaptation strategy or the integration of adaptation measures into relevant existing policies. Mayors Adapt aims to increase support for local adaptation activities, provide a platform for greater awareness and networking, and a heightened attention and public dialogue (EC 2013c, EEA 2012).

Resilience fortifies the immunity of cities against irreversible losses. Resilient cities and communities are the ones that have citizens, businesses, institutions, organisations, assets and flows which can withstand economic, social, environmental and political threats without serious interruptions or irreversible damage. Cities reaching turning points in many regions of the world should invest as early as possible to prevent disasters. European cities have started providing models for the design and organisation of urban transformation processes and the mainstreaming of adaptation measures into spatial planning. The success of the annual global forums on Resilient Cities which attracts experts and decision makers to Bonn, every year since 2010, signals the interest of local governments in adaptation and urban resilience. Handbooks for making cities more resilient contributed to the global campaign 2010–2015 (UNISDR 2012).

The Mayors Adaptation Forum, the leadership segment of the annual global forums on Resilient Cities, fosters close dialogue between local and global leaders, engages in a comprehensive debate of all major strategic issues linked to urban resilience and helps advancing towards concrete political commitments. The 2013 Bonn declaration of Mayors, during the adaptation forum, expressed concern that the historical threshold of 400 ppm CO_2 was crossed in May 2013, underlined the inter-linkage between climate change action and the overall development agenda, and confirmed commitment to globally coordinated local climate action. The declaration welcomed the Durban Adaptation Charter as a powerful mechanism to enhance urban adaptation action by building networks of communication and cooperative action among local governments world-wide, creating a global platform for the dissemination of best practice.

The 2013 Open European Day coincided with the final conference of "EU Cities Adapt", a key event for urban adaptation in Europe. In this framework, European cities, in different stages of adaptation, have worked together towards well-informed and enhanced actions on climate change adaptation. This Open European Day, jointly organised by ICLEI, the European Commission and the European Environment Agency, became a regular annual platform for European cities to exchange experiences on practical challenges and solutions towards local climate resilience. The 2014 event focused on vulnerability and risk assessments, the economics of adaptation and multilevel-governance cooperation and approaches (ICLEI 2014).

In the US, the White House announced measures to help cities adapt in autumn 2014. A task force of US governors, mayors and other leaders focused on recommendations for policies to incorporate climate change and provide tools to help local communities better design their adaptation measures. A Web-based climate resilience toolkit has been designed to help local leaders adopt measures to prepare cities for facing extreme events of rising sea levels, droughts, diseases and other climate impacts. Building a culture of resilience is a shared responsibility among citizens, the private sector, and governments. Sixteen local and tribal communities were selected as Climate Action Champions for their decisive action to cut carbon pollution and build resilience. The selected communities benefit from facilitated peer-to-peer learning and mentorship (White House 2014).

In the emerging world, poverty and informal housing neighbourhoods undermine resilience and social cohesion. Although the poorest metropolitan area in South Africa, Durban strives to be a global leader in climate change adaptation. Rich in its racial, linguistic and cultural mix, including English (49.8 %), Zulu (33.1 %), Xhosa (5.9 %) and Afrikaans (3.6 %), Durban is actively exploring ways to adapt institutions, systems, and processes in order to facilitate integrated, innovative, and flexible planning for a more secure future. This approach can help build the foundation for a more resilient society, economy, and governance.

Present and future buildings and urban infrastructure need huge investments, alongside those for climate proofing, in order to continue supporting urban services even under extreme phenomena. Urban planning and architecture should incorporate lessons from humanitarian emergencies and the creation of temporary shelters for refugees. Social innovations can also unlock multiple opportunities. Climate change adaptation concerns have to be incorporated into building design and standards and eco-retrofitting activities, to ensure that systems can respond under heavier conditions. Adaptation is strongly reinforced through green and blue infrastructure such as parks, wetlands, green walls and roofs, and ocean gardens, wherever feasible and sustainable.

Community resilience is as important as the resilience of the built environment. The US National Academies of Science Koshland Museum celebrated its tenth anniversary in 2014 with the opening of the Idea Lab, a space for visitors to experiment with new approaches to building community resilience, citizen's ability to come together to prepare for, absorb, recover from, and adapt to adverse events. Immersions in interactive games helped in addressing the fundamental questions and creating a model of a resilient coastal city with a jumbo jigsaw puzzle. From disaster simulation and the mapping of risks to alerts and connection to resilience networks, citizens had the opportunity to experiment and engage with uncertainty.

3.5 Blue and Green Urban Eco-Cells, Eco-Organs, Eco-Functions

To lead to sustainable development, cities have to continuously regenerate their bodies, their blood, their vital organs and nerves, but also their minds, with their individual ideas and concepts, and their souls, with their emotions and capacity to wonder. Coastal cities have also to regenerate their marine cells. Recent concepts of innovative cities include the flexible and even instantaneous cities that can adapt instantly, and the frugal or lean city, which enhances innovations to optimise resource use and minimise the generation of waste. Better use of resources in coastal cities is essential for preventing pollutants entering the marine environment. This also links to the concepts of circular or sufficiency economy.

Frugal ideas may inspire innovations for cutting urban costs to the bone, overcoming fragmentation among land and the sea, and leading to flexible modular amphibious products and processes. Frugal urban innovations exploit new possibilities focusing on the epitome of minimalism with all the functions stripped down to their bare essentials. From the initial invention to the final transformation, a chain of interactive processes can bring a thorough change that can further open up an array of opportunities (Haëntjens 2012).

Land, air and water are fundamental resources for all urban settlements. Urbanisation increases pressure on all ecosystems and results in soil, water and air pollution. Forests are being transformed into agricultural land or urban areas. Soil is a living medium which supports human life. It is a vital, natural, multifunctional, largely non-renewable resource that has to be sustainably managed. Soil sealing, the covering of the ground converted into urban land by an impermeable material, is one of the main causes of soil degradation. The expansion of grey infrastructure and urban engineering projects has disastrous consequences for biodiversity. Soil sealing often affects fertile agricultural land and fragile shores and results in the loss of important ecosystem services, such as food production, flood protection and biological diversity. It also increases the risk of flooding and water scarcity and contributes to climate change.

Everywhere in the world, land consumption has grown at more than twice the rate of population growth. In the European Union, an area five times that of Greater London has been consigned to urban sprawl during the last decade of the twentieth century. Comprehensive survey results for the years 1990–2000–2006 suggest that discontinuous peri-urban areas grew far more rapidly than continuous urban areas Although artificial cover accounts for just 4 % of the EU's surface, its intense use and multiple fragmentation and dispersion makes its impacts directly affect more than a quarter of the EU territory (EEA 2006).

The USA experienced large waves of urban sprawl and a surge of suburban satellite developments which sometimes expanded two long fingers on coasts from both sides of a city and some fingers radiating towards the hinterland and across major transport axes. Despite the fact that residential development along the coast affects only 17 % of the total US land, it creates a disproportionately higher impact (Beach 2002).

The extension of urban areas into previously green coastal areas allows some citizens to enjoy more living space in single-family houses, surrounded by gardens and with access to the nature and/or the sea. But it can create serious negative environmental, social and economic problems for society, in particular in the case of low density and scattered urban sprawl. These include increasing travel flows and energy consumption which affect air quality and coastal erosion.

Dispersed low-density residential developments far away from essential places such as work, home, school, and services, and highly-dependent on a car for travel, has increased the fragmentation and degradation of critical ecosystem services in the coastal environment. Important ecosystem services, such as the removal of pollutants from air and water, the mitigation of floods and drought, the protection of coastal shores from erosion and open space and wildlife habitat for recreation, are threatened with disruption and impairment.

Air pollution critically affects humans in both natural and built environments. Poor air quality, both ambient and indoor, can cause respiratory and other serious diseases. The primary source of air pollution is the combustion of fossil fuel in energy generation, industrial processes and transport. Exposure to air pollution is largely a multi-pollutant process, and, in coastal cities, shipping is a major contributor. Nitrogen oxides (NOx), particularly nitrogen dioxide (NO_2), are known to cause specific damage to lung tissues and to contribute to acidification, eutrophication and photochemical smog. The great majority of NOx emissions are produced by energy production and consumption. Despite cuts in nitrogen oxides and non-methane volatile organic compounds, air pollution due to ozone persists. Heat waves can further boost ozone exposure. Particular matter from transport and construction is another major polluter. In USA, more than half of citizens live in areas that do not meet the health-based air quality standards established by the Environment Protection Agency (NOAA 2012). In emerging countries, air quality problems could seriously undermine development and need particular attention from the scientific and policy community (Friends of Europe 2013).

Some emerging cities are notoriously polluted. According to a report by the Ministry of Environmental Protection, only 3 out of 74 Chinese major cities were able to meet national air quality standards in 2013 and 11 of China's most polluted cities were among the 13 cities in the Beijing-Tianjin-Hebei area. Even in the EU, the air pollution problem is still far from being solved, despite falling emission levels and reductions of some air pollutant concentrations in recent decades, and the European cities being among the cleaner in the world. Particulate matter and ground-level ozone continue to be a cause of breathing problems and cardiovascular diseases. Large parts of the population do not live in a healthy environment. Between 2009 and 2011, up to 96 % of city dwellers were exposed to fine particulate matter concentrations above WHO guidelines and up to 98 % were exposed to ozone levels above WHO guidelines. Lower proportions of EU citizens were exposed to levels of these pollutants exceeding the limits or targets set out in EU legislation, which, in some cases, are less strict than WHO guidelines (EEA 2013).

An innovative approach for making citizens highly aware about air quality is by making information available on a platform such as "Dublinked" in the Irish capital. Measurements from five air monitoring sites around Dublin city provide hourly results for Sulphur Dioxide (SO_2), Nitrogen Dioxide, Carbon Monoxide (CO), and particular matter. Black smoke monitoring is also carried out although this has been scaled down since the introduction of the coal ban in heating in the early 1990s (EC 2013b).

In coastal cities, the shipping industry has to become cleaner. Making shipping greener and preventing pollution was one of the original aims of the International Maritime Organisation (IMO), set up in 1948. Environmental rule-making has intensified in recent years as pollution became more serious. The International Maritime Organization asked to limit the sulphur content of maritime fuel, especially in emission-control areas along heavily populated coasts in North America and Europe.

Major port cities cooperate with the maritime industry to reduce air pollution. They have committed themselves to reduce greenhouse gas emissions through the World Port Climate Initiative (WPCI), impacting the sustainability of supply chains, taking into account local circumstances and port management structures. Shipping firms are also under pressure to cut their emissions of carbon dioxide and other greenhouse gases. The IMO reckons that ships cause about 2.7 % of total man-made emissions, a little more than planes but much less than cars and trucks. Ships will have to introduce fuel-economy measures with the aim of reducing their emissions by 20 % by 2020 and 50 % by 2050. The IMO is also pressing on with planned new rules on cleaning up ships' ballast water. An estimated 60,000 ships worldwide would need refitting with one or more cleansing units.

Making shipping greener is subject to technical, organisational, economic and social innovation. The shipping industry is showing increased interest in the use of

Liquefied Natural Gas (LNG) as a cleaner transition fuel towards renewable sources and a decarbonised future. Sulphur and particle emissions would be reduced to almost zero, nitrogen oxide emissions by 85–90 % and net greenhouse gas emissions by 15–20 %. Industrial partners and stakeholders, including port authorities, fuel suppliers and shipping companies, ask port cities to prepare to offer safe storage and bunkering of LNG. Singapore, the world's largest bunkering hub strives to lead on LNG bunkering by 2020.

Voluntary measures are usually preferred by industries that can use them in the most flexible way. The Environmental Ship Index (ESI) is a voluntary instrument designed to improve the environmental performance of vessels. It only includes ships that perform better than obliged by the international legislation defined by the IMO. It assesses the environmental performance of ships regarding air pollutants and CO_2 and scores NO_X and SO_X emissions directly and proportionally. The Environmental Ship Index is intended to be used by ports to reward ships participating in the scheme and can also be used by business associations wishing to share their environmental performance results. The programme is flexible and dynamic and can raise awareness among the global port community and improve the maritime and port environment.

Fresh water, the blue gold, is a vital and scarce natural resource that, unlike oil, cannot be replaced. Access to clean water is regarded as a fundamental human right. Intelligent infrastructure and surveillance systems in order to limit leakage, which often reaches 30 %, are being implemented or planned in many cities. Leakage and risk detection are increasingly parts of integrated management and early warning systems. The Tokyo system for identifying leaks is considered to be exemplary, while Stockholm and Singapore are well known for their proactive water management policies.

Major world cities literally suck water from their surrounding areas to satisfy the needs of their populations. A global survey suggests that large cities occupy only 1 % of the Earth's surface but the watershed providing their water covers 41 % of the land surface. Cities with at least 750,000 inhabitants move 504 billion litres of water each day in an accumulative distance of 27,000 km. Los Angeles, Boston, Mumbai, Karachi, Hong Kong, Tokyo, New York, Tel Aviv, Sydney, and Athens, all coastal cities, top the list of the cities that most suck water from their surrounding regions (McDonald 2014).

Stockholm, the first European Green Capital, is a pioneer city in water protection and management. The Swedish capital stands out for its high quality of environment and life, and demonstrates strong public health performance, high educational attainment and social protection levels. The municipality has done much to improve wastewater treatment and reduce the impact from storm water. The Radically cutting down the discharge of phosphorous and nitrogen has been instrumental for improving water quality. The sensitive Stockholm archipelago has a high water and environmental quality and great recreational value.

Singapore's water and sanitation utility received the Stockholm Industry Water Award for its holistic approach to water resources management. The city-State ranks high in water supply and sanitation in the challenging environment of a densely populated island. Access to water is universal, affordable and efficient. Innovative integrated water management approaches such as the reuse of reclaimed water, the establishment of protected areas in urban rainwater catchments and the use of estuaries as freshwater reservoirs have been introduced along with seawater desalination to reduce dependency on imported water. Singapore's approach does not only rely on physical infrastructure, but also emphasises proper legislation and enforcement, water pricing, public education and heightened awareness, as well as research and development.

Australia is the driest populated continent. Sydney faces particular challenges as an additional 1.3 million people are predicted to move into the city over the next decade. Urban renewal projects are planned around corridors and hubs that have access to public transport, and major job and housing growth in Western Sydney. As a result, Sydney Water is exploring options for future supply models that complement the existing centralised system to provide intelligent and affordable options. A study, led by Arup and Sydney Water, tried to map future water supply in Sydney and other major cities in the face of population growth, increasing scarcity of water and continued budget pressures. The study suggests that water authorities should migrate towards a more hybrid model which incorporates greater decentralisation and autonomous management of water supply, greater participation of additional service providers and smarter management of the water grid (Arup and Sydney Water 2014).

Water is visible and valued in many initiatives. The Canadian Water Sustainable City of the Near Future is built on the expectations of key stakeholders sharing their views on a water-sustainable city. Canada faces a water and wastewater infrastructure deficit which is expected to grow. The Blue Economy Initiative, a national project, seeks to inspire dialogue among Canadian decision-makers and influential actors on the opportunities and benefits of preserving water, and the severe risks of inaction. Responsibility is respected and shared. Four broad areas were identified to make tangible progress towards sustainability, including financial responsibility, progressive regulation and governance, customer-oriented information, and cutting-edge technology. Coastal sustainable cities can incorporate social innovations for the adoption of coastlines and water wise projects.

The water/energy nexus merits special attention. The energy sector already accounts for 15 % of the world's total water use. Its needs are set to grow, making water an increasingly important criterion for assessing the sustainability of energy projects. Expanding power generation could lead to an 85 % increase in the volume of water that is not returned to its source after use, through to 2035 (IEA 2013).

Sustainability asks for water to be respected throughout the life cycle. A US overview of water recycling applications, including drinking water, non-drinking urban and industrial uses, irrigation, groundwater recharge, and ecological enhancement, shed light on various municipal water reuse projects, such as irrigating parks or providing industrial cooling water. Drinkable water projects account for only a small fraction of the water being recycled. However, many drinking water treatment plants draw water from a source that contains wastewater discharged by a community located upstream (NRC 2012).

Wastewater treatment technologies processes are easily adapted to reclamation plants to meet the quality requirements of intended future applications in the water cycle. The concentrations of chemicals and microbial contaminants in reuse projects designed to provide drinking water supplies can be comparable to or lower than those commonly present in many drinking water supplies.

Water reuse projects tend to be more expensive than most water conservation options and less expensive than seawater desalination. Although the costs of reclaimed water are often higher than current water sources, external costs and benefits, for example seasonal peak demands on the drinking water system, can strike a difference. Depending on the specific requirements, recycling and reuse projects could also have a larger or smaller carbon footprint than existing alternatives (NRC 2012).

Effective wastewater management is one of the most widespread challenges to water quality and is linked to climate adaptation policies and urban resilience. Green infrastructure captures and filters pollutants. The green alternative is often less expensive than structural engineering options, and can help reduce costs of flooding. Green roofs, permeable materials, alternative designs for streets and buildings, trees, gardens and parks and rain harvesting systems are increasingly supplementing aquatic or substituting hard infrastructure investments (NRC 2008).

Noise is a local disturbance that seems sometimes strongly linked to port activities and can seriously affect the well-being of residents. Anthropogenic noise in the ocean has also been increasing steadily over the past decades. Shipping is the main source, and the produced noise is often in frequency bands used by many marine mammals for communication. Other significant sources of noise include seismic exploration for the offshore hydrocarbon industry and seabed mining. The impact of noise can be both to disrupt communication between animals and to displace them from their breeding, nursery or feeding grounds, with consequent potential effects on their survival. Cities which care about their marine biodiversity should be very sensitive to this kind of effects.

Apart from their fundamental land, water and air resources, cities impact an amazing array of resources originating from their hinterland, their territorial waters, and, increasingly, the rest of the world. Ecosystem services from surrounding regions provide fresh air, store or drain flood water, as well as drinking water. Differences in urban design and management can greatly impact urban resource efficiency. Eco-efficiency could help decrease the heavy dependency of cities to satisfy their needs in food, water and energy. Urban Eco-Efficiency Labs organised in China and India by the World Resources Forum, searched ecological and smarter solutions for sustainable cities and more efficient and better urban life. Optimising the use of the pedestrian infrastructure, public transport, bicycles, and electric cars can reduce ecological footprint and the loss of biodiversity. Advanced enabling technologies have the potential to cut global carbon emissions by as much as 15 % (WRF 2012).

The Global Partnership on Waste Management, introduced by the UNEP in 2010, suggests that the waste crisis is a very serious one and needs urgent action. Globally, waste production is dangerously mounting. The World Bank highlighted that the annual total amount of 1.3 billion tonnes of generated municipal waste could increase to 2.2 billion tonnes by 2025. Waste management is one of the most complex and cost-intensive public services, absorbing large parts of municipal budgets. Packaging, although only 2 % of the weight of total waste, represents 30 % of the volume, and only 54 % of it is recycled. The threat posed by poor waste management is particularly prominent in low-income countries where waste-collection rates are often below 50 % and garbage concentrates along river and sea banks. Urban population explosion, rapid industrialisation and economic development are generating increasing quantities of waste that are overburdening systems. The vicious circle is aggravated by the advent of a more affluent global middle-class longing for more sophisticated consumption goods and services (UNEP 2010).

Sound waste management is inextricably linked to resource efficiency and the performance of the urban economies. Integrated resource policies take care of the full life-cycle of materials from the extraction of natural resources, through their design, manufacture, assembly, marketing, distribution, sale and use to reuse, repair, recycling and recovery, including energy recovery. It is most important for cities to have a comprehensive picture of the resources required and the waste generated by all urban activities. Business actions and on-line exchange facilities encourage the sharing of resources in order to reduce costs and the environmental footprint. Urban concentrations can enhance multiple-win schemes based on the concept that "the waste from some activity can be a valuable resource for some other activity", orchestrated by cities, business or citizen associations.

Generation of waste mirrors economic growth. Preventing the loss of precious resources is essential. Precious metals discarded as waste also have a global and an ethical dimension. Waste often hides goldmines and the European Environment Agency provided some interesting assessments. Electric household appliances and electronic equipment contain hazardous substances, but also include valuable metals, estimated to contain 450,000 tonnes of copper and 7 tonnes of gold in 2005. At the London Metal Exchange, these metals would be worth €2.8 billion and €328 million respectively, in 2011. However, only a small part of such electronic equipment is collected and reused or recycled. The business case for "urban waste mining" is evident (EEA 2012).

In an era during which the price of raw material is high, coastal cities might be tempted by seabed mining. Conscious cities should insist that such activities not have severe and irreversible effects on marine ecosystems. Transparency, trust and societal acceptance are fundamental issues, as the world has entered uncharted waters and moved towards deep-ocean mining and extracting precious metals from a depth of 1500 m.

Conscious resource management is particularly reflected on eco-buildings which, for residential, commercial or industrial use, are the main cells of all cities. The concept of Eco-habitat, which is a luminous and healthier environment with sustainable use of materials and water and low zero energy requirements, has already endowed many cities with attractive eco-cells. Good integration into a landscape, orientation and adaptation to climatic conditions, recourse to ecological materials and renewable energies, and sustainable management of water and resources, including waste, are critical requirements for eco-performance of an entire built environment.

Bioclimatic architecture and design has made great strides over recent years. Architects, designers, eco-builders and eco-citizens are promoting green constructions and more ecologically sound techniques on coastal environments. Source, design and performance information about the latest low-impact materials and technologies, the best options for the use of renewable energy, water and resources and best practices can help make a difference (Roaf 2007).

Buildings hold the largest potential for cost-effective energy savings. Given the long lifetime of buildings in cities, the largest potential for improving energy performance is in existing buildings. Public buildings and privately owned buildings open to the public can act as pioneers and serve as models for intelligent resource-saving and performance-enhancing buildings. The display of building plans and energy performance certificates and recommended optimal climatic conditions, such as the most favourable indoor temperatures, in all public buildings, can promote awareness about the benefits of eco-performance.

Melbourne offers a prime example with the Council House 2 municipal eco-office, a multi-award winning building which managed to reduce CO_2 emissions by 8 %, electricity consumption by 82 %, gas by 87 % and water by 72 %. It is a ten-storey office building for about 540 staff members of the City of Melbourne. Opened in 2006, the state-of-the-art building also features ground-floor retail spaces and underground parking. The Council House 2 has embraced the best environmental options and local solutions. An in-situ multi-water treatment plant filters out the water and creates clean water suitable for non-drinking uses. Some of the recovered water is used for water cooling and irrigation, and the rest is used in other municipal buildings and public fountains.

The Council House 2 building has been designed to respect ecological principles and enhance the natural 24-h cycle of solar energy, natural light, air and rainwater, to power, heat, cool, and water the building, which purges stale air at night and pulls in pure fresh air during the day. Like a sunflower, the building follows the movement of the sun to collect heat. The north façade of the building has ten dark-coloured air ducts that absorb heat from the sun, then the hot air rises taking the stale air up and out of the building. The south façade has light-coloured ducts that draw in pure fresh air from the roof and distribute it down through the building. The west façade has louvers made from recycled timber and powered by photovoltaic roof panels that move according to the position of the sun.

Schools can be the most instrumental eco-organs of a city and great places for promoting marine environmental awareness and sowing the seeds of a smart and inclusive society. The international award programme "Eco-Schools" has created a movement of schools embedding sustainability principles into school life and action. School children lead the project in all its aspects and help carry out audits to assess and improve the environmental performance of their school. They also learn principles and practices to apply in other spaces including homes and public spaces. Many "cleaning the beach" local action programmes in coastal cities have been initiated by local schools influencing public values.

From the transformation of buildings to the transformation of districts, the scale is of importance. Hamburg embarked on a thorough transformation in an exemplary regeneration project. HafenCity (Harbour City) enhanced the possibilities of abandoned harbour space. At the beginning of the last century, Hamburg's bustling harbour, at the edge of the city centre, hosted the infrastructures and warehouses of the shipping industry. But the advent of bigger ships led to the creation of a new harbour, leaving a vast empty space near the city centre which provided the fundamental resource. By the time HafenCity is finished in 2025, it will stretch 1.5 km between the city centre and the Elbe River (HafenCity Hambourg 2013).

The proximity of the sea is essential for the new district featuring high-quality, high-tech residential and office space, a waterfront promenade, a five-star hotel, a university, and sustainable public transport. HafenCity aims to become a living, breathing part of the city, a place where people want to both work and enjoy their leisure time. Achievements already include a ground-breaking Ecumenical Forum inaugurated in 2010, sponsored by 19 Christian churches in Hamburg and designed according to the strict sustainability standards of the HafenCity Eco-label. In 2011, the district welcomed "Osaka 9", the HafenCity sustainability pavilion with an exhibition space on ecological, sustainable urban development. The jewel in the crown will be a new opera house.

In Malmö, Västra Hamnen, the Western Harbour, provides a fine example of radical transformation of a previously industrial site into an ecological urban area. In 2001, its reconstruction began with 500 residential units, mostly part of an exhibition to develop energy self-sufficient housing units. Among the new buildings is the emblematic Turning Torso, the city's 190 m tall landmark, a residential skyscraper with a twisting design.

Despite Malmö's chilly climate, the beach Ribersborg, stretching along the coastline, hosts open-air baths, opened in the 1890s. The long boardwalk to the Western Harbour has become a new favourite summer hang-out for both citizens and tourists. The harbour is particularly popular with Malmö's vibrant student community.

In Stockholm, Hammarby Sjöstad is an innovative district on the former waterfront industrial land set aside for its ultimately unsuccessful 2004 Olympic bid. The concept includes bioclimatic design, sound surroundings and ecological organisation of local life. The eco-cycle model of Hammarby aims to make the district autonomous throughout its life cycle. In Hanover, the design and construction of the Kronsberg district, built for Expo 2000, incorporated state-of-the-art ecological material and techniques. Construction waste was reduced by 80 % through sorting and recycling measures, but also social and educational models like the waste-free breakfasts for children (Mega 2010).

Stockholm plans to consolidate the urban fabric and brand Stockholm as a leader in green urban planning. The Royal Seaport is the city's latest flagship sustainability project. In 2011, Stockholm Royal Seaport was 1 of 27 projects that received financial support from the Swedish government, through the Delegation for Sustainable Cities. The area has been partly industrial land and royal hunting grounds, and has a particular identity. It is located right by the water and the Royal Urban National Park, making it a very attractive area for citizens and visitors. The Royal Seaport project aims at creating an organic part of the city with 10,000 new apartments and 30,000 new work places. In addition to housing and offices, the area will also host urban parks, cultural spaces and a harbour for cruise ships. The City of Stockholm, owner of the land, is placing high expectations on the project to be at the forefront of urban sustainability. The project aims at being fossil fuel free, through initiatives such as energy efficient transport and generation of biogas from food waste.

Environmental and cultural sustainability are given high attention and much effort is invested in the quality of the natural and built environment, through the planting of oaks for biodiversity and the construction of eco-innovative green buildings. There is also a far-reaching analysis on the effects of the city form on travel patterns and an emphasis on the district´s effect on the marine environment. Social sustainability is also given attention. The Royal Seaport will have a mixed housing composition, with rented flats coexisting with owner-occupied housing and an array of diverse urban functions (City of Stockholm 2014).

SymbioCity, the Swedish trademark for sustainable urban development, launched in 2008, aims at exporting the national know-how on exemplary cities. The essence of the concept is the generation of environmental and economic benefits through unlocking synergies between urban systems. A network of Swedish environmental technology companies and organisations promotes the SymbioCity model in their international exchanges. SymbioCity is scalable, and adaptable to any context. For example, the excessive heat from an industry can warm up a household or the waste from an industry can be useful material for a service. The seven building blocks of the SymbioCity concept include urban functions, industry and buildings, energy, transport, resource management, water supply and sanitation, architecture and landscape design.

Coastal cities can offer a fertile environment for eco-innovation to thrive on the articulated land and space domains and functions and can serve as a magnet for talent, financial capital and entrepreneurship. They can provide a competitive environment to maritime companies and ancillary services and serve as blue green platforms that give partners an unmatched opportunity to develop, test and validate their technologies at a large scale and real-world conditions. Green urban procurement and voluntary practices such as eco-labels are useful instruments.

References

Angel S et al (2011) Making room for a planet of cities. Lincoln Institute of Land Policy, Cambridge, MA

Arup and Sydney Water (2014) The future of urban water: scenarios for urban water utilities in 2040. Arup and Sydney Water, Sydney

Beach D (2002) Coastal sprawl: the effects of urban design on aquatic ecosystems in the United States. Pew Oceans Commission, Arlington

Boston Foundation (2011) The measure of poverty. A Boston indicators special report. Boston Foundation, Boston

C40 Cities and Siemens (2014) City climate leadership awards. C40 Cities and Siemens, New York

CDP (2014a) Protecting our capital. How climate adaptation in cities creates a resilient place for business. Carbon Disclosure Project, London

CDP (2014b) Global corporate use of carbon pricing. Disclosures to investors. CDP, London

Centre for Strategic and International Studies (CSIS) (2013) Nutrition and food security in the city. CSIS, Washington, DC

(The) Chicago Council on Global Affairs (2013) Feeding an urban world: a call to action. (The) Chicago Council on Global Affairs, Emerging leaders perspectives, Chicago
City of Rotterdam (2010) Investing in sustainable growth. Rotterdam programme on sustainability and climate change 2010–2014. City of Rotterdam, Rotterdam
City of Stockholm (2014) Stockholm royal seaport, Stockholm
Convention of Biological Diversity (CBD) (2010) Strategic plan for biodiversity 2011–2020, including Aichi Biodiversity Targets
Ecological Footprint Network (2011) Ecological footprint analysis. Ecological Footprint Network, San Francisco/Oakland/Fremont
Ecological Footprint Network, ESA and WWF (2012) Living planet report 2012. Biodiversity, biocapacity and better choices. Ecological Footprint Network, ESA and WWF, Oakland
Economist (The) (2015) The world in 2015. Economist (The), London
Eriksen et al (2014) Plastic pollution in the world's oceans: more than 5 Trillion plastic pieces weighing over 250,000 tons afloat at sea
European Commission (EC) (2012) The habitats directive. Celebrating 20 years of protecting biodiversity in Europe. EC, Brussels
EC (2013a) Copenhagen. European Green Capital 2014. EC, Brussels
EC (2013b) European Green Capital. Urban environmental good practice and benchmarking report. EC, Brussels
EC (2013c) EU adaptation strategy. EC, Brussels
EC (2013d) 7th Environment action programme. EC, Brussels
EC (2014) Directive of the European Parliament and of the Council establishing a framework for maritime spatial planning and integrated coastal management SWD (2013) 64 final. EC, Brussels
European Environment Agency (EEA) (2006) Urban sprawl in Europe. The ignored challenge. EEA, Copenhagen
EEA (2010) 10 messages for 2010 urban ecosystems. EEA, Copenhagen
EEA (2012) Urban adaptation to climate change in Europe. EEA, Copenhagen
EEA (2013) Air quality in Europe. EEA, Copenhagen
Food and Agriculture Organisation (FAO) (2009) Food for cities. FAO, Rome
FAO (2013) The state of world fisheries and aquaculture. FAO, Rome
FAO (2014) The state of world fisheries and aquaculture. FAO, Rome
Friends of Europe (2013) Global outlook on urban air quality. Friends of Europe, Brussels
Global Ocean Commission (GOC) (2014) From decline to recovery. A rescue package for the global ocean. GOC, Oxford
Gore A (2013) The future: six drivers of global change. Random House, London
Government Office for Science (2014a) The future of cities: what is the global agenda? Government Office for Science, London
Government Office for Science (2014b) The future of cities: living in the city. Government Office for Science, London
Gulbenkian Ocean Initiative (GOI) (2013) Blue natural capital: towards a new economy. GOI, Lisbon
Haëntjens J (2012) The frugal city. Présence/Essai, Paris
HafenCity Hamburg (2013) Essentials. Quartier projects. HafenCity Hamburg
International Energy Agency (IEA) (2013) World energy outlook 2013. IEA, Paris
International Labour Organization (ILO) (2012) Working towards sustainable development. Opportunities for decent work and social inclusion in a green economy. ILO, Geneva
International Council for Local Environmental Initiatives (ICLEI) (2014) Resilient cities 2014. European cities adapt to climate change. Open European Day, Bonn
International Panel of Climate Change (IPCC) (2013) Climate change 2013. The physical science basis. IPCC, Bonn
IPCC (2014) Climate change fifth assessment report. IPCC, Bonn
McDonald R (ed) (2014) World's water stressed cities in global environmental change. The Nature Conservancy, Arlington

McIntyre AD (ed) (2010) Life in the world's oceans. Blackwell, Oxford
Mega V (2010) Sustainable cities for the third millennium. The odyssey of urban excellence. Springer, New York
National Oceanic and Atmospheric Administration (NOAA) (2012) Air quality. State of the art fact sheet. NOAA, Washington, DC
NOAA (2013) Ocean acidification. State of the science. NOAA, Washington, DC
National Research Council (NRC) (2008) Urban stormwater management in the United States. National Academies Press, Washington, DC
NRC (2010) America's climate choices. NAP, Washington, DC
NRC (2012) Water reuse: potential for expanding the Nation's water supply through reuse of municipal wastewater. NAP, Washington, DC
New York City Mayor's Office of Long-Term Planning and Sustainability (2011) Plan NY: a greener, greater New York (Update 2011). PlaNYC, New York
New York City Panel on Climate Change (NPCC) (2015) Building the knowledge base for climate resiliency: New York City. NPCC, New York
Organisation for Economic Co-operation and Development (OECD) (2012) OECD environmental outlook to 2050: the consequences to inaction. OECD, Paris
OECD–FAO (2014) Agriculture outlook 2014–2023. OECD–FAO, Paris
Roaf S (2007) Ecohouse. Elsevier, London
Rockfeller Foundation (The) (2014) Insights. An annual publication exploring emerging issues of consequence. Rockfeller Foundation (The), New York
Sesini M (2011) The garbage patch in the oceans: the problem and possible solutions. Columbia University, New York
Snelgrove PR (2010) Discoveries of the census of marine life: making ocean life count. Cambridge University Press, Cambridge, MA
STOA, European Parliament Research Service (2015) Ten technologies which could change our lives. STOA/European Parliament Research service, Brussels
Tokyo Metropolitan Government (2010) Tokyo cap and trade programme. Tokyo Metropolitan Government, Tokyo
Tokyo Metropolitan Government (2012) Tokyo cap and trade programme. Second year report. Tokyo Metropolitan Government, Tokyo
UN Conference on Sustainable Development Rio+20 (2012) Future perfect. UN Conference on Sustainable Development Rio+20, Rio de Janeiro
UN Convention on Biological Diversity, Stockholm Resilience Institute and ICLEI (2012) Cities and Biodiversity Outlook. Action and Policy. UNEP, Paris
UN Environmental Programme (UNEP) (2010) Global partnership on waste management. UNEP, Osaka
UNEP (2011) Taking steps toward marine and coastal ecosystem-based management—an introductory guide. UNEP, Paris
UNEP / ILO (2012) Working towards sustainable development: opportunities for decent work and social inclusion in a green economy. Nairobi, Geneva
UNEP (2012a) 21 Issues for the 21st century. Results of the UNEP foresight process on emerging environmental issues. Paris, UNEP
UNEP (2012b) Global environmental outlook, 5th edn. UNEP, Paris
UNESCO World Heritage Centre (2013) Safeguarding the crown jewels of the ocean. UNESCO World Heritage Centre, Paris
UNESCO World Heritage Centre (2014) The list. Global strategy. UNESCO World Heritage Centre, Paris
(The) UN Office for Disaster Risk Reduction (UNISDR) (2012) How to make cities more resilient. A handbook for local government leaders. UNISDR, Geneva
van Veenhuizen R (2006) Cities farming for the future. Urban agriculture for green and productive cities. International Institute of Rural Reconstruction and ETC Urban Agriculture, Philippines
White House (2014) 16 U.S. Communities recognized as climate action champions for leadership on climate change. White House, Washington, DC

World Bank (2012) Inclusive green growth. The pathway to sustainable development. World Bank, Washington, DC
World Bank (2013) Turn down the heat: climate extremes, regional impacts, and the case for resilience. World Bank, Washington, DC
World Resources Forum (WRF) (2012) Urban eco-efficiency lab. WRF, Beijing
World Wide Fund (WWF) (2014) Living planet report. Species and spaces, people and places

Selected Internet Links

www.europa.eu.int/comm/environment/urban
www.europa.eu/climateaction
http://ec.europa.eu/environment/europeangreencapital/winning-cities/2014-copenhagen/
http://climate-adapt.eea.europa.eu/
www.calculator.bioregional.com/
www.c40cities.org
www.cbd.int/doc
www.cdproject.net
www.climatesummitformayors.dk
www.eea.eu.int
www.eolss.net
www.footprintnetwork.org
www.iclei.org
www.ipcc.org
www.kk.dk
www.london.gov.uk
www.myclimate.org
www.unfccc.int
www.unhabitat.org
www.unep.org
www.un.org/esa/sustdev
www.panda.org/livingplanet
www.cityofboston.gov/climate/bostonsplan/
www.dublinked.ie
www.hamburg.de/stockholm.se
www.worldwatchinstitute.org
www.acrplus.org
www.oneplanetliving.net
www.stockholmresilience.org
www.surfrider.org
www.symbioCity.com
www.wpci.nl
www.thebluecarboninitiative.org
www.whitehouse.gov/share/climate-action-plan
http://marinedebris.noaa.gov/hawaii-nets-energy-program

Watercolour 4
Venice: A Sinking Eternity?

Chapter 4
Sustainable Energy and Transport Systems

Abstract Responsible coastal cities have to ensure, for all inhabitants and visitors, uninterrupted access to clean, secure, competitive and affordable energy and transport. Offshore energy generation can extend possibilities of all those benefits, allowing cities, producers and consumers of green energy to become energy autonomous. Investments in energy efficiency and renewable energy sources can only be beneficial. This chapter offers an insight into the components and vectors of sustainable energy production and consumption in cities and the efforts made at the local level to overcome national and supranational energy policy targets. Blue green electricity from offshore wind, ocean thermal, tides and waves, is expected to be among the energy vectors of the future.

Coastal cities could be among the first post-carbon cities. Renewable energies have conquered Europe and the world. They have become major players in the energy market and a significant generator of employment. However, the performance of the highly energy-intensive transport sector, the most intractable contributor to global warming, has to be radically improved. Sustainable mobility and accessibility are fundamental in coastal cities for linking the first and the last mile of all journeys to terrestrial and maritime transport networks and all parts of the city with a port and a seafront through integrated, smarter and cleaner services.

4.1 Sustainable Energy Options: The Rise of Renewable Energy Sources

Responsible coastal cities, as all cities, have to ensure continuous access for all citizens to secure, clean and affordable energy. Port cities have been critical for the transport of fossil fuels. They increasingly become instrumental for cleaner transport uses and the connection points with renewable energy produced offshore. Energy from the sea enters the policy debate and promises to contribute to clean and abundant energy and the reduction of greenhouse gas emissions.

The UN has described sustainable energy as the golden thread which connects economic growth, social equality and healthy environment. The 2012 International

V.P. Mega, *Conscious Coastal Cities*, DOI 10.1007/978-3-319-20218-1_4

Year of Sustainable Energy for All tried to raise awareness of the world's 1.3 billion energy poor and to generate action for eliminating energy poverty by 2030. There seems to be a shared world consensus on the key challenge of affordable and inclusive energy security for an increasing world population and drastically reduced greenhouse gas emissions. The equipment of emerging world cities with energy infrastructures and the replacement of ageing infrastructures in industrialised countries represent unique opportunities for the decarbonisation of cities and the world in the most efficient way.

In the European Union, the strategy known as Europe 2020 aims at a reduction of greenhouse gas emissions of at least 20 % below 1990 levels, 20 % of EU energy consumption from renewable resources and a 20 % reduction in primary energy use through improved energy efficiency. More ambitious targets for emissions reduction for 2030 have already been agreed. They include a binding EU target of at least 40 % reduction of greenhouse gas emissions by 2030, a binding target of at least 27 % of renewable energy used at EU level, an energy efficiency increase of at least 27 %, to be reviewed by 2020, having in mind an EU level of 30 % for 2030, and the completion of an internal energy market by reaching an electricity interconnection target of 15 % between Members States.

Cities are concentrated energy demand centres and have to participate in all decisions on the future of energy, together with other levels of governance. They should discuss with citizens the dimensions of a reliable, flexible and diverse energy supply to satisfy their needs and co-decide the energy investments of the future, especially in the context of climate change. The IEA warned that an increase of global energy demand by a third by 2035 may lead to a "long-term average global temperature increase of 3.6 °C". If countries do not respect their commitments to reduce emissions, energy demand could be even greater (IEA 2012).

Science and innovation have an essential role to play in improving energy efficiency and in exploring and capitalising on the potential of all energy options. Industrial production patterns have largely been adapted to environmental requirements. They are more concentrated in space and easier to target and change. A real change in consumer and society behaviour is still to be accomplished even in the most advanced cities and countries. Taxation measures can steer demand towards energy options that are more environment-friendly and generate revenues to be invested in sustainable development policies.

Energy options depend on the macroeconomic environment, geopolitics, international trade, market dynamics, citizen preferences and environmental concerns and global governance especially in relation to climate change. In the EU, energy demand increasingly outstripped indigenous production, and, since 2004, the Union imports more than half of the energy that it needs. Energy dependency can have serious consequences, such as supply uncertainty, higher and unpredictable energy prices and exposure to political instability of exporting world regions (EC 2011).

Sustainable coastal cities have to ensure that energy systems are economically robust, socially inclusive and environmentally sound and that energy services satisfy present citizens' demands and respect the rights of future generations. They can extend possibilities by enhancing renewable marine energy resources. Local governments play an important role in leading communities into the post-carbon age with investments in renewable energies and energy efficiency as major cornerstones. A bold vision for cities could be to become net producers instead of net consumers of energy. Urban buildings and districts could be converted into active energy generators (Mega 2013).

The spectacular rise of renewable energies is probably the most auspicious sign of the transition to a sustainable world order. However, fossil fuels continue dominating the global energy mix. Supported by subsidies, which, in 2013, were four times more important than the subsidies to renewables, fossil fuels form the backbone of the world energy primary sources (IEA 2014). Coal is the dominant primary energy source. China accounted in 2011 for half of the global coal consumption and became the world's largest coal importer. Oil has still the lion's share in the primary energy mix in many world regions. Transport is the economic sector most notoriously dependent on oil, although electricity generation in some small island states such as Malta and Cyprus is also nearly exclusively dependent on oil.

Gas is often seen as the transition fuel towards sustainable development. Natural gas became a polyvalent source of energy and gained ground in all sectors, including power generation, heat and transport. Global gas markets are undergoing a profound change, notably because of developments with shale gas in North America. With Liquefied Natural Gas, trade has become increasingly global since transport has become more independent from pipelines. Unconventional gas is seen as a promising new source of energy supply. However, due to the early stage of exploration, uncertainties abound. Shale gas could be a bridge fuel from coal to renewables, but the needed investments are enormous and could divert money from renewables (IEA 2012; Gore 2013).

The management of fossil fuels throughout the value chain has crucial environmental implications. During the extraction phase, methane (CH_4) and carbon dioxide can be released when, for example, natural gas is flared from oil wells. Natural gas may also be released during the extraction and processing of coal. The fortuitous release of fuels during transportation is extremely hazardous. Accidents occurring to oil tankers and releasing devastating hydrocarbons into the oceans have serious consequences to marine environments. Oil spills into the marine ecosystems are particularly dangerous and often unfolded into major disasters, reminding the world of the multiple unbearable and unacceptable risks on the marine environments from offshore and coastal activities. The investigation of the Macondo Well Blowout Disaster in 2010, following an explosion of the Deepwater Horizon, which caused a seafloor gusher and spilled oil into the Gulf of Mexico, provided many world lessons about the fragility of marine resources but also the challenges for global governance to bring a decisive end to such disasters.

During the combustion of fossil fuels, in addition to energy production, a high number of by-products are also generated, including carbon dioxide, carbon monoxide, methane and other hydrocarbons, nitrous oxide and nitrogen oxides, particulate matter and certain metals and radionuclides. These by-products are crucial pollutants and contributors to local and regional air pollution and greenhouse emissions.

One third of the oil and a quarter of the natural gas consumed in the world come from underwater areas. Offshore oil and gas drilling took off in the 1970s and it gradually moved further and deeper offshore. It is growing in the Arctic, which possesses 13 % of oil and 30 % of gas respectively, the Mediterranean and East Africa. This represents multiple threats, especially for the extreme fragile Arctic environment. Marine ecosystems can be seriously disturbed from seismic surveys and spills are hard to address in deep and remote waters. Binding safety and environmental standards are suggested for the offshore energy industry including risk management and capacity building (GOC 2014).

The rise of unconventional gas could revolutionise the future energy landscape. Hydraulic fracturing, or fracking, a technology for extracting natural gas and oil from shale rock formations, has generated heated debate around the world, raising questions on the amounts of water consumed during the process and even pollution of groundwater supplies. Even though the environmental controversy is far from settled, the surge in shale gas production has caused a great shock to natural gas markets. Fracking is taking off in the United States owing to bountiful shale deposits, the development of effective extraction techniques, and the availability of a high-capacity gas pipeline infrastructure. Politics is, also a big factor in the equation in some EU countries, especially in France. The interactions between geopolitics and markets in relation to the ethical dimensions of the energy options are crucial for the European Union. It is important for policy makers to appraise the opportunity costs of each energy option and technology and discuss them with all stakeholders.

Geopolitics is affected by the trajectory of gas markets as much as it shapes that trajectory. Russia and the United States could dominate the future gas market. The most optimistic scenario suggests an extensive global shale gas revolution, which, along with extremely liberalised markets, foresees a manufacturing boom in the United States spurred by falling natural gas prices. Argentina could be motivated to exploit its massive shale gas reserves. The US transport fleet shifts increasingly toward natural gas-fuelled vehicles, while the Japanese industry prepares such vehicles for the export market. In the least optimistic scenario, the US shale industry is dealt a blow when fatal accidents are attributed to poisoned well water, leading to bans on fracking (Belfer Centre and Rice University 2012).

The spectrum of possibilities is expanding. A study by the World Energy Council suggests that global energy supply by 2050 is set to increase between 27 and 61 %. Two contrasting policy scenarios, the more consumer-driven Jazz scenario and the more voter-driven Symphony scenario, suggested that renewable energy sources are expected to mark the highest growth and increase 20–30 %, but fossil fuels will still dominate the energy landscape supplying 59–77 % of the global energy mix. In terms of GHG emissions, both scenarios suggest an increase in excess of the universally agreed 450 ppm CO_2 target (WEC 2013).

The IEA 2014 World Energy Outlook suggests that the world might need 37 % more energy by 2040, almost all of the increase coming from non-OECD countries. An encouraging sign is that energy demand in China, which accounted 50 % of the increasing demand in recent years, is stabilising, as its population levels off and its economic growth slows down. India takes over as the leading engine of energy demand. In 2040, the world energy supply mix looks almost equally divided into oil, gas, coal and low-carbon sources. Each part faces particular challenges. Oil supply until 2025 mainly comes from non-OPEC countries, United States, Canada and Brazil. But by the mid-2020s, total non-OPEC oil supply starts to fall back, increasing the call on major resource-holding Middle East countries. Gas is the fastest growing fossil fuel especially in liquefied form and offers some protection against supply disruptions. This clearly seems to be the transition fuel, but it still is 2.1 times more expensive than coal. Growth of coal demand is constrained by air pollution and climate policies mainly in the United States and China. Coal use continues to grow briskly in India. China, India, Indonesia and Australia alone could account for over 70 % of global coal output by 2040, underscoring Asia's importance in global coal trade and pricing (IEA 2014).

Renewables have rapidly gained ground, and the EU has sent a green wave to the world. Solar, wind, biomass and biofuels, geothermal, ocean and tidal energy are fundamental vectors towards a sustainable energy future. Starting from a rather marginal role in the previous century, renewable energy dramatically increased its contribution. Enthusiasm about renewable energy is escalating, barriers are being removed, and infrastructure is expanding. Many trajectories of renewable energy growth progressed more than initially expected and the promises for the future abound.

A rising numbers of cities, regions and countries strive to transition to renewable energies. China, the United States, Brazil, Canada, and Germany remained the top countries for total installed renewable power capacity. In China, for the first time in 2013, new renewable power capacity surpassed new fossil fuel and nuclear capacity. Denmark, Germany, Portugal, Spain, Sweden, and Austria are leaders in per capita renewable power capacity installed by the end of 2013. In many places, municipal governments have emerged as some of the most active and ambitious drivers of renewable energy growth, while civil society organisations also played a major role (IEA 2014).

Renewable energy technologies are entering the policy and market mainstream. The rapid growth of the sector is increasingly being led by emerging nations. In 2005, only 15 emerging countries had renewable support policies. In 2013, 95 emerging countries had support policies for renewables and formed the majority of the 138 countries with such policies in the world. The rise of support in the emerging world contrasts with declining support and policy uncertainty in some industrialised economies (REN21 2014).

Renewable energy markets have been growing strongly over the last years, as renewable energy technologies become mature and investment costs decline. Solar, wind and biomass are the technologies progressing most rapidly. Solar and wind develop for electricity generation. Biomass, solar and geothermal energy have a huge potential for the heating and cooling sector. Onshore wind and photovoltaic (PV) systems are moving steadily towards competitiveness. The EU strategy on renewables states that renewable energy should be gradually integrated into the market with reduced or no support and should, over time, contribute to the stability and security of the grid on a level ground with other electricity generators (EC 2012a).

The future of energy seems irreversibly leaning towards renewable energy sources, the world's second energy source. Renewable power capacity jumped more than 8 % in 2013, accounting for the majority of net additions. Their share in the primary world energy mix was 13 % in 2011 and could rise to 18 % in 2035 (IEA 2012). The IEA forecasts that renewables will account for one third of the total electricity output by 2035. More than 35 GW of wind power capacity were installed in 2013. Among wind technologies, conflicts on land make the deployment of offshore wind the most rapidly growing option. In addition to power generation technologies, heating and cooling from renewable sources, such as modern biomass, solar, and geothermal, account for a gradually rising share of final global heat demand (IEA 2014).

Solar energy has the highest theoretical potential for energy production. The sun is the universal primary source of energy. Solar energy systems can harness solar rays and produce electricity and heat. Solar energy represents a real chance for distributed energy in sunny cities, also in the emerging world, where micro-power is often cheaper than extending the grid. Photovoltaics use solar cells to convert light directly into electricity. The produced energy goes directly to the grid or gets stored in batteries. During the period 2005–2010, average photovoltaic system costs have declined by 48 % and module costs by 41 %. Industry expects costs to fall further driven by current government support policies, structural reforms and removal of market barriers (EC 2012a).

Wind is the fastest growing world option, followed by hydropower and solar. Wind power deployment has more than doubled during the years 2008–2013, approaching 300 GW of cumulative installed capacity, led by China, the United States and Germany. Wind power provides 2.5 % of global electricity demand and up to 30 % in Denmark, 20 % in Portugal and 18 % in Spain. Policy support has been instrumental in stimulating this growth. Onshore wind investment costs fell by 10 % between 2008 and 2012. Conflicts with neighbouring communities led to a lower allocation of local permits and the search of new spaces. Terrestrial wind power generation can already be competitive where wind resources are strong and financing conditions are favourable, but it still requires support in most countries.

The best wind resources are offshore and are largely undeveloped. Offshore wind technology costs levelled off after a decade-long increase, but are still higher than that of terrestrial developments. Offshore wind farms seem to cause more limited scenic and acoustic impact and less tension with local communities. The precautionary principle has however to be cautiously adopted and the impacts on the marine ecosystems have to be carefully examined.

The geographical pattern of deployment of wind energy is rapidly changing. While OECD countries led early wind development, since 2010, non-OECD countries are installing more wind turbines. After 2030, emerging countries will have more than 50 % of global installed capacity. A policy roadmap developed by the International Energy Agency assumes that the cost of energy from wind will decrease by as much as 25 % for land-based and 45 % for offshore by 2050 on the condition of strong research and development to improve design, materials, technology and reliability, optimise performance and reduce uncertainties (IEA 2013b).

Bio-energy sources of energy include residential organic, agricultural, forest residues and algae. Biomass is versatile and can generate electricity, heat, and/or transport fuel. Furthermore, it can transform waste into an energy asset. At the world level, biomass is the fourth largest energy source. Second generation biofuels, made out of agricultural, aquatic and forestry waste, offer a promising avenue for creating an asset out of liabilities. Food security crises reminded the world that biofuels should not be in conflict with food safety (NRC 2012).

Producing energy from waste, including marine waste, is the most-efficient potential energy generation. Second and third generation biofuels are expected to provide multiple benefits to the economy, the society and the environment. Many cities use biogas for their operations. For some years, Amsterdam has been using biomass from municipal waste to generate green electricity and heat. The metro and trams operate on green electricity. The Waste and Energy Company is the larger producer of sustainable energy. More than half the incinerated waste of non-fossil origin is used as biomass (Amsterdam Climate Office 2008).

Many coastal cities invest in fuel cells producing energy from hydrogen and oxygen in an intrinsically clean and efficient way. Unlike batteries, fuel cells do not store energy, but support a continuous flow process. Fuel cells may replace, in the medium term, a large part of the current combustion systems in industry, buildings and road transport. In the long term, fuel cells and hydrogen are expected to form an integral part of renewable energy supply and lead to a significant international market for fuel cells in transport and industry.

Many cities chose hydrogen as the privileged fuel for their public transport. Hydrogen is a key energy carrier for a future sustainable energy economy. It is abundant and perfectly clean. It provides a unique pathway for gradually increasing the contribution of renewable energy sources. Hydrogen can be used in fuel cells for all final, stationary and mobile, applications. Cost-effective transport, distribution and storage of hydrogen are major issues, together with the creation of an appropriate infrastructure. Amsterdam is one of the leaders of the hydrogen fuelled public transport including buses and boats. In cooperation with other urban regions, the city provides a test site for large scale sustainable transport operations (Amsterdam Climate Office 2008).

Global geothermal resources can contribute more than 2 % to the global electricity production. Ocean, wave and tidal energies represent some of the most plentiful sources and the EU invests much to harness their power (EC 2014a). Ocean cities can be pioneers in engaging in projects for harnessing of power from the waves that wash their shores. They can be true strongholds of blue green energy and interfaces between local generation and national grids.

Nuclear energy faces a very uncertain future, especially after the Fukushima accident, despite the sector's negligible contribution to GHG emissions. Plans for nuclear energy could scale back as some countries have reviewed policies, but capacity could rise in China, Korea, India and Russia (IEA 2013a). However, the issues of nuclear proliferation and waste management have posed serious ethical questions for humanity.

Centralised electricity generation is inherently inefficient. An essential change in electricity supply is the transition towards networks of smaller decentralised power plants nearer consumers. Decentralised micro-generation from renewable energy sources is advancing fast despite an overall respite linked to the economic crisis. Micro-generation implies that every home and district can become a mini power station. Decentralised energy can be much more efficient, as it allows the energy losses and financial costs associated with the long-distance national transmission to be radically reduced (Greenpeace 2005; UNEP/ICLEI 2015).

Cities are impactful partners in moving towards decentralised energy production able to fuel homes, offices, shops, factories, public and private activities and transport. The integration of decentralised energy resources and renewable energy into the main electrical grid is expected to change the energy paradigm of urban societies, with electricity generated in large power plants and delivered to consumers through a passive distribution infrastructure. The involvement of all stakeholders, including utilities, independent producers, central and local governments, regulators, industry, port authorities and consumers, is a major precondition for achieving the transition.

Energy consumption has almost stabilised in developed countries over the two last decades. Energy demand patterns of various cities present key differences but also many common points. They depend on socio-economic conditions, climatic and cultural features, population density and urban form and functions. The global crisis and advancing climate change have increased uncertainty. Northern cities with colder climates and low population density are huge energy consumers. The severity of winters is mirrored on energy consumption peaks, while warmer summer conditions are reflected on a growing number of cooling systems.

Industrial energy consumption is being stabilised, mainly as a result of the transition to a digital, knowledge-based and service-oriented economy. Investment in technology and innovation has enabled industry to become more eco-efficient. Heavy industry such as the automobile sector has demonstrated significant reductions in energy consumption during the last years. Environmental regulation and enforcement have been key driving forces for the development of eco-innovations.

Electricity has been the symbol and the quintessence of industrial societies and economies. Energy consumption by the transport sector, depending almost entirely upon oil, has increased steeply, even though the global crisis has slightly delayed growth. This is mainly due to continuing growth of road transport, passenger and freight. Air transport is also increasing dramatically, due to the rise of low-cost leisure trips.

Coastal and offshore sources of renewable energy provide coastal cities the possibility of sustainable energy. Micro-power generation is expected to continue emerging gradually alongside the grids, and increasingly incorporate renewable energies. Photovoltaic, wind, biomass and geothermal power and combined heat and power systems can significantly reduce the need for power from the grid, for households, office and industrial buildings. As consumers become "procumers" (producers-consumers), they also gain a stronger sense of ownership and control over their energy use.

Emboldening consumers as micro-producers is crucial for removing barriers to renewable energy growth. Many benefits are expected to come from the combination of micro-generation and smart electronics. Intelligent equipment and policy measures can increase consumers' awareness on the amount and the price of the electricity services and encourage them to improve their consumption patterns. Smart products responding to price signals could further induce consumers to change.

Heat production is predominantly decentralised, and includes individual heating systems and dedicated heat stations with their associated local networks. Needs range from office and household heating and cooling to steam production for industrial port uses. Combined heat and power generation can substantially increase energy efficiency. The co-generation process uses waste energy from electricity production for heating, and, at the same time, helps to avoid the environmental impacts from additional heat generation. The overall system can reach very high efficiencies due to the inherent characteristics of the process. The produced heat can be used locally for residential district heating. Tri-generation, combined heat and power with additional production of cooling, holds potential for ever-higher efficiencies.

Empowering producers and consumers and increasing cooperation and confidence in renewables are essential for cities. In some cases, changes to support schemes have lacked transparency, have been introduced without proper consultation and at times have even been imposed retroactively. Moreover, diverging national support schemes and local approaches based on differing conditions can create barriers and prevent market operators from deploying cross-border models, possibly impeding business opportunities. Transnational cooperation among coastal cities in the context of regional sea basins can be decisive for overcoming frontiers.

Local authorities can boost green electricity through innovative public procurement and high standards, invigorating partnerships with citizens and associations, and raising access to adequate infrastructure and awareness. They can remove constraints in the value chain and foster accessibility to renewable energy sources. All municipal buildings should be equipped with renewable energy and demonstrate energy-efficient spaces fuelled only by local smart distributive systems. Coastal cities should give priority to energy from the sea for all their municipal needs, including all public services and transport fleets. They should serve as lighthouses to the road of decarbonisation.

Public acceptance of renewable energy projects is fundamental for further development of options and technologies. Technological limitations, in relation, for instance, to energy transmission and storage could be a bottleneck for revolution of energy systems. Market failures also abound, as the level playground of various energy options is notoriously uneven. The non-inclusion of external costs, such as the effects of air pollution on human health or the harm of sea pollution on ecosystems, in energy prices prevents renewable energy options from competing with fossil fuels on equal terms. At the dawn of the millennium, the EC-supported ExternE project highlighted that, if all costs were taken into consideration, coal should be 2–15 times more expensive, oil 3–11 times its price and gas 1–3 times more expensive. At the other end of the spectrum, PV could be 0.6 times cheaper and wind energy 0.05–0.25 times less expensive (Mega 2005).

4.2 Energy Efficiency, Decarbonisation and Offshore Prospects

Most international studies and projects highlight the urgency to act at the earliest possible stage for effective decarbonisation of world energy systems. Energy infrastructures are long-lived and energy investments take time to bring results. Present decisions impact already the energy reality of 2050. The beginning of a new investment cycle should spearhead energy transition. A deferment of investments would result in more costly solutions and greater disruption in the longer term (EC 2011; IEA 2013a; HKS 2011).

Energy efficiency is the most certain multi-beneficial action of all future energy options and a major driver on the demand side. The residential, transport and tertiary sectors hold the highest potential for improvements and deserve enhanced attention. Cities can play a major role in increasing awareness through appropriate regulation, public procurement policies and economic signals. Increased action on efficiency can serve as a unifying energy policy foundation that brings multiple benefits, including substantial energy security and economic savings. Greater efforts on energy efficiency could cut the growth in global energy demand by half (IEA 2012, 2013a).

The EU Energy Roadmap 2050, adopted in 2011, proposed a broad long-term vision and possible directions for the EU to achieve the goal of full carbonisation, i.e. reducing greenhouse gas emissions to 80–95 % below 1990 levels by 2050. The analysis of a set of different pathways highlighted the urgency for the EU to act united as early as possible and intensify efforts beyond 2020. The exercise intended to help policy makers and industrial actors to make appropriate energy choices and create a stable climate for longer-term private investments (EC 2011).

The EU policies and measures to achieve the Europe 2020 strategy are expected to continue to bring results beyond 2020 and help reduce emissions by about 40 % by 2050, only less than half of the aimed reduction by 2050. Effective energy transition would demand more radical structural reform and social change.

The post-2020 strategies will be critical for decarbonisation. Four main decarbonisation routes for the energy sector have been identified, including energy efficiency impacting mostly on the demand side and, on the supply side, renewable energy sources, nuclear and carbon capture and storage. The scenarios proposed different combinations of these four interacting decarbonisation paths.

The EU roadmap suggests that decarbonisation is possible but all pathways entail radical changes in policy, technology, society and the markets. They all allow at least 80 % reduction in greenhouse gas emissions implying some 85 % decline of energy-related CO_2 emissions including from transport. The long time horizon associated with these developments brings much uncertainty, given also the fact that the underlying assumptions are also shifting. Social and behavioural changes will have significant impact on the energy system (EC 2011).

Decarbonisation scenarios get to their target with no major differences in terms of overall costs or implications for security of supply. The total energy system cost, including fuel, electricity and capital investment in equipment and energy efficient products, could represent about 14.6 % of European GDP in 2050, compared to 10.5 % in 2005. Decarbonisation scenarios suggest an import dependency of 35–45 % in 2050.

All decarbonisation scenarios highlight a transition from the current model, with high fuel and operational costs, to an energy system based on higher capital expenditure and lower fuel costs. The average capital costs of the energy system, comprising investments in power plants and grids, including connection of offshore systems, industrial energy equipment, heating and cooling systems, smart metering, insulation, more efficient and low carbon vehicles, and devices for exploiting local renewable energy sources, such as solar heat and photovoltaic, could increase substantially. The impact on the urban economies and jobs in manufacturing, services, construction, and transport can be very important.

The High Energy Efficiency scenario, which requires political commitment to very high energy savings, includes for example more stringent minimum requirements for appliances and new buildings, high renovation rates of existing buildings and energy savings obligations on energy utilities. This could lead to a decrease in energy demand of 41 % by 2050 as compared to the peaks in 2005–2006.

Electricity is expected to increase its part in the future, almost doubling its share in final energy demand to 36–39 % in 2050. Final electricity demand increases even in the high energy efficiency scenario. Electricity could provide around 65 % of energy demand by passenger cars and light duty vehicles. The power generation system would have to undergo structural change and achieve a significant level of decarbonisation as soon as 2030 in order to reach 96–99 % in 2050. This highlights the magnitude of the transition and the necessity of signals to minimise investments in carbon-intensive assets.

Energy efficiency should increase substantially in all decarbonisation scenarios. Achieving significant energy savings will require a stronger decoupling of economic growth and energy consumption in all cities, regions and economic sectors. Primary energy demand drops to 16–20 % by 2030 and 32–41 % by 2050 as compared to peaks in 2005–2006 (EC 2011).

Cities are in general ready to go for renewables given the significant local benefits for public health and the integrity of ecosystems. Coastal cities should pioneer energy from the sea. The share of renewable energy rises substantially in all scenarios, achieving at least 55 % in gross final energy consumption in 2050. The share of RES in electricity consumption reaches 64 % in a High Energy Efficiency scenario and 97 % in a High Renewables Scenario that includes significant electricity storage to accommodate fluctuating RES supply.

Decarbonisation can be an advantage for cities moving early to renewable energy and contributes to reduction of import dependency and exposure to the volatility of fossil fuel prices. Centralised large-scale systems, such as nuclear and gas power plants, and decentralised renewable energy systems, will increasingly have to work together, as will have to work together terrestrial and offshore systems. The future energy architecture of decentralised and centralised, terrestrial and offshore systems needs to build on the complementary advantages of all scales.

Higher energy efficiency in new and existing buildings is vital and feasible. Nearly zero energy buildings should become the norm. Buildings, including homes, could produce more energy than they use. Materials, equipment and appliances will have to fulfil highest energy efficiency standards. In transport, efficient vehicles and incentives for behavioural change are required. Citizens will gain better information on technological complex issues and more predictable energy bills. With smart technologies, such as home automation, consumers will be better equipped to transform their consumption patterns.

The role of cities and local organisations is expected to be much greater in energy systems of the future. Energy optimisation should be an essential design parameter incorporated into all urban functions. Households and companies will have to invest in the transition to a better energy system. Greater access to capital for consumers and innovative business models is crucial. Energy efficient smart systems have to be incorporated in a wide range of activities, from, for example, digital health services to standards for consumer appliances.

Urban coastal and spatial planning can contribute to saving energy in the medium and long term, with all land and marine uses making a complementary contribution. Local authorities should train advisers to help citizens optimise energy options and conditions, such as insulating buildings, reducing heating and cooling, and should review the necessary infrastructure and networks to systematically use the waste heat of electricity generation in combined heat and power plants.

In 2030, all decarbonisation scenarios suggest growing shares of renewables in the EU of around 30 % in gross final energy consumption. The challenge is to enable market actors to bring down the costs of renewable energy through technological development and more efficient innovation policies and support schemes. Renewables move from small scale to larger scale, integrating local and more remote sources, and increasingly coastal and offshore renewable energy sources. As the scope for expansion of land-based renewable energy generation becomes constrained, the marine space offers a potential solution to public acceptance issues related to visual impact and local disturbance, which may hinder terrestrial renewable energy developments.

In the near future, wind energy from the Northern Seas and the Atlantic can supply substantial quantities of electricity with declining costs. By 2050, wind power will provide more electricity than any other technology in the High Renewables scenario. In the medium term, the contribution of ocean energy can provide an important contribution to electricity supply. Wind and solar power from the Mediterranean countries could deliver a substantial part of electricity. Many renewable energy options need further development to become competitive and penetrate the market. Greater efficiencies and competitive costs require improved infrastructure for integration. Sufficient interconnection capacity and a smarter grid could help manage the variations of wind and solar power in coastal cities.

Heating and cooling from renewable energy sources are vital to decarbonisation. Decarbonisation will require a large quantity of biomass for heat, electricity and transport. In transport, a mix of several alternative fuels will be needed to replace oil. Biofuels will probably be a main option for aviation, maritime transport, long-distance road transport, and rail. The market uptake of second and third generation biofuels based on fish waste, and residues, and algae respectively, should continue to be promoted.

Gas is expected to substitute for coal and oil in the short to medium term and help to reduce emissions with existing technologies until at least 2030. Although gas demand in the residential sector, for example, could drop by a quarter by 2030, due to energy efficiency measures, it may stay high in other sectors, such as the power sector, over a longer period.

Carbon capture and storage (CCS) could play a pivotal role in system transformation. Large-scale application of CCS could help gas to become a low-carbon option, otherwise the long term role of gas could be limited to a flexible back-up for renewable energy. For all fossil fuels, the decarbonisation objectives demand CCS to be applied from around 2030 onwards. CCS is also an important option for decarbonisation of some heavy industries. However, the future of CCS depends on public acceptance and adequate carbon prices. It needs to be sufficiently demonstrated on a large scale by 2020, and then further deployed, in order to be feasible for widespread use by 2030 (EC 2011). Many world thinkers suggest that investing in CCS would divert investments from renewables (Gore 2013).

Enabling information and communication technologies in energy and transport and for smart urban applications are expected to play a major role. The digital infrastructure can make a city more intelligent in all its functions including energy consumption for electricity, heat and transport. The increasing contribution of intermittent renewable generation from terrestrial and offshore sources demands more flexible power systems.

Electricity from wind and solar has low marginal costs and, as its penetration in the market increases, prices could decrease and remain low for longer time periods. Revenues would then be reduced for all generators, including those needed to ensure sufficient capacity to meet demand when wind or solar are not available. This could bring concerns about price volatility and the ability of investors to recover capital and fixed operating costs.

Local energy production from renewable sources asks for a more intelligent grid to deal with variable generation from many distributed sources. With more decentralised generation, from terrestrial and offshore sources, smart grids, new network users, such as electric vehicles and ships running on biofuels, there is a greater need for a more integrated approach on transmission, distribution and storage. Coastal cities should be among the key stakeholders to organise a dialogue with all actors and make sure that the voice of citizens is heard.

Carbon pricing can provide an incentive for the deployment of efficient, low-carbon technologies. The EU Emissions Trading Scheme, the central pillar of European climate policy, is designed to be technology neutral, cost-effective and fully compatible with the internal energy market. It will have to play an increased role. The scenarios show that carbon pricing can coexist with instruments designed to achieve better energy efficiency and development of renewables.

A higher carbon price creates stronger incentives for investment in low-carbon technologies, but may increase the risk of carbon leakage. Such carbon leakage is a particular concern for industry sectors subject to global competition and price patterns. A well-functioning carbon pricing system should continue to include mechanisms such as incentivising cost-effective emission reductions and free allowances based on benchmarks to prevent significant risks of carbon leakage.

Local coastal authorities should invest in discussing the various options and fully involving citizens in their technological choices and the pricing mechanisms and incentives which have to remain transparent and understandable to final consumers. Vulnerable customers could need specific support in the transitional period. Citizens should be engaged in the decision-making process, while technological choices have to take into account the local environment, including also offshore possibilities. Institutional investors could become greater players in the financing of promising costly offshore energy developments.

To drive decarbonisation, coastal cities should deploy numerous efforts on all strands of urban policy. Energy efficiency and renewable energy are considered to only have positive impacts. Irrespective of technological choices, higher energy efficiency and shares of renewables can only be safe beneficial choices. There is no path to decarbonisation without investment in these two prominent policy paths. Improving energy efficiency and incorporating renewable energy infrastructure, including its offshore components, can provide multiple local benefits.

Strategic planning and transparent implementation are needed for renewable energy policy measures and support schemes. Research and development programmes and initiatives and process and social innovation are necessary to speed up developments and guarantee higher and better penetration of renewables in the energy market. Twin coastal cities schemes could support offshore renewable energy developments and help to ensure fair trade and open access to emerging technologies and international energy markets. International cooperation can be further enhanced through joint investments in renewable energy schemes and the creation of new enterprises and jobs.

Important developments at the European front include the SET-Plan, introduced in 2008 as the technology pillar of the EU's energy and climate policy. The Plan aims at fostering excellent science, technology transfer and up-take, industrial leadership on low-carbon energy technologies, and contributing to the worldwide transition to a low-carbon economy by 2050. The SET-Plan has the ambition to transform the entire energy landscape. The implementation of the plan advances through European Industrial Initiatives which bring together industry, the research community, Member States and the European Commission in risk-sharing, public-private partnerships aimed at the rapid development of key energy technologies. In parallel, the European Energy Research Alliance has been working to align research and innovation activities and the SET-Plan priorities, and to establish a common EU framework. An update of the SET-Plan, in early 2015, emphasised a more integrated EU energy system and innovation to speed the advent of a post-carbon economy. Special attention is given to energy efficiency through intelligent technologies that give consumers more control over consumption and storage, and encourage them to transfer surplus renewable energy to public grids.

In 2012, the European Commission launched the Smart Cities and Communities Innovation Partnership to stimulate Europe's capacity for research and innovation to enhance the urban environment. The partnership brings together cities, academia, industry, civil society and citizens to improve urban life through more sustainable integrated urban projects. A Strategic Implementation Plan provides a wealth of detailed examples for smart urban initiatives including better sustainable planning, higher energy efficiencies, better transport responses and a smarter use of digital infrastructure (EC 2012b).

Selected high-impact lighthouse projects bring competent industrial consortia together with cities to demonstrate technologies. A transformation agenda provides strategic orientation for identifying bottlenecks, shapes incentives and defines actions to be undertaken. The replication of successful urban applications and the mainstreaming of innovations are facilitated through the Smart Cities and Communities Stakeholder Platform, bringing city authorities, industry, academia and civil society organisations together, to exchange and capitalise on experiences.

Smart buildings and neighbourhood projects, integrating for example local and renewable energy sources and expanding the use of highly efficient heating and cooling, are the components of future smart cities. The construction of nearly zero energy buildings and positive energy built environment can be scaled-up to energy producing cities. Smart supply and demand service projects can provide data and information to citizens on energy and mobility options and services.

Urban mobility projects could include electric public transport vehicles that are able to exchange surplus energy, braking and accelerating energy, with the rest of the system. Smart technologies would help to manage energy flows or hydrogen as an energy carrier for storing energy and to balance the total urban demand. Sustainable digital infrastructures could help reducing the carbon footprint of the Internet, in particular data centres and equipment, and advance towards intelligent heating and more and better cooling and lighting solutions (EC 2012b).

Such initiatives can be of great interest for coastal cities as they promote industrial technology tests to better assess developments on the ground, with reasonable costs and clear advantages for citizens and the community. During a brokerage 2013 event, Greek and Cypriot cities suggested some lighthouse projects in coastal cities. Near-to-market technologies which intersect energy, transport and ICT have been specifically designed for coastal localities.

Energy from the sea is essential for coastal cities and for maritime transport. Coastal cities should partner with governments and businesses towards enhancing coastal and offshore energy sources. The global ocean energy resource far exceeds all present and projected future energy needs. Exploiting this indigenous resource would help countries and cities to mitigate dependence on fossil fuels and enhance energy security. This may be particularly important for island nations and regions, where ocean energy can contribute to energy self-sufficiency and replace expensive imported energy.

The offshore wind energy is expected to increase. The EU is the global leader in offshore wind energy with more than 90 % of the world's installed capacity. In 2012, offshore wind represented 10 % of the annual wind capacity installed in the EU. Offshore wind capacity grew by 33 %, a faster rate of growth than the onshore wind sector. Two landmark projects reached financial close in 2013. Twelve offshore projects currently under construction will bring cumulative capacity to 9.4 GW. The European Wind Energy Association expects the percentage of new offshore installations to rise to 60 % by 2030 (EWEA 2014).

A further promising concept is offshore floating wind power. A floating platform anchored to the seabed could lead to more cost-effective technological and policy options. Europe already counts two offshore wind floating demonstration projects in operation, in Portugal and Norway. Nearby cities should be involved in the decision-making process, ensure that impacts on marine ecosystems have been properly addressed, and examine synergies with other activities.

Ocean energy can be instrumental for fuelling coastal urban economies and societies. Many governments invest in research and development for wave and tidal energy converters. Most of the existing technologies are still in the demonstration phase. The UK Carbon Trust estimated that the global wave and tidal energy market could be worth up to €535 billion by 2050. Ocean energy has also the potential to create new, high-quality jobs in technology development, manufacturing and operations (EC 2014a).

Ocean energy could be a valuable asset in urban coastal energy portfolios and could help to balance out other renewable energy sources to ensure a steady aggregate supply of clean energy to the grid. Ocean energy devices, entirely or partially submerged, have a low visual impact, and do not disturb local communities. As for offshore wind energy, the potential impact on marine ecosystems has to be carefully studied. Supporting policies could draw from the early measures to support offshore wind power development in the 1980s and 1990s. Since that time, the wind sector, including offshore wind, has grown exponentially having benefited from targeted policy support. Ocean thermal energy conversion technology has a strong potential in tropical regions, displaying the highest difference of temperature between surface and deep waters. Local deployment can provide for the needs of the coastal populations and contribute to their independence from imported energy.

Biofuels from marine waste and algae is a major energy avenue for transport in coastal cities. Second generation biofuels from fish residues can transform a liability into an asset, as can do third generation biofuels from algae. Algae have been already used as food, feed and fertilisers for centuries. Their size ranges from micrometers of unicellular micro-algae to macro-algal seaweeds of tens of metres. Algae contain several high-value molecules, such as lipids, proteins and carbohydrates, and present a growing interest for multiple uses. Approximately 200 species are used worldwide in different sectors, from pharmaceutical to cosmetics industry and the production of ethanol or biodiesel, while research is ongoing on genetic engineering of micro-algae (EC 2014b).

4.3 Urban Leadership and Action to Surpass National Energy Targets

The governance architecture is decisive for the decarbonisation of the world. Governments, at all levels, have to lead by example and invest in post-carbon, energy-efficient systems and buildings, goods and services. Urban governments will have to drastically reduce the energy consumption related to their spaces and functions by making the required investments and the necessary audits to continuously monitor energy performance. They also have to weave the links among all levels of governance, stakeholders and the networks with other cities.

Local authorities and industrial actors can help with the energy audits to increase energy efficiency in businesses and especially small and medium-sized enterprises. It is also crucial to introduce schemes for the effective recovery of heat losses from electricity and industrial production, and to enhance cogeneration of heat and power. Port authorities can also contribute much. The "WaterNet" Waste Water Treatment Company at the Port of Amsterdam is operating a successful monitoring system for electricity and combined heat and power, acting as a buffer facility for energy. This approach leads to a significant reduction of GHG emissions, higher energy efficiency and energy costs reduction.

The energy efficiency of the built residential environment is crucial for all urban governments. The great majority of the housing stock of European cities is old and its energy performance has high potential for improvement. In Amsterdam, almost half of the housing units were built before the second war. The Amsterdam Climate Office collaborates with stakeholders, private landlords, owners and residents, to promote insulation, double glazing, efficient heating installations, and the use of renewable energy. The much better energy performance of new houses has however been counteracted by an increase in the number of housing units and especially by an increase in electricity consumption.

Housing associations, which own more than 50 % of the housing units in Amsterdam, are important players along with tenant organisations. They invested for many years in better insulation and higher efficiency. Since 2007, the municipality of Amsterdam has headed the alliance of leaders in which the housing associations are challenged to participate in concrete initiatives and achieve the best performance. Model houses to reduce emissions were the joint results of common efforts. It was estimated that if all housing associations undertake serious renovation by 2025, a CO_2 reduction of 37 % could be achieved (Amsterdam Climate Office 2008).

The lion's share of electricity consumption in the United States occurs in buildings where a very diverse array of cost-effective efficiency investments is possible. For example, replacing appliances such as air conditioners, refrigerators, freezers, and hot water heaters with more efficient models could reduce energy use by 30 %. The savings from attaining full deployment of energy-efficient technologies in buildings could eliminate the need for new electricity generation capacity through 2030.

In the US, the Better Buildings Initiative was announced as part of the plan to ensure that America invests in innovative clean energy technologies and doubles the share of electricity from clean energy sources by 2035. In parallel, the Better Buildings Initiative aims at making commercial and industrial buildings 20 % more energy efficient by 2020 and accelerates private sector investment in energy efficiency. Through a variety of efficiency improvements, e.g. improved lighting, greater insulation, more efficient heating and cooling, clear information and access to financing, buildings can become more energy efficient and better places to live and work, while generating employment and fuelling growth.

The White House Better Buildings Challenge is inspiring public and private sector leaders to commit their organisations in saving energy and money, and to demonstrate the results of the best strategies. Partners commit to an energy savings pledge, a showcase building, and the presentation of their achievements. They enjoy public recognition, technical assistance, and sharing of best practices through a network of peers. Their leadership can provide real models for emulation and continuous overcoming. The US Conference of Mayors and the American Institute of Architects recognise that creating energy-efficient, high performance buildings is critical to address climate change and promote integrated, sustainable design with a goal of reaching carbon neutral buildings by 2030.

Although there is great potential, many barriers exist to widespread adoption of energy efficiency technologies. The upfront expenditure can be high and can deter investment despite the prospect of long-term cost savings. Volatile energy prices can cause delays in the purchasing of more efficient technology due to a lack of confidence concerning the return on investments. Public authorities closest to the citizens must inform on local investments in energy-efficient infrastructures, which can impact patterns of energy use for decades. They are responsible for providing trustworthy information for consumers wishing to learn about the comparative advantages in terms of performance and costs of energy-efficient technology alternatives. Driving change will require significant public and private support, and sustained effort.

Successful energy efficiency initiatives include the US Department of Energy and Environmental Protection Agency (EPA)'s Energy Star labelling scheme. The average Energy Star building uses 35 % less energy and produces 35 % less carbon dioxide emissions than typical buildings. EPA lists the US cities with the most Energy Star certified buildings that meet strict energy performance standards and produce limited greenhouse gasses. By the end of 2011, 16,500 buildings had earned an Energy Star, saving the US economy an estimated $2.3 billion in annual utility bills.

Fifteen types of commercial buildings can earn the Energy Star, including office buildings, schools and retail stores. Energy used in commercial buildings accounts for nearly 20 % of greenhouse gas emissions. With 659 Energy Star buildings, Los Angeles has, by far, the most energy efficient buildings in 2011. Major coastal metropolises, including San Francisco, New York and Boston, also feature at the top of the energy star champions' pyramid.

New York City introduced the Greener, Greater Buildings Plan (GGBP), within the city's ambitious plan for 2030. Improving the energy efficiency of existing large buildings is crucial for achieving New York's goal of reducing greenhouse gas emissions 30 % by 2030 compared to 2005 levels. The emphasis on large existing buildings is driven by three facts. First, 75 % of GHG emissions in New York City come from energy used in buildings. This is almost double the proportion in the US as a whole. Second, almost half of New York City's GHG emissions are generated by the city's largest buildings, which constitute only 2 % of the city's properties, but half of the built space in the city. Third, 85 % of the buildings that will form the 2030 horizon are already built.

New York's largest buildings have a huge potential for energy savings. On the cusp of an information revolution concerning energy use in building stocks, Local Law 84 of 2009 mandates that all privately-owned properties, with individual buildings over 50,000 square feet or multiple buildings with a combined surface of over 100,000 square feet, annually measure and report their energy and water use (New York City Mayor's Office of Long-Term Planning and Sustainability 2012).

In 2011, the first benchmarking report analysing a year of energy and water use marked a first milestone in increasing knowledge about buildings' energy and water management and highlights opportunities for property owners to make savings through higher buildings efficiency. The benchmarking documented the state of consumption and performance in large buildings in New York City. Using the EPA's Energy Star Portfolio Manager, an interactive energy management tool that allows tracking and assessing energy and water consumption across the entire range of buildings, nearly 1.8 billion square feet of built space was benchmarked. This is the largest collection of performance data gathered for a single jurisdiction and can help identify opportunities for further efficiencies.

Energy efficiency in the heating and cooling of buildings can reduce greenhouse gas emissions. Landmark buildings pave the way. New York's Empire State Building was retrofitted for energy efficiency, reducing its energy consumption by 38 %. The benchmarking report highlights that energy use varies greatly between property types, uses, functions and locations, with some properties using 3–5 times more energy than buildings with similar volumes and uses.

In the EU, the 2010 Energy Performance of Buildings Directive (Recast) asked for all new buildings to be (net) nearly zero energy buildings by 2020. The Energy Performance of Buildings Directive (EPBD), issued in 2002, had introduced a common framework of harmonised measures for the development of integrated energy performance standards, to be applied to new and existing buildings when renovated. An integrated method addressing all energy aspects could facilitate the most effective and efficient concert of measures and offer a basis for designers and builders to recognise and promote high standards.

Coastal cities often place outstanding buildings at the waterfront. From the Aegean island of Syros to the Tokyo port, public built environments looking into a sea include a very broad spectrum of buildings which can mark transition to the new energy order. Energy Performance Certificates can be displayed in public buildings providing information and increasing public awareness on energy conditions and objectives. Public authorities are responsible to implement the recommendations included in the certificate, monitor their performance and communicate their results to citizens. They can also interact with citizens, involve them in life cycle assessments and modify conditions and objectives according to public expectations.

Illuminated waterfronts hold pride of place and can magnify coastal cities. Public lighting has often been a subject of controversy, as it has been perceived both as a factor of public security and local identity and as a driver of high electricity consumption. Ample lighting has a beneficial effect on the atmosphere of a city. It has proved to help reduce both road accidents and insecurity incidents, while the illumination of monuments and public buildings can redesign a city. Light designers were frontrunners in a contest to dramatically brighten up the city of Montreal. An Iceberg show consisting of illuminated arcs and the day of Eight Suns with projections on the façades of the buildings were some of the innovations, parts of a truly artistic energy adventure.

Coastal cities can make sure that blue green electricity is given high priority for all uses. Lighting typically accounts for about 20 % of a city's total electricity consumption. Intelligent lighting can optimise illumination, increase comfort and urban safety and reduce costs. The latest generation street lighting is robust, weather-proof and economical, often lasting 3–5 times longer than traditional street lamps. In a report by Siemens focusing on New York, as "the city where the future comes to rehearse", performance contracting is proposed as an essential driver (Siemens 2012).

Responsible public lighting, especially on the waterfront, would benefit from lighting adapted to the degree of darkness. The municipality of Amsterdam replaced older lamps with more efficient ones and entirely switched to electronic components. In 2007, the City Hall of Amsterdam was already fully equipped with energy-efficient lamps, achieving 45 % energy savings. Furthermore, new generation streetlights have been installed and provide huge energy savings throughout the city (Amsterdam Climate Office 2008).

Cooperation among all cities and their elected representatives is critical for global change. The EU Covenant of Mayors is an ambitious initiative associating cities and citizens in the energy future. It was launched by the European Commission, with the support of the Committee of the Regions and the European Parliament, during the 2008 European Sustainable Energy Week, as the mainstream initiative involving local and regional authorities in the fight over climate change and towards a more sustainable energy future. The EU Covenant of Mayors is a voluntary commitment by local elected leaders to meet and exceed the EU 20 % emissions reduction objective through increased energy efficiency and the development of renewable energy sources. With 5709 signatories in March 2015, this movement brings together the most pioneer cities to exchange good practices on energy efficiency and promote low-carbon socio-economic development.

The Covenant, open to cities of all size or form, aims to support the efforts made by local authorities to implement sustainable energy policies. When local and regional authorities engage with the Covenant, they commit to establish a Baseline Emission Inventory and a Sustainable Energy Action Plan and submit them to the European Commission within the year following their adhesion. The Action Plan includes the Covenants' signatories foreseen actions to reach the CO_2 objectives by 2020, the measures to be taken to reach their targets and the responsible actors and stakeholders. Regions usually play a key role as territorial coordinators and co-investors in the energy future.

The EU Covenant has generated many creative ideas and impactful alliances, including with over 1000 American cities working within the US Mayor Climate Protection Agreement, via a memorandum for cooperation between the Committee of the Regions, bringing together all cities and regions of the European Union, and the US Conference of Mayors, bringing together all 1200 American cities with a population of at least 30,000 inhabitants.

In China the engagement of local authorities has also been vivid. In September 2012, the first EU-China Mayors Forum brought together the EU Covenant of Mayors and the Chinese Mayors' Association for an exchange on the challenges and opportunities in the field of sustainable urban development. The exchange led to the signing of the EU-China Mayors' Charter which commits the signatories to share experience and know-how in sustainable energy management.

In USA, the day that the Kyoto Protocol became international law, the mayor of Seattle launched the US Conference of Mayors Climate Protection Agreement to advance the goals of the Kyoto Protocol through leadership and action. In signing the agreement, the Mayors committed themselves to initiatives which accelerate local efforts, including energy efficiency and conservation block grants.

The US Conference of Mayors Climate Protection Agreement 2009 Awards and Best Practice models recognised the efforts of some pioneering cities. Honolulu introduced the concept Ahupua, as its overarching brand for sustainability and climate protection. "Ahupua'a" is the Hawaiian term for the sustainable resource management system that have allowed living in balance with nature for over 1800 years. Enhancing the legacy of the Polynesian ancestors, the city tries to project an important message and combine traditional culture with technological innovations to accelerate efforts towards self-sufficiency and sustainable development. The twenty-first century Ahupua'a has accomplished several major milestones, including the first sustainability plan, greenhouse gas inventory processes for city operations, authored public awareness programmes and partnerships with the private sector and academia to promote sustainability and climate protection practices (US Conference of Mayors 2009).

The Seattle Climate Partnership is a voluntary agreement among Seattle-area employers to assess and reduce their carbon footprint, and to help meet community-wide goals for reducing climate pollution. The Partnership has been joined by many organisations, from a variety of sectors including health care, finance, engineering, manufacturing, and education. These organisations have committed to assess and reduce their carbon footprint, consider the strategic impacts of climate change on their organisation, and help forward the community reduction goal of 7 % below 1990 levels. In 2008, the Partnership has launched a recruitment effort aimed at the largest ship owners, expanded its technical assistance programme to assist partners in reducing emissions related to fleets and fuels, and created a networking and recognition programme.

American cities expect to have important benefits from sustainable energy. According to a 2011 survey, 70 % of them wish to develop a greener economy. Eighty-three per cent of cities deployed new energy technologies with the support of the Energy Efficiency and Conservation Block Grants Programme offering grants for energy technology investments. The main technologies funded include energy-efficient lighting, new building technologies and photovoltaics. Financial constraints are highlighted as the most important problem for the adoption of low/zero carbon technologies. Adaptation to climate change is already integrated in the policies of many cities and some of them have set targets for the use of renewable energy. Mayors suggest that there is a great potential for non-governmental cooperation on clean energy and public-private partnerships (US Conference of Mayors 2011).

From the EU and US, the movement expanded to the world connecting leaders and accelerating action. Mayors adopted the Global Cities Covenant on Climate, which aims to ensure transparency, accountability and comparability of local climate action. The initiative encouraged signatories to report commitments and actions to the carbon Cities Climate Registry (cCCR). Cities announcing commitments have to report their baseline and latest available GHG inventory, in order to monitor their progress towards achieving their targets. The cCCR, operated by the International Council for Local Environmental Initiatives, acts as a global online platform for local governments to report climate and energy commitments and accomplishments.

Another interesting global initiative comes from the World Wide Fund. The Earth Hour City Challenge has been designed to mobilise action and support from cities in the transition towards a climate friendly, one-planet future, and to stimulate the development and dissemination of best practices for climate mitigation and adaptation. The initiative is a global challenge for cities to present ambitious, holistic, inspiring and trustworthy plans for low carbon development and for dramatically increasing the use of efficient renewable energy in the next decades. The primary focus of the challenge was on city actions that help move investment flows from fossil fuels to sustainable energy solutions.

The Earth Hour City Challenge has awarded Cape Town the title of Global Earth Hour Capital 2014. In the final round of the global competition, Cape Town stood out as a role model for the global South, with a portfolio of innovative actions that other cities can replicate. The city has taken bold steps towards energy transition. A solar water heating programme has been instrumental for promoting renewable energy. Community engagement on sustainability issues is high and has brought important energy savings, especially through a large-scale building retrofitting programme.

4.4 Optimising Sustainable Mobility Patterns for Coastal Cities

Conscious coastal cities have to offer competitive and efficient, affordable, clean and safe transport to citizens and visitors. Transport infrastructures are the arteries for circulation of socio-economic vitality in cities. Urban transport is a socio-technical system including interacting paradigms and visions, cultural and mobility patterns, technologies and infrastructures, business models, and policies responding to citizen desiderata. As most journeys begin and end in cities, increasing demand for urban mobility has often created an unsustainable situation characterised by severe congestion, poor air quality, noise and high levels of CO_2 emissions and citizen dissatisfaction (EC 2013b).

The transport sector is responsible for the fastest growing source of greenhouse gas emissions and presents the greatest problems for decoupling economic prosperity from pollutant release. Despite the important technological improvements that the transport sector witnessed over the last years, the increasing demand outweighed the eco-efficiency gains.

The key strategies try to reduce the carbon intensity of transport, improve the modal split and support sustainable travel. A sustainable urban mobility plan gives priority to more sustainable transport modes, i.e. walking, cycling and public transport, while fostering a balanced citizen-friendly multimodal system able to flexibly respond to all needs. Plans usually include integrated sets of technical, political and social measures to improve performance and cost-effectiveness of innovative technologies. As all urban sustainability plans, a sustainable mobility plan embraces a transparent and participatory approach and is closely discussed and monitored by all major stakeholders. In particular in coastal cities, the mobility plan has to ensure that all transport hubs linking the maritime, aerial, rail and road networks are effectively linked and flows are optimised.

Barriers and success factors for sustainable urban transport abound and are often linked to the local context. The main categories of barriers include resource barriers, including physical, financial and institutional hurdles, often related to uncoordinated actions, legal barriers linked for example to the regulations providing the framework conditions, social and cultural barriers related to public acceptance and side effects. The formulation of a well-informed consensual vision and the development of a smart, sustainable and equitable portfolio of policy measures, public participation, accountable governance and enlightened research and innovation are among the promising avenues.

Prospective studies are essential in order to formulate a mobilising vision for the future of a city and its transport models. The mobility projections by the International Transport Forum suggest that global passenger transport volumes in 2050 could be up to 2.5 times as large as in 2010, and freight volumes could grow by a factor of 4. The strongest growth is expected from emerging economies. Car use levelled off in OECD countries and passenger transport grows more slowly than the GDP, but this trend could be transitory. In 2000, half of global passenger kilometres were driven in OECD countries but this share could decline to 20 % in 2050 and follow the Eastward shift of economic gravity. Emissions of CO_2 grow more slowly because of increasing energy efficiency, but could nevertheless more than double (ITF 2012).

The future of mobility demand in cities depends on topography and urban form, culture and lifestyles, local economy, social landscape and governance. A new culture of sustainable urban mobility asks for less car-dependent lifestyles and the promotion of better alternative forms of mobility. Increasing the part of walking, cycling and public transport has never been more urgent. Optimal urban sustainability policy portfolios include legislation, the promotion of lower-consumption vehicles and innovative fuels and propulsion technologies, demand-management schemes, such as parking and access restrictions, fair and efficient pricing, teleworking and land use and planning. Mix of compatible land uses is highly important, as it can lead to mutually enriching functions and ensuring the accessibility of places (OECD–ECMT 1994; ITF 2012).

The transition to a more sustainable urban transport system asks for the right policy measures and technologies to influence citizen behaviour. The car fleet may continue in the medium term to be dominated by internal combustion engine technology, but it is expected to display smaller and lighter vehicles, more electric vehicles and higher levels of hybridisation, and also autonomy and interconnectedness. Electric city cars could be foldable or adapted to the space requirements and better served by an expanding infrastructure. Most fundamental changes are likely to have occurred in the organisation of transport and innovative inter-modal mobility services. As the sharing economy expands, car and bike sharing are expected to grow in some urban areas and this could lead to the reduction of urban vehicle stocks and flows. Freight logistics could become optimised, thanks to much more nimble ICT applications.

Governance strategies should deal with the transport system as an integrated organic whole. Many studies conclude that just using cleaner fuels and propulsion technologies and optimising the efficiency of transport flows is not enough. Successful pathways do not only require transformational change in one of the elements of the system, but in several or in all of them, and at the same time (STOA 2012, 2014).

The search of a seamless transport system in the urban world helps to identify win-win investment options that provide sustainable value. The upward pressure on mobility demand from larger populations, including a rising middle class with higher income in the emerging world, is crucial. Cities in the emerging world start investing in integrated less unsustainable transport networks, smarter and greener infrastructures and alignment of mobility objectives with socio-economic and environmental policy objectives. Multi-dividend practices should be highly encouraged, especially in times of crisis and restricted public and private funding.

Access to places, goods and services is the ultimate objective of transport. Planning and design for sustainable urban mobility argues that the development of transport systems requires a conceptual leap. The purpose of sustainable mobility is to offer access to destinations and activities, socio-economic and environmental services and goods. Urban design should focus on bringing people and places together in healthy environments. Urban form and the functionality of the city therefore merit major attention, and this highlights the importance of integrated land-use and transport planning (OECD–ECMT 1994; UN-HABITAT 2013).

In 2013, the United Nations' World Habitat Day celebrated urban mobility because access to goods and services is essential to the efficient functioning of expanding cities and the well-being of their citizens. Accessible cities encourage a shift towards more sustainable modes of transport and draw more and more travellers out of cars and onto trains, boats, buses, bike paths, and pedestrian areas. Mobility is a service depending on far more elements than just the mode of transport. Urban planning and design should focus on creating conditions for optimal accessibility rather than simply expanding urban transport infrastructure.

A comprehensive integrated and coherent approach is a sine qua non condition. The International Transport Forum suggested that the lack of cross-industry cooperation, conflicting policies and missing standards, and the inherent risks of hyperconnectivity are the main barriers for seamless travel and transport. It suggests that these barriers can be overcome through a game-changer combining intermodal travel, a future traffic management system for megacities, and the optimisation of the logistic systems (ITF 2012).

Sustainable seamless passenger travel in urban areas asks for local and regional authorities to reassess the boundaries of geographical and administrative structures. These will have to be more flexible and less mode-oriented and jurisdiction-bound to allow for imaginative new mobility services. Infrastructure investment across all modes and innovative funding arrangements are very important. Business processes must be much more open to mode-neutral mobility and common standards are crucial for interoperability. Sharing risk and liability with public–private partnerships could be crucial and facilitate for instance greater uptake of road congestion charging or urban tolls. Transparent and clear pricing rules fully aligned with sustainability objectives are fundamental pre-conditions. Public acceptance depends on informed stakeholder involvement and the clear allocation of revenues from pricing schemes to sustainability projects.

Congestion is a huge burden on transport. In the US, Urban Mobility Reports published by the Texas A&M Transportation Institute underlined the magnitude and impacts of traffic congestion. Unreliable travel has a high cost for commuters and businesses and emits more greenhouse gas emissions. Rankings of the most congested cities include repeat performers. The list of the top ten is composed of Washington DC, Los Angeles, San Francisco-Oakland, New York-Newark and Boston, followed by Atlanta, Chicago, Philadelphia and Seattle. Highlights suggest that the delay endured by the average commuter was 38 h in 2011, up from 16 h in 1982, and that the cost of congestion is more than $120 billion, nearly $820 for every commuter in the US. Congestion is becoming a more general problem outside traditional rush hours, with about 40 % of the delay occurring in the mid-day and overnight hours. As traffic congestion gets worse and the time for a given trip becomes more unpredictable, the 2012 report proposed a Planning Time Index as a measure of time reliability illustrating the additional time needed. The most sustainable policy options involve public transport, combined with traffic and demand management strategies, such as signal coordination, teleworking and flexible work hours (Texas Transportation Institute 2011, 2012).

Freight transport is pivotal for the socio-economic functioning of cities but the efficiency of the sector has a high potential for improvement. Construction works and retail are responsible for much of freight transport. The sector accounts for about 10–12 % of vehicle traffic in cities but causes disproportionately higher impacts on congestion and the environment. The importance of urban logistics could become more intense in OECD cities, given the rise of e-commerce and the needs of an ageing population. Many cities impose vehicle size or weight restrictions, or limit access in certain areas. A balance has to be struck between access requirements for operators and tolerable impact to local populations.

Intelligent traffic management is a growing challenge for world megacities. Seamless transport is being promoted as a smart investment in advanced transport systems that minimise obstacles to interconnection and access to information and services. The Condition-based Megacity Traffic Management System integrates and processes information from vehicles, travel infrastructure, citizens and the social environment. Using real-time data collection and analytics, it helps to reduce congestion, greenhouse gas emissions and fuel consumption and accidents. The total net benefits of the system can range from $2 billion to $10 billion per city (ITF 2014).

In the European Union, transport infrastructures cover 10–15 % of the urban space. Household spending in the transport budget amounts to 13 % of the total budget. On average, a European citizen makes 1000 trips per year and half of these are less than 5 km long (EC 2009). According to a 2013 Eurobarometre survey, half of all Europeans use a car every day, which is more than those cycling and using public transport (16 %) combined. Across the EU, only a third of respondents use public transport at least once a week. Around four in ten Europeans have problems when travelling within cities and a substantial majority of Europeans believe that air pollution, road congestion, travelling costs, accidents and noise pollution are serious problems within cities. The cost of urban trips is important as more than half of Europeans believe that better public transport and lower prices for public transport would be the best ways to improve urban travel. Less than a quarter of Europeans believe that the urban traffic situation will improve in the future (EC 2013b).

Some cities display dedicated places for electric cars equipped with charging battery infrastructure, such as for Zen cars in Brussels. Despite some encouraging signs, conventionally-fuelled cars are by far the dominant urban transport mode, contributing about 75 % of kilometres travelled in EU conurbations. Increased car use has not only been the cause of environmental problems, but also underinvestment in public transport. More than one third of fatal road accidents happen in cities and half of the victims include pedestrians and cyclists. Improvements have to address safety conditions of vehicles and infrastructures, enforcement of road rules, education and training, and protection of vulnerable populations including post-accident emergency services.

Codes of the street ask for more attention to the rights of pedestrians, cyclists, children, the elderly and the handicapped. The Belgian code requires drivers to respect 1 m minimum distance from pedestrians crossing the streets. The concept of crossing curbs can promote urban safety and quality of life. The EU aims at halving the number of casualties by 2020 and bringing it near zero by 2050. Progress has been slower in cities than in the rest of the Union.

A comprehensive EU strategy for resource-conscious competitive urban transport systems, the Transport 2050 Roadmap aims to dramatically decrease Europe's dependency on imported oil and reduce carbon emissions by transport by removing major barriers and bottlenecks in key fields across transport infrastructure, investment, and innovation. The roadmap proposes 40 measures for different types of journey, including within cities, between cities, and long distance. They include a major commitment for phasing out conventionally-fuelled cars in cities and a 50 % shift of medium distance intercity passenger and freight journeys from road to rail and waterborne transport. By 2050, the majority of medium-distance passenger transport, about 300 km and beyond, should go by rail and by 2030, 30 % of road freight over 300 km should shift to other modes such as rail or waterborne transport, and more than 50 % by 2050. Together with a 40 % use of low-carbon fuels in aviation and at least a 40 % cut in shipping emissions, these measures were assessed to contribute to a 60 % reduction of transport emissions.

The manifesto on sustainable mobility for European regions, towns and municipalities by Council of European Municipalities and Regions (CEMR), bringing together approximately 100,000 local and regional authorities, suggested some key areas for further reflection and action at all levels. These focus on the barriers to limit the use of public transport, the promotion of public transport and alternatives to car use, incentives to manage demand for road space, quality and safety issues, privatisation, public procurement, concessions, and public service obligations, economic instruments, and synergies between urban planning and transport at all levels of governance (CEMR 2005).

The European Union has assisted cities to implement and evaluate a broad array of innovations to promote sustainable urban transport. Since 2002, the CIVITAS (CIty-VITAlity-Sustainability) initiative facilitated the sharing of best practice among more than 200 European cities forming a network Forum, open to all cities engaged to a sustainable modal shift in their urban transport systems. By signing the CIVITAS Declaration, a non-binding voluntary commitment, cities communicate and exchange intensively in a joint effort to raise the bar of excellence in sustainable urban mobility. An early rental bike system was one of the achievements of the initiative.

Multilevel governance is very important as urban, national and international networks link together in cities, especially in port cities having also an airport. The EU Action Plan on urban mobility (2010–2012) implemented 20 key actions and proposed measures to encourage local, regional and national authorities in achieving their sustainable urban mobility objectives. In 2013, the European Commission adopted the Urban Mobility Package, reinforcing its supporting measures in the area of urban transport by sharing experiences and best practices, and fostering cooperation, research and innovation on new and better solutions for urban mobility (EC 2013b).

Pioneer coastal cities have often advanced issues of sustainable mobility. Many coastal cities have introduced policy measures for intermodal mobility, linkages to national and supranational transport networks and support to walking, cycling and public transport. A cardinal element in their approaches is the integration of the waterfront and harbour area in the mobility network and the intelligent articulation to regional, national and international transport hubs.

The Athens urban mobility system, finalised with the occasion of the 2004 Olympic Games, integrated the airport and the port of Piraeus. The new metro, equipped with the latest technology, incorporates the former Athens-Piraeus Electric Railways, a conventional steam railway opened in 1869 and electrified in 1904. During the construction of the metro tunnels, archaeological artefacts were discovered and are exposed in galleries of the underground system.

The concept of sustainable urban mobility plans promoted by the European Commission in 2013 suggests that the ultimate aim is improving accessibility of urban areas and providing sustainable mobility services to, through and within the city. It considers the "functioning city" and its hinterland rather than a municipal administrative area. The plan has to be an organic part of a long-term strategy for the future development of the city and its transport and mobility infrastructure and services and contribute to the economic, social and environmental urban sustainability. The development of a sustainable urban mobility plan should start with a careful assessment of the present performance of the urban transport system and forward analysis of the possible and wished demand (EC 2013b).

Noble pedestrian streets and districts have expanded throughout the world and became the hallmarks of many cities. "Is the "car-free city" a utopia or a reality? A research study by the European Commission, in early 1990s, suggested that the city could be re-conceived and redesigned to give priority to pedestrians. A car-free city would be a polycentric one, made up of small islands on a human scale which would be connected by high-speed means of transport. It seems that the car-free city would benefit not only the environment but also the economy, since it appeared to be 2–5 times less expensive, dependent on density. Amsterdam, which had gone through a recent referendum on the restriction of the private car, in early 1990s, organised an emblematic conference "Car-Free Cities?" The question mark is not negligible but many negative perceptions and reserves have long been overcome (Municipality of Amsterdam 1994).

Italian cities were among the first in protecting their splendid historic centres from private cars and introducing peripheral parking and ride systems. Some urban mobility plans made an ode to pedestrian areas and streets as shared dignified public space. Their role is well highlighted in the European Urban Charter issued by the Council of Europe in 1992 and reinvigorated in 2008 (Council of Europe 1992, 2008).

The transformation of European urban hearts into pedestrian areas with cycling paths and human-centred infrastructure has celebrated five decades of astonishing results. Despite difficult beginnings due to strong opposition from commercial lobbies, the schemes were adopted by many cities in the world. Copenhagen was a pioneer city in recognising the sustainable value of central pedestrian streets. The creation of pedestrian precincts evolved in parallel with public transport and a bicycle network, with parking policy and elimination of 2–3 % of the parking spaces per year in the city centre. Urban space was freed from traffic and gained by citizens, initiating a tide of return from anonymous peripheries to the historic heart of the city. The pedestrian Stroget area attracted civic architecture, sculptures, fountains, musical and cultural events and became the backbone of an archetypal Northern pedestrian precinct (EEA 2010).

Pedestrian vitality of main streets also abounded in US coastal cities like New York, Boston and San Francisco with increasingly safe and inviting streets for pedestrians. The litmus test of pedestrian friendly areas includes the high attraction of citizens outside office hours and round the clock, a continued expansion of public transit to flexibly respond to the movement of pedestrians in longer trips, restricted parking, and the ending of mega-garages of the 1950s and 1960s. Widening sidewalks and reorienting and broadening pedestrian walkways across downtown were the first signs of a new era in urban transport (Boston Foundation and The Citistates Group 2004).

Bicycles are the only sustainable transport means besides walking. Policy measures and infrastructures to promote cycling activities have expanded in most European cities. The modal split of cycling has grown significantly in European cities during recent years. Amsterdam and Copenhagen are the capital cities endowed with most elaborate bicycle network, complementing the road and canal routes. Copenhagen, selected by the International Cycling Union as the world's first Bike City, is registering 41 % of urban trips made by bicycle. Both cities developed successful public cycling plans.

Is the bicycle the future of urban mobility? Copenhagen counts 150,000 citizens cycling every day. The city plans to accommodate an additional 60,000 cyclists by 2025 by developing a network of Super Cycle Highways in the area around Copenhagen connecting residential areas with places where people work or study. A Super Cycle Highway is a route where commuters' needs have the highest priority. For instance, cyclists will have a green wave of traffic lights during rush hour. The highways offer fast, comfortable and safe service, and transit areas for public transport. They are as direct as possible, cutting down on the number of times cyclists have to stop. All municipalities were enthusiastic about cooperating in this innovative project and jointly developed a conceptual strategy and a plan for the network of Super Cycle Highways, in total 28 routes of about 500 km.

The fundamental policy measures, infrastructure and conditions to support cycling are expanding and many coastal cities provide examples. Helsinki has more than a total of 1500 km of bicycle paths, Stockholm and Hanover about 750 km. Smaller cities like Turku, Aalborg, Tampere and Aarhus, featuring a cycling network of about 300 km, managed to convince about thousands of citizens to use the bicycle also for commuting to work. The bikes of La Rochelle can be hired to discover the city. As in many cities, public bikes are free of charge during the first 2 h. Dublin deploys many of its 220 km of cycling routes along the coast (EC 2013a).

Towards the last decades of the twentieth century, bicycles started being valued as components of everyday urban transport and recreation patterns. The "Velib" bike sharing service has been successful in many European cities and disseminated extensively in the world. Velo-city events and conferences have been held in Copenhagen, Dublin, Barcelona, Perth, and Montreal to promote pedal power in bicycle-friendly coastal cities. Each event offers participants a chance to share best practices for cycling-friendly cities and involve debates on health, economics and the environment with experts and citizens. More cities associate cycling with sightseeing. Bangkok created a cycling path surrounding its historic Rattanakosin island.

More than 500 bike sharing systems in cities involve a global fleet of more than 500,000 bicycles. The scheme is expanding very quickly and the number of initiatives has doubled during the last 5 years. Each city introduces various innovations. Buenos Aires suggested the bicycle's manual, while Hangzhou, one of the most prosperous cities in China, organised a perception survey to understand drivers and barriers to bike sharing. The results suggested that almost one third of the users had incorporated bike-sharing into their everyday practices and, most frequently, used a bike sharing station closest to either home or work. The modal shifts indicated that bike sharing acted both as a competitor and a complement to existing public transport. Recommendations for improving bike sharing included expanding infrastructure and parking services, improving bike maintenance and locking mechanisms, and extending operational hours.

Cycling has critical links with education, health and safety. In 2012, Boston added 11 miles of bike lanes and more than 100 bike racks including two new on-street bike parking sites. The New Balance Hubway bike share system expanded within the city and spread into Cambridge, Brookline and Somerville. Hub on Wheels, Bike Week, and Bike Fridays were larger than ever, and partnered with

Circle the City to offer three Open Streets events. Some of the major objectives concern children and their access to safe pedestrian walkways and bikeways, and arts facilities. After extensive collaboration with other city departments including the Boston Police Department and Boston Public Health Commissioner, a substantive safety report has been launched, together with a comprehensive safety and education programme (Boston Bikes 2013).

The promotion of walking and cycling has often been complemented with car restriction policies, such as road pricing. Is it a coincidence that the cities which have introduced urban tolls are coastal? Could this indicate a higher complexity and difficulty in dealing with urban transport? In Europe, the experience of Oslo, Stockholm and London, charging motorists entering their central districts, offers many lessons on urban tolls design and execution, public consultation and acceptance, capital and operating costs and pricing, information provision, technological aspects and environmental impact. In London, the congestion charging scheme introduced in 2003 represented an investment of £100 million in electronic equipment. The annual revenues have been estimated 20 higher than the annual cost of the system. Good communication with the citizens and potential users has been crucial.

Public transport can only be a worthy alternative to the private car, if safe, clean, reliable, fast, frequent, noiseless, flexible, easily accessible, well-designed, environment friendly, and affordable. It plays a major role in the larger and denser cities, in which it carries 2.5–3 times as many people as private transport. Public transport is also important for the important share of households without a car, which reaches 40 % in the European Union.

Highly populated world cities like Tokyo, where only 1 % of commuters use their private car, can provide many inspiring lessons. The 2000 urban rail development master plan identified five major policy targets to address the challenges in Tokyo's urban rail market. The first policy target is congestion. The second target relates to the Tokyo Metropolitan Area's decentralised land-use policy, which has resulted in satellite business sub-centres. Measures addressed the need for efficient rail connections between sub-centres and reduced commuting time from residential areas to business districts.

The third target also relates to the redevelopment of seaside areas near Tokyo Bay and aims to increase rail capacity, particularly in the central business district of Tokyo. Since the 1990s, a number of high-rise buildings were built for both business and residential use, due to the redevelopment of seaside areas near Tokyo Bay for business and residential use, and the gradual change in preferences of younger generations starting moving from suburban areas to the city centre. The fourth target relates to the globalisation of business and tourism markets. The government has also implemented a policy that includes deregulation of the air transportation market and the promotion of tourism in Japan. The improvement of rail access both to and from airports, and of high-speed rail, is critical for better business and tourism conditions. Finally, the last target reflects Japan's rapidly ageing society and the introduction of new devices and applications for older adults or disabled passengers (ITF 2014).

Eco-ride gains ground on all public systems of the world obliged to become more efficient, smart, green and inclusive. Many metropolitan underground systems, already more than 100 years old, go through transformational innovations. Driverless operations, state-of-the-art information, ticketing and security devices, mobile communication capabilities and wireless connections are among the main innovations.

Since the 1980s, tramways returned in many European cities, where they had often been abolished in the 1960s. The tram, abolished in 1960 in Athens, returned to the city in summer 2004, just in time for the Olympic Games. It is a valuable silent transport mode connecting the city centre to the seaside suburbs. In Marseille, trams continued to operate even through the 1950s and beyond. The original tram system remained operational until the closure of the last line 68, in 2004. Three years later, a new generation of trams started serving the city.

Marseille's new tram has been compared to a boat on rails. In 2007, the first phase of the new tram network opened. More recent developments ensured links to the rapid train and the old port of Marseille. Customised Bombardier Flexity Outlook trams composed of articulated sections, introduced a new element to the urban landscape. The exterior appearance seems like the hull of a ship, and the driver's cabin resembles the bow. A lighted circle displays the colour of the line the tram is on and the blue colour dominates the interior design.

Valencia advertised its new tramway as a tramway named desire. The Valencia Metro has been developed out of the former regional narrow gauge network. A tunnel through the city centre, 100 years after the inauguration of the first narrow gauge line, gave birth to the Metro de Valencia, unveiled in 1988.

Buses, including electric and hybrid buses, biofuel and hydrogen buses, and intelligent transport management systems develop continuously, in parallel with smart transport passes, often linking transport to other public and cultural services and spaces. Many local authorities have chosen to contract out public transport services, through tendering processes which incorporate sustainability criteria in the public service obligations. Stockholm managed to achieve a bus fleet that uses 100 % renewable fuels. The city promoted the use of clean vehicles running on biogas, ethanol or hybrids. It also offered incentives to clean vehicles, including free residential parking in the congestion charging zone (CEMR 2005).

Regulation is an important driver of change. In California, the Zero Emission Bus Rules were instrumental for inventing better buses. The Proterra EcoRide transit bus was the first full-size fuel-cell transit bus to enter service in the US and meet California's rules. Proterra already introduced a new generation of 100 % electric buses in 2014, building on EcoRide's leading design and engineering. The second-generation bus measures 40-ft, it has 30% fewer parts and is built from lightweight, durable composite that utilises fiberglass and balsa wood. Proterra offers clean, quiet and user and neighbourhood-friendly vehicles that meet government regulations and local mandates. The company aims at achieving exceptional fuel savings and dramatically reduce maintenance costs.

The 2013 edition of the International Association of Public Transport (UITP) "Grow with Public Transport International Awards" rewarded the public transport of Dubai for its outstanding innovation in supporting the growth and attractiveness of urban public transport services. These awards are expected to act as a key milestone every two years until 2025, by which UITP hopes to fulfil the ambition of doubling the worldwide market share of public transport. The initiative "Youth for Public Transport" (Y4PT) works collectively with youth to educate, disseminate, innovate and improve the urban public transport and the cities of the world. The Y4PT began in 2006, when UITP asked young people to report about their experiences and interactions with public transport. Y4PT acts as the voice of young people in urban transport policy making and mentors young citizens wishing to start a career in urban transport services.

In 2012, the International Association of Public Transport presented the results of a European project on urban buses, aimed to develop a new generation of urban bus systems adapted to the specificities of European cities and to achieve key breakthroughs in improving the perception of bus transport. The project highlighted the importance of addressing the bus system as a whole, not only vehicles, but also infrastructure and operations. Innovations in digital systems, vehicle design, driver cabins and bus stations were tested in real city conditions (UITP 2012).

Many cities have adopted fast and clean, green lanes exclusively reserved to public transport and, often, priority vehicles or users. Special lanes restricted to all traffic except buses, taxis, emergency vehicles and cyclists have been adapted in various ways to the local conditions of coastal cities. Priority lanes can also be used by private cars for a price. Dublin introduced high frequency bus services operated by environment-friendly buses, featuring well trained staff, high quality shelters and furniture, kerb alignments and traffic signals. The system provides a faster, more reliable, safer and cleaner bus service, adequate loading and parking facilities for businesses and improved safety (CEMR 2005).

Thessaloniki, in the middle of a deep economic recession, shaped a Sustainable Urban Mobility Plan. In partnership with all key stakeholders, the Thessaloniki Public Transport Authority devised the mobility priorities for the entire metropolitan area, which consists of ten municipalities and over one million inhabitants. The authority benefited from its participation in the EU "Attractive Urban Public Transport for Accessible Cities" project involving urban areas of eight EU countries. The mobilisation and commitment of a wide range of stakeholders, including central government, regional authorities, municipalities, trade unions, passenger and cycling associations, transport institutes and the university, and an emphasis on the better use of public transport were key elements of the planning process.

A seaborne transport system in the Thermaikos Gulf was one of the services, also including an integrated ticketing system, rapid bus transit, bus priority at traffic lights and awareness campaigns to discourage private cars and promote sustainable mobility modes. The system favours intermodality among the metro, trams and buses, flexible transit systems including the restructuring of taxi services and integrating parking policy, and a controlled street-parking system, a bike sharing scheme and a congestion charge and access to the city centre.

Network emulation design can reverse a vicious circle in public transport, in which declining demand translates into poor service, into a virtuous circle, in which greater demand leads to higher requirements and more seamless service. Smart ticketing with integrated tariff structures can improve user service, reduce fraud, increase capacity and bring considerable benefits to providers and citizens. The London Oyster card illustrates a low-cost alternative to the urgency of expanding station entrance areas. Contactless smart cards can considerably speed up boarding on public transport (ITF 2012).

Public waterborne transport, once the primary mode of urban transport in cities, is being rediscovered. Waterways in London, Gothenburg, Oslo and Hamburg regain importance. Waterborne transport can become a fully integrated mode of travel that complements other modes in cities. The renewed interest in waterfront development and urban rejuvenation, and also the increased congestion on land-based transport services, has provided an opportunity for waterborne transport to play a significant role in integrated urban mobility schemes. Until recently, many waterborne systems operated independently as complementary commuter or leisure service, but started to be designed as organic parts of local urban sustainable planning and mobility schemes.

The waterborne transport also offers relatively fixed and reliable travelling time and scenic ways of the urban waterfronts. It is a unique mode of travel that is often more pleasant than other transport modes. Urban cruises are among the sought after services by sightseers in cities like Venice and Amsterdam. Innovative design of boats and stops with dedicated spaces for children, passengers with reduced mobility and bikes, can far increase its attractiveness. In cities where waterborne transport is part of the urban environment, ferry stops are often located in close proximity to land-based transport nodes and city landmarks, encouraging connectivity between the transport modes and inviting tourists to use the public system.

By focusing on service quality and integration with other modes of transport and waterfront development projects, waterborne transport has a bright future in the overall mobility options for coastal cities. Hybrid and solar boats are being more commonly used to reduce emissions, as are fuel cell boats. La Rochelle, one of France's foremost seaports, displays innovative public transport and promotes intermodal mobility. The Sea Boat is a regular boat service between the Old Port and Les Minimes, a district of La Rochelle built on reclaimed land towards the end of the previous century, with one of the largest marinas in the country. "Yeló" runs an integrated network of public buses, boats and bikes.

The capacity of urban waterborne transport is not negligible. Venice ranks first with around 110 million passengers per year, the great majority of which are residents. The waterborne system in Lisbon transports around 23 million and in London about seven million passenger per year. Waterborne transport is also well established in a number of cities, such as Amsterdam, Istanbul, Stockholm, Gothenburg, Hong Kong and New York. Other cities with much smaller systems also use waterborne transport to promote the attractiveness and efficiency of city mobility. Many more coastal cities with waterways are currently considering adopting a waterborne system.

Maritime transport and urban waterborne mobility have to be well integrated with the urban transport network and the regional and national systems. Since seaborne trade accounts for almost 90 % of global trade in terms of volume, the support to maritime connexions is critical for port cities. Barcelona provides a prime example. Port Vell is promoting a productive mix of functions (OECD 2013), including those typically antagonistic, like port functions and bathing spaces at relatively close distance. Barcelona receives 20 million visitors per year, including 2.5 million cruise passengers. Good connections with the airport are crucial. More cities promote sea connections well integrated into the transport intermodal system. Istanbul is investing in improving maritime connections between Yenicapy, the main modal interchange in Istanbul, and Mudanya, on the southern coast of the Sea of Marmara.

Rio de Janeiro is enhancing some exceptional events of the 2014 World Cup and the 2016 Olympic Games as a catalyst to make significant improvements to its transport design. One of Rio's flagship projects is the Morar Carioca initiative, which has the ambition to improve accessibility, health, education, and the environment in the city's informal settlements, the favelas. Favelas, home to the city's poorest residents, are often cut off from city services. This often creates a vicious circle of exclusion as the populations that most need social services have no access to the services most needed by them.

Effective links and access to public transport should ensure that residents move and have access to jobs and opportunities within the city. A household travel survey revealed that 90 % of residents travel on foot while inside favelas, yet 70 % use public transport when outside their settlement. Morar Carioca is the one of the first favela improvement initiatives in Rio which emphasised connecting favela communities to the city via new public lines like the TransCarioca. Effective public transport is expected to provide vital access to jobs and opportunities in the city centre and help ensure that favela residents are connected to the full range of opportunities offered by urban life.

Many more coastal cities try to innovate in matters of urban mobility. Gothenburg proposes itself as a testing ground for autonomous drivels cars to speed up developments. Oslo is often described as the electric vehicle capital of the world. The experience of the city suggests that to succeed, electric vehicles must bring a clear advantage to the consumer, from buying to using. Therefore, good sale, charging and parking conditions must be established. The main challenge for the municipality has been to facilitate citizens wishing to own an electric vehicle and to use it as for daily commuter. During the planning phase, it was crucial to secure a wide political consensus, involving all strata of government in order to make the long-term investment in charging stations possible. Charging stations must be fully integrated within the urban fabric and easily accessed. This requires active interaction and cooperation with the Electric Vehicle Association and private property owners.

Charging infrastructure is a major issue for all cities investing in electric vehicles. Oslo established 400 free charging points with reserved parking for electric vehicles in the city over the period 2008–2011. The number of public charging stations soon exceeded 500 and could soon double this number. There are also 758 privately owned charging stations, of which 323 have been implemented with subsidies granted by the Oslo municipality. Trends in electric car sales illustrate the enthusiasm of the population. Between 2012 and 2013, an increase of 93 % of the number of electric vehicles on the road took place and the trends continued in 2014.

In 2013, the City of Oslo decided that all municipal cars must be emission free by 2015 and allocated $8.3 billion to support agencies with an interest-free loan when replacing vehicles. More than a quarter, 27 % of Oslo's municipal fleet, is projected to be electric vehicles. Oslo intends to cut emissions from transport by 50 % by 2030 compared to 1991, despite the expected population growth, and to achieve emission-free transport by 2050.

The European Commission's Sustainable Urban Mobility Campaign sheds light on many innovative alternative mobility initiatives. Since 2002, hundreds of European coastal cities participate, every autumn, in the European Mobility Week, evolving into a global movement inviting citizens to sustainable mobility. The week encourages local authorities and citizens to propose better ways to travel and contribute to making cities more healthy and pleasant places to live in. The car-free day is the highlight of the week, with the challenge of organising it during a working day! Increasingly popular, mobility weeks have supported cities in creating a more civilised environment with reduced traffic congestion and pollution and more sustainable transport means, including boats. Lasting impact is particularly valued and participating cities are encouraged to launch at least one permanent practical measure beyond the events of this exceptional week.

Larnaka invited a wide range of partners in the organisation of events during the European Mobility Week 2011. Its comprehensive promotional activities included a free bus day, a hybrid car exhibition, an Environmental Café and awareness raising events on sustainable mobility. The city showed its commitment to space reallocation by transforming one of the roads in the central business district into a pedestrian street. The city also created numerous bicycle parking spaces and set up a "Bicycle bank" to facilitate maintenance and reuse of bicycles.

Many cities invested in raising public awareness and presenting urban mobility initiatives as part of an effort to improve urban life and not constraining freedom of movement. Porto developed a cutting-edge mobile phone application "Move me", providing real-time travel information to public transport users. The system offers users a recommended travel route based on real-real time data on the services by the various public transport providers and intermodal options. The system has already been extended in Portugal, and was praised for its replication potential.

Smart ticketing, offering seamless travel between different operators and modes, and access to other services, is a key choice. Transport for London, the Association of Train Operating Companies and Cubic, took London's Oyster smart ticketing scheme to a new level. Since 2010, customers have enjoyed the benefits of travelling on all major public transport modes with one flexible ticket, covering 250 rail destinations within Greater London. Within 2 years since launch, usage has grown fivefold, from 0.5 to 2.5 million journeys per week, on surface railways alone. The project demonstrated that well-designed alliances among public and private stakeholders can create a more integrated seamless transport system.

It seems that there is a considerable potential for integrated eticketing in Europe, as public transport is attractive for financial service providers and telecommunications operators, but also for museums and cultural events. A European survey on eco-efficient transport, involving a consultation of stakeholders and scientists from the transport sector, suggested that respondents see a very high degree of desirability of interoperable electronic ticketing applications for public transport. Still, additional service improvements are necessary to sustain the modal shift. Tourists would particularly profit from integrated ticketing schemes, as they usually value local experiences, and urban mobility, especially by the sea (STOA 2014).

Sustainable urban transport policies should not forget the importance of beautiful urban transport places. The "Art in Transit" workshops, organised by the UITP, highlight artistic achievements in public transport and encourage reflection about the impact of quality transport environments on everyday lives. That the integration of art, architecture and engineering can uplift transport spaces is a key question. A most notable example is the Fulton Centre in New York opened in 2014 as part of the rebuilding of lower Manhattan after 9/11. The creation of the Fulton Centre gave the opportunity to artists, architects and engineers to create projects in vast covered spaces with view corridors. It completes an extraordinary underground museum of contemporary art that spans the whole city. The Metropolitan Transportation Authority network suggested that artworks should participate in a unified vision with the environment and strive for integration and orientation, rather than decoration (UITP 2014; Bloodworth and Ayres 2014).

References

Amsterdam Climate Office (2008) New Amsterdam climate—summary of plans and on-going projects. Amsterdam Climate Office, Amsterdam

Belfer Center and Rice University (2012) The geopolitics of natural gas. Belfer Center and Rice University's Baker Institute Energy Forum, Cambridge

Bloodworth S, Ayres W (2014) New York's underground art museum, MTA arts & design. Monacelli, New York

Boston Bikes (2013) State of the Hub. Boston Bikes 2012 Update

Boston Foundation and The Citistates Group (2004) Boston unbound Tapping greater Boston's assets and talents to create a world leading citistate. Boston Foundation & The Citistates Group, Boston

Council of Europe (1992) The European urban charter. Council of Europe, Strasbourg

Council of Europe (2008) La Charte urbaine européenne II—Manifeste pour une nouvelle urbanité. Council of Europe, Strasbourg

Council of European Municipalities and Regions (2005) CEMR Manifesto achieving sustainable mobility in Europe's towns and municipalities. Council of European Municipalities and Regions, Paris/Brussels

European Commission (EC) (2009) Action plan on urban mobility. EC, Brussels

EC (2011) Energy 2050 roadmap. EC, Brussels

EC (2012a) Communication on renewable energy: a major player in the European energy market. EC, Brussels

EC (2012b) Smart cities and communities. European innovation partnership. EC, Brussels

EC (2013a) European Green Capital. Urban environmental good practice and benchmarking report. EC, Brussels

EC (2014a) Blue energy. Action needed to deliver on the potential of ocean energy in European seas and oceans by 2020 and beyond. EC, Brussels

EC (2014b) Micro-algae-based products for the food and feed sector: an outlook on Europe. JRC Science and policy papers. IPTS, Seville

European Environment Agency (2010) 10 messages for 2010 urban ecosystems. European Environment Agency, Copenhagen
European Wind Energy Association (EWEA) (2014) The European offshore wind industry—key trends and statistics 2013. EWEA, Brussels
Global Ocean Commission (GOC) (2014) From decline to recovery. A rescue package for the global ocean. GOC, Oxford
Gore A (2013) The future: six drivers of global change. Random House, London
Greenpeace (2005) Decentralising power: an energy revolution for the 21st century. Greenpeace, London
Harvard Kennedy School (HKS) (2011) Transforming US energy innovation. Harvard Kennedy School, Belfer Center, Cambridge, MA
International Energy Agency (IEA) (2012) World energy outlook 2012. IEA, Paris
IEA (2013a) World energy outlook 2013. IEA, Paris
IEA (2013b) Technology roadmap: wind energy—2013. IEA, Paris
IEA (2014) World energy outlook. IEA, Paris
International Transport Forum (ITF) (2012) Transport outlook. Seamless transport for greener growth. ITF, Paris
ITF (2014) Urban rail development in Tokyo from 2000 to 2010. ITF, Paris
Mega V (2005) Sustainable development, energy and the city. Springer, New York
Mega V (2013) Quintessential cities, accountable to the future. Innovation, sustainability and leadership. Springer, New York
Municipality of Amsterdam (1994) A car-free city? Municipality of Amsterdam, Amsterdam
National Research Council (2012) A sustainability challenge: food security for all: report of two workshops. NAP, Washington, DC
New York City Mayor's Office of Long-Term Planning and Sustainability (2012) New York City Local Law 84 Benchmarking Report, New York
Organisation for Economic Co-operation and Development (OECD) (2013) The competitiveness of global port cities. Synthesis report. OECD, Paris
OECD–ECMT (1994) Urban travel and sustainable development. OECD–ECMT, Paris
Renewable Energy Policy Network for the 21st Century (REN21) (2014) Renewables global status report. REN21, Paris
Siemens (2012) Smarter neighbourhoods, smarter cities. Siemens, New York
STOA, European Parliament (2012) Urban transport technology options. STOA/European Parliament, Brussels
STOA, European Parliament (2014) Integrated urban e-ticketing for public transport and touristic sites. STOA/European Parliament, Brussels
Texas Transportation Institute (2011) Annual urban mobility report. Texas Transportation Institute, Texas
Texas Transportation Institute (2012) Annual urban mobility report. Texas Transportation Institute, Texas
Union Internationales des Transports Publics (UITP) (2012) European bus system of the future—Paving the way for a bus revolution. UITP, Brussels
UITP (2014) Art in transit workshop. UITP, London
UN Environmental Programme (UNEP)/International Council for Local Environmental Initiatives (ICLEI) (2015) District energy in cities: unlocking the potential of energy efficiency and renewable energy. UNEP/ICLEI, Paris
UN-HABITAT (2013) Planning and design for sustainable urban mobility—Global report on human settlements. UN-HABITAT, Nairobi
US Conference of Mayors (The) (2009) US mayors and climate protection best practices. US Conference of Mayors (The), Washington, DC
US Conference of Mayors (The) (2011) Clean energy solutions for America's cities. US Conference of Mayors (The), Washington, DC
World Energy Council (WEC) (2013) World energy scenarios: composing energy futures to 2050. World Energy Council, London

Selected Internet Links

www.europa.eu/energy
www.eumayors.eu
www.erec.org
www.energie-cites.org
www.ere-renewables.org
www.ewea.org
www.ises.org/estif.org
www.fuelcell-info.com
www.managenergy.net
www.renewables2004.de/en/2004
www.sustenergy.org
www.savemorethanfuel.eu
www.totnes.transitionnetwork.org
www.windday.eu
www.greenpeace.drg
www.nyc.gov/plany
www.usmayors.org/cleanenergy
www.energystar.gov
www.worldenergy.org
www.ec.europa.eu/transport
www.internationaltransportforum.org
www.civitas-initiative.eu
www.niches-transport.org
www.fubicy.org
www.uitp.com
www.velo-city2012.com
www.velocityclub.be
www.velib.fr
www.gart.org
www.tfl.gov.uk
www.citypass.com
www.gart.org
www.smilesprogramme.eu
www.zencars.eu
www.seattle.gov/climate/partnership.htm
www.ec.europa.eu/transport/modes/maritime/
www.bostonbikes.org
www.cityofboston.org
www.proterra.com

Watercolour 5
London, Leading European Global Coastal City

Chapter 5
Cities and the Creative Blue Green Economy

Abstract Coastal cities are privileged crossroads of exchanges on a globalising chessboard, the very places where many diverse value chains on land and sea interact and economic flows can be converted and intensified to better satisfy the expectations of citizens, business and organisations. Since 90 % of the total volume of the world trade is done by sea, port cities by definition are the cities through which most of the world's wealth moves. However, they are not just "passage" cities. Brainports and well-functioning harbours can reinforce global value chains and can generate sustainable value and employment. Their competitiveness depends on the strength of their functions, their maritime links and their hinterland connections. Sustainable infrastructures can help reconcile ports and cities with the oceans and the planet.

This chapter highlights the capacity and achievements of coastal cities as strongholds of the coastal, blue and green economies. In times of uncertainty, multiple dividend innovations are essential for creating new assets, often out of liabilities, and capturing synergies. Industrial symbiosis in ports is critical for the creation of thriving maritime clusters and high value services. Eco-responsible blue green businesses have a key role to play in creating sustainable value out of values and leading corporate action for responsible cities and global ocean governance.

5.1 Port Cities, Competitiveness and the Polyphony of Value Chains

Globalisation and socio-economic waves of change have placed cities and metropolitan areas at the forefront. It seems that cities will lead and governments will follow in grasping the noble opportunities of the twenty-first century (Katz and Bradley 2013). Cities are not only the stages of socio-economic operations and environmental impacts, but genuine players in the theatre of nations and the journey to sustainability. Coastal cities are very intense places where many diverse value chains interact and precious synergies can be generated and harvested through the mutual reinforcement of activities which cluster together on land and sea.

V.P. Mega, *Conscious Coastal Cities*, DOI 10.1007/978-3-319-20218-1_5

Cities could be seen as the brains of the global economy, which provides many more actors the opportunity of becoming parts of a world conglomeration. Strong urban infrastructures and leading-edge institutions can greatly improve the position of cities in an equitable globalisation horizon. Conversely, shortcomings in urban hardware and software are important constraints, whose cumulative effects could drain the potential for sustainable development.

The World Economic Forum (WEF) highlighted the leadership role that cities are taking in stimulating competitiveness and proposed a four-fold taxonomy for city competitiveness, including institutions, policies and regulation of the business environment, hard connectivity and soft connectivity. The WEF Future of Urban Development initiative has developed a strong brand. It focuses on the transformation of cities to enhance their chances. In the first 2 years (2012–2014), the initiative worked hand-in-hand with the Cities of Tianjin, Dalian and Zhangjiakou to address the most pressing urban challenges. During the third year, the initiative is cooperating with the Government of India in the development of its planned 100 Smart Cities.

A city cannot thrive if it cannot offer citizens and investors convincing security and credibility, infrastructure and services, and quality of work and life. Cities seem on orbit to continue to be essential to the future growth and competitiveness of territories, especially in emerging economies. Key drivers of change include demographics megatrends, urbanisation and the emerging middle class, inequality, strategic resource scarcity, deficit of global democratic governance and declining trust in public authorities (Gore 2013; WEF 2014).

Democratic cities are stepping up to re-conceptualise sustainable development and create a resilient recession-proof economy and society in balance with the environment. Networking with their stakeholders and among themselves is of the utmost importance. Especially in times of global democratic deficit, visions, ethics and actions for sustainability offer cities the opportunity to perform as sustainable democratic spaces between the world macro-regulations and the micro-regulations of the local communities (Gore 2013; Mega 2013).

Cities have long been the world's economic turbines. The collective GDP of the world's top ten cities exceeded the total GDP of 162 countries combined, with Tokyo alone generating a larger GDP than Canada. Tokyo is the world's largest urban economy, followed by New York, generating more wealth than India or Mexico. National economies become networks of metropolitan economies which produce disproportionally more wealth than their population share. The top 50 metropolitan areas in the EU host 36 % of its citizens, but generate 43 % of its GDP. In the US, 388 metropolitan areas host 84 % of the population and generate 91 % of the GDP (Katz and Bradley 2013).

Many emerging metropolitan areas are dynamic hubs of production, consumption, and trade, even in an uncertain period for the global economy. An analysis of per capita income and employment changes during the year 2010–2011 for 200 of the world's largest metropolitan economies, which account for 14 % of global population but almost one-half of global output, reveals that 90 % of the fastest-growing metropolitan economies among the 200 were located outside North America and the European Union. In almost all regions, metropolitan areas generated disproportionately larger shares of wealth and employment. Most of them significantly outperformed the national average on income growth, while several others significantly underperformed on employment growth (Brookings Institution 2012).

As emerging world cities enter the stage of intensive growth, their contribution to the global wealth is impressive. Through a combination of investment in physical and built capital and consumption, rising cities could inject up to $30 trillion a year into the world economy by 2025, also integrating growing parts of informal markets. The 600 cities making the largest contribution to a higher global GDP, the "City 600", including over 440 cities in emerging economies, the "Emerging 440", could generate nearly 65 % of world economic growth by 2025. The Emerging 440 could account for close to half of overall growth (MGI 2011, 2012).

A crucial part (61 %) of the world's total gross national product comes from areas within 100 km of the coastline (GPO 2013). Maritimisation is a prime form of globalisation which intensifies exchanges among world citizens and socio-economic actors. According to estimates by the United Nations Conference on Trade and Development (UNCTAD), owners from five countries (Greece, Japan, China, Germany and the Republic of Korea) accounted for 53 % of the world tonnage in 2013. Among the top 35 ship-owning countries and territories, 17 are in Asia, 14 in Europe, and 4 in the Americas. The multiple interactions between local and global flows and dynamics, which led to the term "glocalisation", rely on cities as privileged chain links. The strength of the future world will, however, only be as strong as the weakest link.

Coastal cities are essential drivers of national and regional economies and account for proportionally more wealth than their population share. Many coastal cities and especially ports have historically been strongly linked to global value chains. Port cities operate on a global horizon, competing with other world players for investment and opportunities. Their competitiveness depends on a multitude of endogenous and exogenous factors, including the macroeconomic environment, openness to trade, markets and investment, education, training and skills, ability to create and innovate, flexibility of the labour market, capacity of physical and digital infrastructure, governance and leadership.

Given the concentration of population, economic activities and infrastructure along the shores, coastal cities have to lead the preservation of the integrity of those natural and marine resources that increase the value of these places as being highly-desirable areas in which to live, work and visit. From leisure activities to coastal and marine fisheries and non-conventional security issues of food, water, and energy, ports support an array of activities and are often located in the heart of sensitive coastal and marine ecosystems. Understanding and managing the interdependencies among safe maritime transport, efficient port operations, and coastal and marine stewardship are critical for coastal economies and societies.

The strength of port operations, their maritime links and their hinterland connections can boost the strategic profile of a city. Harbour functions can increase industrial output and trade, generate value and employment, promote innovation and attract leading-edge companies and clusters. Many regional economic benefits associated with well-performing ports often spill over into the larger area or even into other regions. Doubling the port efficiency of two countries is found to raise their bilateral trade volume by 32 % (OECD 2013).

The economy of coastal cities also greatly benefits from efficient links with other sectors operating outside the port region. Sometimes, only a limited part of the linkages with suppliers takes place in the port or the port-region, with a larger share being located in the main economic centre of the country, which could be relatively far from the port, e.g., Greater Paris from the ports of Le Havre and Marseille. In some particular cases, the economic capital city and the main harbour form a bipolar conglomeration, like Athens and Piraeus, with well-defined distinct functions and roles.

The integration of ports in metropolitan areas and the organic inclusion of their functions in sustainable economies are crucial for the mitigation of negative effects, mostly related to the environment, land use and traffic congestion. These impacts can be very substantial in extended ports, such as the port of Antwerp, which occupies more than a third of the surface of the city. In very dense, concentrated ports, like Hong Kong, the majority of SO_2 emissions are related to shipping. Most of the negative environmental impacts take place close to the port area, but some types of pollution, like air emissions, expand on a regional scale. Noise from ships and port operations can also produce a kind of "acoustic smog" expanding beyond the port, while congestion patterns depend much on the local and regional transport policy (OECD 2013).

Many port cities went into years of transition. Usually, port operations include industry and refineries, shipbuilding and repairing facilities, supporting services and passenger and freight transport and transit. They increasingly yield more value as environmental policies, and incentive schemes have reduced a variety of impacts, while transport policies in and around ports address negative impacts, including air pollution, emissions, noise and congestion. Port relocations, often combined with waterfront developments, have freed up central space in many coastal cities for other public and private, residential and cultural functions. Seafront development has frequently managed to capitalise on port and maritime heritage and transform this into a source of urban prosperity.

Maritime industry and services are increasingly organised in formal and informal clusters. Maritime clusters try to attract high value added services related to the maritime industry, such as maritime finance, consulting, law and engineering services. Industrial development related to traditional ports attracts industries highly dependent on imported resources and consumer markets. Port activities attract small and medium-sized enterprises, considered to be propellants of innovation and job creation, and dynamic actors of many urban maritime webs. Their incubation and growth are crucial for the blue green economy. Intelligent micro-enterprises are associated with various goods and services supporting the maritime economy, for example, equipment for offshore energy, yachting and ICT services. Smart and flexible new structures and networks, created in many port cities, provide eco-services for blue sectors.

Clusters can catalyse concentrations of economic activities, stimulate innovations, enable grasping of new opportunities and promote a port city as key anchor in various value chains. They can contribute to a rapid diffusion of best practices and provide incentives to improve performance vis-à-vis local competitors. Clusters can facilitate efficient access to specialised services, skills, information, institutions, training and other public goods, and ensure smooth transactions and exchanges across firms. The presence of multiple suppliers and institutions assists in knowledge creation and dissemination. The proximity of competing companies encourages strategic differentiation. Last but not least, clusters offer opportunities for new related companies, especially SMEs. Spin-offs and start-ups are encouraged by the presence of other related companies, available skills, commercial exchanges, and value chains (Ecorys et al. 2014).

Policy portfolios for sustainable port cities include a range of policy instruments, such as incentives and free zones, training and education, platform organisations and knowledge transfer and sharing schemes to attract high added value companies that could make the city an international maritime services centre. Singapore is a good example of integrated policies to attract investors. Rotterdam positioned itself as a site for industrial sustainability, and Bremerhaven is trying to become a global leader in renewable energy. Adequate sustainable infrastructure is most necessary to support the sound ecological development of port operations and the prevention and minimisation of waste and pollution. Spatial planning and financial mechanisms for sustainable redevelopment have been applied to multiple waterfronts in order to create a functional mix, including high quality residential places, cultural services and noble public spaces.

Metropolitan port cities with high-quality infrastructure, green and public spaces, a clean marine environment and healthy residential areas have great potential to attract investors, as well as highly qualified professionals and tourists. Effective governance depends on leadership at all levels, from the national government, a government at the metro-regional level, and local authorities and networks that mobilise non-governmental actors, civil society and businesses. To balance the financial needs of harbours with those of other activities, cities can use public-private partnerships to raise funds for public projects that enhance aquatic environmental assets (OECD 2006).

Conscious port cities need a sustainable infrastructure and smart grids to underpin the blue green economy. Public policies and public-private partnerships can be effective in increasing "port-city" sustainability performance. Several ports have also started to monitor environmental trends and impacts and promote balanced environmental and economic development through dedicated sustainability agendas. Transport and land-use policies and city-university-business cooperation can be very influential.

Coastal urban partnerships usually bring together, often in forms of dynamic helices, universities, enterprises and municipal institutions. Cities must lead in defining long-term objectives, businesses must inject vigour, and educational institutions must instil a culture of striving for excellence. Triple helix schemes can better address the demand for natural resources and water, and for new businesses and housing, office buildings and infrastructures.

Maritime clusters have already existed for many years in some industrialised maritime countries. In Belgium, the Flanders Maritime Cluster brings together 80 companies, including port authorities and major dredging companies, as well as the Flemish Institute for the Sea. Four complementary sub-clusters are organised around the four major ports, including the more established ones of Antwerp and Oostende, the growing port of Zeebrugge and the port of Ghent for inland waterways.

The Port of Amsterdam is an example of smart symbioses of different services and utilities which can reduce energy and costs. Industrial symbiosis is especially focused on synergies between material production, infrastructure, transport and energy. The Port Authority acts as an infrastructure manager for all industries located in the harbour. This approach would allow for a win-win situation since the different sectors operating at the Port could concentrate their activities on core manufacturing activities supported by a specialised infrastructure for the optimisation of incoming and outgoing materials, feedstock, and energy. This would add a new dimension to existing cluster concepts, in particular by encouraging new forms of cross-sectorial cooperation and sharing infrastructure for transport and energy.

Across the Mediterranean and Black Seas, some 600,000–700,000 people are estimated as potentially employed in maritime clusters. A study which analysed the clusters of the region suggests that almost half of the surveyed clusters can be considered growing, while 29 % are mature, and 17 % are emerging. The EU clusters present a more balanced ratio between mature, growing and emerging clusters, while two out of three of the non-EU clusters are classified as growing. This could indicate that much maritime growth in the Mediterranean and Black Sea regions could come from the non-EU side. Innovative maritime sectors are underrepresented in the surveyed clusters which present a concentration typical for the Mediterranean and Black Sea regions around rather traditional activities, such as coastal tourism, cruise tourism, shipbuilding and ship repair, shipping, passenger ferry services, and fishing. Differences among the sub-regions include the particular role of the shipbuilding and oil and gas sectors in the Black Sea area. Tourism, either coastal or cruise, is by far the most common activity of East Mediterranean clusters. The Adriatic-Ionian sub-region demonstrates a greater heterogeneity, with a higher diversification of activities.

Maritime clusters involve a very diverse array of actors and can pursue smart specialisation strategies. Mainstream cluster policy tends to focus on specialisation, while diversification appears to be critical for maritime cluster development. Maritime cluster development touches upon an array of policies, ranging from transport, economic policy, environmental policy and physical planning to skills development, education, employment and safety regulations. Proximity underpins triple helix approaches and business-to-business and research cooperation, but cluster activities can further enhance this process and build new value chains of products and services (Ecorys et al. 2014).

Brest hosts France's largest maritime cluster, the "Pôle de competitivité mer", a network of 300 partners, including large companies, public and private laboratories, and universities located in Brittany. With over 2700 km of coastline and 95 % of the population living less than 60 km from the sea, regional culture, identity, and prospects have a crucial marine dimension. A long-term vision for sustainable development led to regional strength in sea-related activities and created 40,000 jobs, including for 2900 researchers. The cluster intends to respond, through innovation, to the requirements for safety and sustainable development and make the difference in international markets. A certification scheme allows projects to attract capital for innovation. Since 2005, the supported cooperative projects focused on six areas of excellence, including maritime safety and security, naval and water activities, fossil and renewable energy resources, marine biology resources, fishing and aquaculture, and port, infrastructure and maritime transport.

The shipbuilding and repairing sector is the fourth largest industrial sector in Brittany, and involves the construction and repair of navy, military and civilian vessels for fishing, cruising, oceanography and passenger transport. Leading companies anchor the region with world value chains in naval design and construction, equipment and integration of systems, testing facilities and associated services. The sector has experienced many waves of world crises and is searching for a new future in terms of location, resources, functions and markets.

Shipbuilding is being reinvented. Globally, new shipbuilding activity increased in 2014, continuing the trend that began the previous year, after 5 years of decline. New orders for all types of vessel have increased and bulk carriers have tripled in demand, followed by gas tankers and other cargo carriers. Demand for non-cargo carriers has also increased, with more sustainable designs complying with new international and European regulations for safer and cleaner maritime operations (SEA Europe Shipbuilding 2014).

Far Eastern countries continue leading the shipbuilding market thanks to governmental policies, cheaper production costs and availability of financing. China has become one of the most influential shipbuilding countries in the world. In the coming years, the restructuring and upgrading of China's shipbuilding industry is expected to focus on accelerating innovation, promoting high-end products and enhancing the industry's international market share. European shipyards are increasing their activity and maintain a good position in terms of value. The European maritime technology industry continues at the forefront of innovation and excellence, attracting demand for high technology vessels and equipment. European marine equipment providers are market leaders, with 43 % of global shares.

Another extraordinary location for this sector is Gibraltar. Its shipyard and repairing facilities are among the most renowned and extensive. The very location of the rock between the Mediterranean Sea and the Atlantic is an important factor of success. The climate is ideal for year-round work and many shipping lanes organise fast stop-over repairs. Gibraltar claims a broad range of expertise and activities to ensure quality repairs and services.

New generation shipyards for niche markets have emerged in Europe. In the Baltic region, Helsinki, for example, specialises in steamships, car ferries and icebreakers, Gdansk in ship hulls and the Kiel region in container vessels, submarines and luxury yachts. Arctech, Helsinki's shipyard, focuses on vessels equipped to work in ice conditions. Arctech has been using Russian facilities to build most of the blocks before the vessels are assembled, painted and fitted out in Helsinki. Baltika, the oblique icebreaker and oil spill response vessel, is known for its unique aesthetics and its lack of symmetry. Baltika handily combines aesthetics and functionality, as one side of the vessel is an oblique icebreaker, creating 50 m wide channels, while the other side has oil recovery equipment for open seas.

Last but not least, amongst the array of coastal cities, many have to promote regional economies. An example is the city of Heraklion, in Greece, which is the stronghold of the Cretan economy. The surrounding economy exhibits strong specialisations highly dependent on tourism, which accounts for 31 % of employment, and agro-products, 59 % of the exports being primary products. In the agri-food sector, the Region of Crete has recognisable quality agricultural products, both for domestic and international markets. The Cretan diet, which is a model of the Mediterranean diet, is an "intangible heritage" with a strong potential for exports. The main weaknesses are found on the small entrepreneurial scale, limited standardisation and marketing of products, and little systematic penetration of local products into tourism businesses. Opportunities reside primarily in renewable energies and the gradual transformation of the building stock through bioclimatic architecture. Finally, but still importantly, the rich local marine biodiversity provides many opportunities for new products, services and applications. The Hellenic Centre for Marine Research and its local facilities are investing much in research and awareness projects.

5.2 The Sea as the New Frontier: Emerging Blue Green Activities

Strong sustainability demands the preservation of all urban capital, natural and physical, including marine capital, human, social and cultural, as well as man-made and financial capital. The evaluation of the impacts on all distinct but interlinked capitals highlights the potential benefits of green growth as a shortcut to sustainable development. Investments in port infrastructure may increase manufactured capital, part of the man-made capital, but can undermine overall wealth if negatively impacting marine ecosystems, part of the natural capital, or damaging public health through traffic congestion, part of the human capital (EEA 2012).

Along with green growth, the blue growth concept is also attracting attention. Since the marine environment is already under much pressure, the blue economy needs to be solidly anchored in sustainable development, otherwise it undermines its very foundations. Blue green growth should focus on multi-dividend options beneficial to the economy and the environment, equally profitable for present and future generations. In the long term, economic and social development without sustainable management of natural resources, including from the sea, is brittle. Policy makers should set the ground and establish the framework conditions to allow new breakthroughs to emerge and overcome institutional or socio-economic inertia. Investments in education and innovation are most important and global governance is necessary to prevent the tragedy of the commons.

Blue green growth means fostering economic growth and development while ensuring that natural and marine assets are wisely managed to provide the goods and services on which human well-being depends. To achieve this, a city must catalyse the investment and innovation which underpin sustained growth and give rise to new opportunities. The 2014 European Green Capital provides a good case for green growth. Copenhagen's push to be at the vanguard of sustainable development has yielded many economic benefits. Blue green growth is seen by Copenhagen as an opportunity to empower and resource cities.

The acceptable balance is a fundamental challenge for the twenty-first century political process. More and more citizens want economic growth, but not at an unacceptable cost to the blue planet. Business as usual is unwise and ultimately unsustainable, involving risks that could impose constraints on further growth and development. Coastal cities are ideal hotbeds for blue green policies, not only through new technologies but also new partnerships and governance models. Risks include increased resource bottlenecks, air, water and soil pollution, and irreversible climate change and even disasters, accompanied by a myriad of ripple effects. The way forward requires a balanced policy portfolio to overcome deeply entrenched behaviours, and develop the financing instruments and indicators to monitor progress (World Bank 2012a).

Blue growth aims to increase food, energy and water security, support sustainable management of aquatic resources and achieve global, regional and national impact. In 2012, the European Commission adopted a Communication on blue growth which aims to develop the potential of the EU's seas, oceans and coasts, create sustainable wealth and employment and promote innovation. Extended arrays of marine and maritime economic activities have been identified as potential drivers of blue growth (EC 2012d).

Most drivers for sustainable development are linked to awareness of climate change and the need for the preservation of the integrity of the ocean ecosystems and habitats. Obstacles mainly involve social and political inertia and a lack of financing instruments. Fiscal instruments, such as environmental taxes, pollution charges, subsidies for renewable energy technologies, including for offshore developments, and tax incentives can play a crucial role in promoting a blue green economy. The political acceptance often depends on effective measures, addressing, in particular, societal impacts.

Blue green growth strategies can help coastal cities to become more resilient as they strive to meet demands for food, water, housing, energy and transport. Sustainability-oriented policies can help mitigate the impacts of adverse shocks by reducing the intensity of resource consumption and environmental impacts, while alleviating pressure on commodity prices. Pioneer cities, which create a culture of change and commit to policy innovations, usually act at the earliest possible stage. The market for blue green goods and services is fast expanding and can offer the multiple benefits of environmental quality and job creation (World Bank 2012b).

Coastal cities are the gates to the seas. They started weaving their nets to grasp new and better opportunities. The blue economy represents 5 % of global wealth and promises much more if well aligned with sustainable development and as ocean science and technology development advance. There is an increasing awareness of the equilibrium needed between resource extraction and resource preservation, be it in sea mining or managing fish stocks. Scarcity of raw materials and energy and the search for new opportunity spaces are a global phenomenon. While matter and energy are abundant on earth and in the universe, scarcities are geostrategic, and they can disrupt, for example, local lives and trajectories.

Crises have the power to stimulate and cross-fertilise ideas and markets and generate growth. Developing new sources of wealth will depend on the intellectual assets needed to conceive, create, promote, diffuse and adopt innovative initiatives. There is no single blue green growth model but various possibilities for reconciling the short with the long term. Strategies will vary across countries and cities, reflecting local contexts and preferences, but all countries, rich and poor, have chances for growing cleaner without growing slower. There is a growing consensus that green growth should focus on actions in the next 5–10 years to prevent getting locked into unsustainable paths and to generate immediate, local benefits (OECD 2011a; World Bank 2012a).

A blue green growth strategy incorporates a longer time horizon and takes into account the full value of natural capital and terrestrial and marine ecosystems. Diverse port-related activities could create propitious conditions for the development of lean blue green innovative SMEs. The essence of coastal urbanisation, the concentration of people and activities close to the sea, offers a range of diverse resources needed by the green maritime industry and sustainability conscious consumers. The features of a city favourable to green SMEs include access to precious human and natural resources, the availability of other high-quality services, complementary producers and a market for green maritime services. Port cities can also provide the necessary knowledge basis and advanced services and support the generation or integration of innovative green SMEs in maritime clusters.

The EC's blue growth strategy has been developed in the framework of the EU's Integrated Maritime Policy and in support of the overarching European strategy 2020. The strategy aims to harness the potential of the blue economy to create new jobs, foster innovation and contribute to sustainable economic recovery. It underlines the importance of coordinated efforts from the EU, national and regional authorities, business, and civil society on issues such as research and development, access to financing, and promotion of education and innovation. A framework directive adopted in 2014 aims to introduce Maritime Spatial Planning and Integrated Coastal Zone Management to promote the optimal location and distribution of marine activities throughout the blue economy, in a coordinated and coherent way. Spatial planning provides investors and operators with more predictability about future directions and the impact of maritime activities (EC 2012b, 2014e).

Fisheries are the oldest blue activity on earth and have contributed substantially to food security. They continue to be at the epicentre of many world concerns linked to overfishing, described by many as a crime (GOC 2014). The restoration of the fishery stocks through sensible optimal sustainable yield policy options is a must (GPO 2013). Strong international governance systems should ensure that stocks are restored. In the European Union, the new fisheries management framework should redress the overfishing that led to many stocks being depleted. The number of stocks overfished has fallen from 94 % in 2005 to 39 % in 2013, although the situation continues to be critical in the Mediterranean and Black Seas where only a few stocks are assessed.

Aquaculture provides half of the consumed global seafood, and organisations suggest that, by 2030, it could surpass two thirds (FAO 2014d; World Bank et al. 2014). The main challenges for aquaculture include lack of available space, feed, breeding technology, competition in the global market and administrative procedures and constraints. Sustainable aquaculture must also consider potential impacts on wild fish stocks and water quality and ethical considerations addressing, for example, gender issues and the creation of decent jobs. More space for marine aquaculture could be freed up by the development of offshore production systems.

Progress in feed technology could achieve a higher proportion of feed originating from plants, algae, and, if possible, waste. Technical developments suggest that wide-spread automation and control techniques can be instrumental for realising aquaculture potential (OECD–FAO 2014).

Cities have often had to address conflicts among tourism and aquaculture. Coastal cities are gateways to the recreational economy which, if sustainably managed, can provide precious blue green jobs. Traditionally second to shipping, the maritime and coastal tourism has become the largest single maritime economic activity in the EU, the first world destination (EC 2014d).

A healthy land-sea environment is fundamental to blue tourism and favours the growth potential of new forms of sustainable and green tourism in and around coastal cities. Accessible high quality bathing waters and pristine coastal and marine habitats have a high recreation value and contribute to the attractiveness of cities and the potential for activities as varied as nautical tourism and sports, and dolphin, whale or coral reef watching. The sheer variety of blue journeys provides cities with an opportunity to enhance their natural and cultural assets and boost their local economy in partnerships with their seas and their surrounding regions.

Over the past six decades, tourism has experienced continued expansion and diversification, becoming one of the largest and fastest growing economic sectors in the world. Many new destinations have emerged which challenged old patterns and practices. In spite of occasional shocks, linked, for instance, to accidents, economic crises or terrorist events, international tourism experienced considerable growth, with a steadily rising share of the world's emerging regions (UNWTO 2014; EC 2014d).

Sustainable tourism is expanding on all continents. The European Union attracts the majority of tourists. The USA, France, Spain, Italy, and China, the five first world destinations, have exceptional cities that serve as cultural gates for sustainable tourism. Sea basins and coastal regions provide a unique resource of natural and cultural wealth. Worldwide, international tourism rebounded strongly, and the increase more than offset the decline caused by the economic downturn. Recovery was particularly strong in emerging countries, where arrivals grew faster than in advanced ones (UNWTO 2014).

Destination management is important for all cities wishing to attract visitors and create healthy experience economies. There are multiple excellent examples of sustainable tourism activities. Boston is a cultural heritage-oriented destination and is encouraging the tourism industry to improve its operations and services to visitors, including visits to the waterfront and the New England Aquarium. A handful of docking facilities in the Boston Harbour are maintained by private interests and some wharves have been converted into residential or tourist facilities. A multifaceted programme has been established after consultation with all stakeholders promoting Boston as a preeminent destination for eco-tourists (Boston Foundation 2009).

Cruise tourism is developing in most seas. It is a global business, which depends on geopolitical stability and uncertainty. Each cruise is a multi-destination experience which competes with "stationary" tourism on shore and can impact other blue economies, but also threatens the carrying capacity of biodiversity hotspots and sensitive sites such as Venice and Dubrovnik. There are 421 cruise ships world-wide, carrying from 400 to over 4000 passengers. Despite misfortunes, the sector has weathered crises well and, even in the Mediterranean, 2014 was the only questionable year since 1995. The average growth rate of Caribbean cruising, approaching 3 %, can be reached in the Mediterranean, which accounts for 85 cruise ports in 2015 versus 40 in 2000.

Yachting tourism has also developed for super yachts, of lengths over 30 m. Yachting is expected to grow by 2–3 % a year. EU shipyards and repairing centres have been successful in serving this specialised market, both with large cruise ships and small leisure vessels. According to the European Cruise Council, there were 41 lines domiciled in Europe in 2011, operating 120 cruise ships. Furthermore, 25 non-European cruise lines, operating 76 ships, shared the facilities of 250 European ports. The number of marinas in Europe is estimated at 4500. The Mediterranean is the most attractive region in this regard, with more than 200 super yacht marinas. Cruise terminals try to attract consumers through immersion in the local economy and culture (EC 2014d).

The share of returns from cruise and coastal tourism with local communities is a thorny question. Congestion is the main burden for local communities and time management of ports as "chronotopes" is a key issue. Local authorities typically accuse the cruise lines of unfair sharing of benefits. The data provided by the cruise lines, suggesting that the average cruise passenger spent 75€ per day in each port of call in the Mediterranean in 2015, are considered overestimated by local authorities, who, nonetheless, do everything they can to attract cruise ships.

Energy activity in ports highlights a move towards renewable energies. Bremerhaven port reinvented itself to address the needs of manufacturers and suppliers in the offshore wind industry. Maritime activities were enriched with related cultural and ecological projects, such as the German Emigration Centre and the German Maritime Museum. The Bremerhaven Zoo features Arctic wildlife, both terrestrial and marine. The Lloyd-Werft shipyard is renowned for building and renovating large cruise liners. Every 5 years, the Sail Bremerhaven, a large sailing convention, attracts tall ships from all over the world.

Blue energies have the potential to enhance renewable energy resources, minimise land-use requirements and reduce greenhouse gas emissions. Renewable energy targets and incentives for investments, such as green certificates, have begun to expand rapidly. Wind is the fastest growing world coastal option. Among wind technologies, offshore wind is the most rapidly growing sector, and synergies with other offshore activities are increasingly explored (IEA 2014).

Other offshore renewable energy technologies are still at an early stage of development. Wave and tidal energy appear far more costly to harness. The challenge is to accelerate the commercialisation of ocean energy through technology development and cost reduction. Geographic and oceanographic conditions suit different technologies, which can offer a more predictable base-load supply of electricity and back-up the fluctuating supply from wind. Tidal barrages are used to capture energy from moving water. Wave power devices are currently being demonstrated and underwater turbines driven by currents are close to commercialisation. Ocean thermal energy conversion, which enhances the temperature difference between cooler deep and warmer surface waters, could be another option for tropical territories (EC 2012a, 2014a).

Coastal cities are privileged interfaces between the energy producers offshore, the grid operators onshore and the final consumers. The commercial exploitation of blue energy may require costly investments in grid connections and transmission capacity. Many local partnerships with small businesses and the local society could promote new emerging technologies and smart devices. Good governance architecture is essential for the articulation of needs and solutions at local, regional and national levels of responsibility.

Among the emerging blue activities which will have to pass through the filter of public scrutiny and acceptance, sea mining is boosted by the increase in price of many non-energy raw materials, mainly in emerging economies and the related scarcity risks. Under the 1982 UN Convention on the Law of the Sea, a country has exclusive rights to explore its continental shelf and exploit the natural resources, including mineral resources. Offshore mining is currently confined to shallow-water coastal regions. About 75 % of the world's tin, 11 % of gold, and 13 % of platinum are extracted from near the surface of the coastal seabed. Aggregates, including sand, coral, gravel, and shell, are also important. The United Kingdom, the world's largest producer of marine aggregates, extracts approximately 20 million tonnes per year, meeting around 20 % of the domestic demand. Deep-sea mining operations can also be carried out in the international seabed, outside jurisdictional marine areas. In these areas, the International Seabed Authority (ISA) is responsible for organising and monitoring activities. The world has moved closer to deep ocean mining in 2014, as a Canadian company concluded an agreement with Papua New Guinea to start digging up and extracting precious metals from a depth of 1500.

Seabed mining could have a disproportionate and even irreversible impact on coastal and marine ecosystems and habitats. Present commercial scale seabed mining operations, limited to shallow waters, mainly provide for aggregates, sand and gravel, for the construction industry and the creation of artificial beaches. The technologies most used include dredging, vacuum pumps and remotely operated vehicles. Aggregate extraction has a large environmental footprint and could irreversibly change the seabed in shallow waters and affect other sectors and stakeholders. Social acceptance issues become extremely important and the precaution principle has to prevail when moving into uncharted waters.

Blue biotechnology is another promising blue growth activity with great potential for new products and processes. The biological sciences have demonstrated enormous advances beneficial for health, medicine, food and agriculture. The unexplored nature of much of the underwater world means that the capacity of marine organisms to nurture blue economic activities is just beginning to be discovered, partly through new gene sequencing technologies for living organisms. Exploration of marine biodiversity could help us understand, for example, organisms that can withstand extremes of light, temperature and pressure and can assist us in developing new industrial enzymes or pharmaceuticals.

In the immediate future, blue biotechnology could emerge as a niche market focused on high-value products for the health, cosmetic and industrial bio-materials sectors. By 2020, it could grow as a medium-sized market, expanding towards the production of compounds as inputs for the food, feed and chemical industries. In the longer-run and subject to technological breakthroughs, the blue biotechnology sector could provide a broad and diverse range of specialised high-added value products.

Coastal cities can also boost the production of food, feed and fuels from algae. The global market for micro-algae-based food and feed supplements is well developed and has a great potential for growth. Micro-algae are currently used both as dried whole algae and for the extraction of high-value food/feed supplements and colorants. Although the total production volumes and market size of food and feed supplements derived from micro-algae are still relatively small, they have increased fivefold since the beginning of the century, and some applications already have a long tradition (EC 2014b).

Bio-based products and sustainable food production systems can help create new markets. Cities could do much to facilitate procurement for bio-based products by adopting specific standards and labels and introducing training for public procurers and incentives and mutual learning mechanisms for improved resource efficiency. The creation of innovative urban markets can help accelerate the commercialisation of bio-inventions. The procurement of bio-based products could also create jobs in rural areas surrounding cities and help foster urban-rural partnerships.

Cities could also play a major role in managing and enhancing the transit of activities taking place offshore. Multi-purpose offshore platforms could act as innovation spearheads for sustainable maritime economies. Some offshore platform consortia are developing innovative design and new business models to become key players in the future cross-sector offshore economy. The governance architecture is trying to generate and capture synergies among diverse economic chains, such as aquaculture and offshore wind energy, facilitated by robotic technologies.

The public and private sectors should work in concert to promote the sustainable blue economy, starting with precompetitive associations, in which competitors pool investments for resources, knowledge, and expertise to learn from successes and failures, to invent, deploy, and scale the cutting-edge technologies that may spark new breakthroughs. This offers the opportunity for coastal cities to organise the conditions for new actors to get into the innovation chains and invest creativity and vigour in alliances to drive emerging and changing blue green growth activities.

Effective public–private partnerships require a convincing policy framework, a trusted leadership and a large number of both public and private actors in cooperation. Policy makers should carefully map and identify relevant local maritime clusters, and put in place instruments and tools to support their emerging phase. A coherent longer-term policy framework is vital for the development of the cross-maritime basis. It is also crucial that governance levels are aligned and that efforts are made towards well-articulated multi-level governance. Equally, policies need to take into account sub-regional specificities.

Maritime partnerships and clusters can be a powerful resource for policy makers, as they are a unique platform for business, education, research and government to connect and engage in substantial dialogue on the future of blue activities. Well-functioning clusters are often engaged in horizon scanning to serve the longer term needs of their members and partners. The blue green maritime sectors are interlinked with competing economic activities, as well as the marine environment. Benefits from clusters depend strongly on strategies to strike the right balance between local and international activities and care for coastal marine ecosystems. Competency, skills and research are prominent areas for international collaboration (Ecorys et al. 2014).

5.3 Business-Friendly Cities and Ocean Corporate Responsibility

Sustainability and commercial excellence are not mutually exclusive. Reconciling economic objectives with longer term political goals is at the heart of strong sustainability. Urban sustainability implies constant climbing up the scale of values. Innovation and entrepreneurship create sustainable value when moving from low-yield to higher-yield renewable resources. Large enterprises are important for pioneering cities and SMEs have high potential for innovation and for revitalising urban fabrics, including harbours. Coastal cities can create enlightened business environments and establish specialised enterprise areas, start-ups, seed and capital risk companies, innovative clusters and micro-financing mechanisms. They can also partner with their businesses and engage with citizens in ocean corporate responsibility.

Businesses have a substantial role to play in creating substantial sustainable wealth. Socially responsible investing has been gaining ground as decision makers and investors seek to embed sustainability ethos in policies and create collective long-term value out of sustainability values. Various recognitions and awards, like the Sustainable Shipping awards, shed light on best practices and the dedication of business to helping, for example, shipping reduce its carbon footprint, and to encourage other operators to take the sustainability journey.

Public recognition is essential for highlighting the link between sound business ethics, environmentally sustainable operations and positive long-term business performance. But to combine the three requires a great deal of innovative thinking and bold decision making. Before the turn of the last century, the Bremen declaration “Business and Municipality: New Partnerships for the 21st Century” focused on local, national, regional and international frameworks for the sustainable development of communities and economies. It suggested that businesses ask cities for favourable locations and fees, preferential treatment and social recognition as a significant wealth and job generator. Municipalities seek employment for citizens, tax revenues and investment in the local economy from businesses. At the crossroads of these requirements, cross-fertilised investments can be beneficial for citizens and the city (Mega 2010).

Climate change has added a major and most powerful dimension into the social responsibility interactions between cities, business and citizens. Climate change in cities affects the array of blue economy sectors, from shipping and food production to tourism and service industries. The Carbon Disclosure Project suggests that cities recognise and report climate change threats to business. Businesses and cities seem aligned in their recognition of climate change risks, and most of the severe risks from climate change that businesses disclose are also recognised by the hosting cities. City adaptation actions contribute to business resilience. Cities are providing information, incentives and regulations, as well as investments in infrastructure, that assist businesses in being more resilient to climate change. Other policy actions and services support better resilience for businesses and the wider community (CDP 2014a, b).

A more resource-conscious economy advocates for new ways of designing business, responding to the fundamental needs of all and respecting the availability of terrestrial and marine resources. Scarcity can be turned into plenty when the waste of one product or service becomes the resource providing the input to a new process in the circular economy. Cascading nutrients and energy can become a beneficial process which generates jobs, builds social capital and increases income for all. Coastal cities are best situated to orchestrate such partnerships and processes.

Communication technologies, education and knowledge-sharing, transportation and urban migration are transforming world dynamics and the relative advantages of cities and enterprises. A most interesting world "Vision for 2050", conceived by the World Business Council for Sustainable Development (WBCSD), based on country dialogues with several hundred companies and experts, highlights that humankind can achieve a low-carbon or even zero-carbon society, but all stakeholders have to change their eco-behaviour radically, particularly their interactions with economic and ecological processes (WBCSD 2010a).

This vision of genuine sustainability for 2050 highlights a future world with nine billion people living well and within the limits of the planet. More than two-thirds of humanity lives in cities. Education, cooperation and empowerment of citizens, and especially women, bring more awareness about the socio-economic challenges. The vision for 2050 underlines that conflicts and disasters may not have disappeared, but societies are resilient, able to withstand disruption and recover quickly.

A redefinition of notions and values is a precondition for the success of Vision 2050. Its implementation asks that society redefine the notion of prosperity and well-being and that economic growth be decoupled from energy and material use and ecosystem destruction. Markets have to redefine values, costs and benefits. In a complex, yet interconnected world, nations have to engage in responsible global governance to effectively manage international systems and resources, like climate, water or space.

Humankind can progress to a low-carbon society with a secure and sufficient supply of low-carbon energy and transport. This vision is also shared in the European Union's vision for 2050, "living well within the limits of the planet" enshrined in the Seventh Environment Action Programme (EC 2013). The circular and sharing economy concepts advocate for optimal benefit to all, with zero particles of waste, all of it having been thoroughly turned into resource. The WBCSD suggests that, in concretising this vision, humanity will be using the resources regenerated by just over one planet, versus the impossible equivalent of 2.3 planets needed according to the business as usual scenarios (WBCSD 2010a).

Coastal cities require new or upgraded infrastructure, which may vary among regions and purposes. Cities need to construct floor space and equipment for residential and commercial purposes. The capacity of ports to handle container traffic needs may have to rise considerably. China may hold a share of nearly 40 % of growth in global demand for urban building floor space up to 2025. Africa and the Middle East will account for almost 14 % of the global rise in municipal water demand in large cities, almost twice their share of urban GDP growth. But cities also have to care about green infrastructures and resource and energy-efficient developments.

Businesses are essential for the development of urban coastal infrastructures, which provide the "hardware", increasingly incorporating "software" for a better performance which depends much of the local context. The main infrastructure, technology, and services to reinforce the sustainable visions and strategies of cities are predominantly developed and implemented by the private sector. The WBCSD's Urban Infrastructure Initiative brought together a diverse group of companies from key sectors, including energy, materials, water, equipment, and support services to help urban authorities develop pragmatic and cost-effective plans for sustainable infrastructures. The case for action can be compelling both for cities and for companies. A sustainable city can use resources more efficiently, thrive economically and create an inclusive community. Urban markets offer companies the opportunity to provide system solutions, products and services for sustainable buildings, energy, infrastructure, and resource and waste management. Working with local authorities, the WBCSD helps translate the identified needs into landscape solutions for sustainable urban development (WBCSD 2010b).

Another related inspiring initiative is the ICLEI Global Town Hall at Metropolitan Solutions, held as part of the 2014 Hannover Fair. This event provided a paramount platform for city-enterprise dialogues on innovative and sustainable urban infrastructure, bringing together businesses, local governments, researchers, and developers for 5 days of face-to-face conversation, trust-building and information exchange. The Global Town Hall provided an excellent opportunity to launch the final report of the WBCSD Urban Infrastructure Initiative and to promote further dialogue on effective city-business partnerships. It seems that there is a shared win-win opportunity to drive transformational global action on climate change through interconnected city and business leadership.

Cities win by getting practical, cost-effective solutions to realise their ambitious sustainability visions. Leading businesses win through growth in the markets for innovative new solutions that are essential for transformational change. The World Business Council on Sustainable Development suggested some key actions for businesses to drive progress towards sustainable development through the market. Its platform, Action2020, helps businesses to define and shape action for sustainable development. Through collaboration toward common shared goals, business can address some of the critical environmental and social challenges while strengthening resilience. Action2020 member companies developed the societal targets, based on a review of the latest scientific consensus led by the Stockholm Resilience Centre. Business solutions must be scalable and replicable by many companies, in multiple sectors, cities and countries, in order to make a global difference (WBCSD 2014).

Once more, evidence confirms that global in nature, grand societal challenges can only be addressed by business, government and society working together. Enlightened leaders in cities and metropolitan areas will only succeed in reshaping their economies, and thereby the economy of the nations, in concert with companies, universities, citizen associations, labour unions, environmental groups, cultural institutions and philanthropies, working in often overlapping and interlocking networks.

The social responsibility of enterprises is essential for the blue green economy. Cities and enterprises embracing and supporting the sustainability ethos should make optimal use of marine resources and technologies. They have to ensure that the best green products and services come to the market and ultimately improve citizens' and consumers' lives. Enterprises have to deploy their inventiveness and bring to the market blue technologies, products and services.

Improving social and environmental performance of business and disclosure of information on their results is capital. Ethical citizens and vigilant consumers and organisations call for coastal cities and companies to design and prove their contribution to sustainable blue growth. Awareness campaigns try to overcome information deficits and increase consumers' knowledge on the ecological impacts and potential benefits of alternative patterns. Citizen associations and the media tend to create a climate of trust, surrounding sustainable cities and businesses. Companies and cities without a declared commitment to and action for sustainable development could face citizen and consumer boycotts, attacks on fixed assets, failure to attract forward-looking stakeholders, stockholders and employees, restrictions on port operations and obstacles in infrastructure development. Pro-action is a must, since damaged reputation, impaired licenses, disillusioned shareholders and disappointed citizens may impact future investments.

A growing number of cities and companies make public their intentions, plans, actions and results in preparing for the post-carbon age. Corporate codes of conduct have gained momentum. Businesses have adopted a broad range of approaches, from non-reporting to social reporting to increase confidence in their performance. Assessment and public reporting have to be organically integrated components of the business process. The voluntary non-binding nature of most codes is often related to the absence of independent auditing, even if codes spell out the necessity for monitoring, life cycle assessment and reporting.

Eco-responsible businesses are instrumental in creating value out of values and can be precious allies to conscious coastal cities. They increasingly integrate social, environmental, ethical and human rights concerns and principles into their strategies and business operations in close collaboration with their stakeholders. The aim is to achieve results which are beneficial both to society and the economy, and to minimise and prevent negative impacts on the environment, including marine habitats and ecosystems.

Social responsibility can be expressed or mobilised through many diverse schemes. The "Social Return on Investment" initiative in Amsterdam, designed to include social obligations in public contracts, started as a pilot project in 2007 in Amsterdam's South-East District. Contractors are asked to translate profits into employment opportunities for youths usually excluded from the labour market, especially ethnic minorities. The project included the development of guidelines and monitoring instruments (Eurocities 2010).

Corporate engagement for sustainable development is highly promoted by many investors, environmental and public interest organisations and advocacy groups. The Coalition of Environmentally Responsible Economies and Societies (CERES), founded by a small group of investors largely in response to the Exxon Valdez oil spill that occurred in March 1989, wished to bring together environmentalists and capitalists to forge a new sustainable business ethos. The oil spill in Alaska's Prince William Sound devastated one of the world's most pristine habitats. Suddenly and painfully, businesses had to account for the environmental and social impacts of their operations. The CERES brought forward a bold vision of a world in which business and the capital markets jointly promote the well-being of human society and the protection of the precious natural environment. At the heart of the vision, the coalition's Principles proposed a ten-point code of corporate conduct to be publicly endorsed by companies as a sustainability ethic.

The coalition has introduced many innovative instruments to weave environmental and social challenges into company and investor decision-making and the capital markets. The global reporting initiative, launched in 1997, had proposed a harmonised public disclosure to deliver a steady flow of structured verifiable and comparable information. The coalition's sustainability reports enabled organisations to "walk the talk" and serve as a model for other businesses trying to improve the transparent reporting of their actions and outcomes against commitments. Recognising the need to set higher expectations for corporate water management, CERES developed the Aqua Gauge, a free water stewardship assessment tool backed by investors managing $2 trillion in assets. Hundreds of companies in dozens of sectors are using this aqua gauge to evaluate their water performance and set ambitious goals to reduce their impacts.

In 2010, CERES further raised the bar of excellence for corporate sustainability performance with the Roadmap for Sustainability, which outlines meaningful environmental and social improvements that companies have to undertake in order to succeed in the resource-constrained twenty-first century. The Roadmap was intended to bring investors, companies and other key economic players together to find scalable solutions for integrating sustainability across capital markets and the economy. The Roadmap for Sustainability contains 20 specific expectations for corporate performance, broadly divided into four areas of activity, including governance, stakeholder engagement, performance, monitoring and disclosure (CERES 2010). In 2014, CERES launched the Clean Trillion, highlighting the annual global investments needed to limit global warming to 2° C and ensure a clean energy future. Meeting this goal is very challenging, given the current less-than-adequate amount of $300,000 billion in clean investments, and will only be possible if businesses, investors and policy makers join forces (CERES 2014).

Corporate social responsibility is crucial for mainstreaming ethical principles in business activities and catalysing actions in all sectors. It could link different activities of the blue economy and diffuse to a larger scale of initiatives, such as the UN Global Compact, the largest voluntary corporate responsibility initiative in the world. This strategic policy initiative engages businesses which align their operations and strategies with ten universally accepted principles in the areas of human rights, labour, environment and anti-corruption. A commitment to transparency and disclosure is critical.

The Compact is a conceptual yet practical framework for the development, implementation, and disclosure of sustainability policies and practices. It calls for businesses to support a precautionary approach to environmental challenges, undertake initiatives to promote greater environmental responsibility, and encourage the development and diffusion of environmentally friendly technologies. The Compact is global and local at the same time, private and public, voluntary yet accountable. It seeks to combine the moral authority and convening power of the UN with the private sector's vigour and the contributions of a range of key stakeholders.

Benefits include the sharing of a globally recognised policy framework and governance, the exchange of best practices and world linkages among business units and subsidiaries across the value chain through the Global Compact's Local Networks, including in emerging markets. The Compact also has an accountability policy component, the Communication on Progress. The annual communication is an important demonstration of a participant's commitment to the UN Global Compact and its principles. Failure to communicate may result in the loss of participant status and possible expulsion.

Corporate social responsibility for blue green growth and sustainable development is promoted through the Global Partnership for Oceans (GPO), a growing alliance of governments, international organisations, civil society groups, and private sector interests committed to the health, productivity and resilience of the ocean. The alliance has targeted the problems of overfishing, pollution, and habitat loss which are contributing to the depletion of the natural resource that provides nutrition, livelihoods and vital ecosystem services. The GPO has been established as a platform that can facilitate access to finance and expertise (GPO 2013).

Convened by the World Bank to advise the Global Partnership for Oceans, the Blue Ribbon panel, bringing together CEOs of some of the largest seafood companies in the world, governments and prominent marine ecologists, emphasises that, without action to restore the declining health of the ocean, the consequences for economies, communities and ecosystems will be irreversible. Improving ocean health is a complex process that requires leadership and participation across a broad sphere of communities, industries and governments and the concentrated effort of society, government and science. Fragmented approaches that fail to consider social, political, economic and ecological relationships cannot meet the complex challenges facing ocean health. The panel calls for an integrated approach to ocean investment and emphasises the essential role of public-private partnerships.

Policy options must be multidimensional and dynamic and integrate all aspects of the socio-ecological system. The Blue Ribbon panel's principle-based strategy provides an approach that prioritises action with high impact. The panel recommends five principles to be incorporated into all levels of ocean policy reform, from fishery management to incentives and pollution reduction to habitat restoration. The five interlinked principles include sustainable livelihoods, social equity and food security, a healthy ocean, effective governance systems and long-term viability and capacity building and innovation (GPO 2013).

5.4 Quality of Life in Marine Cities, Magnets of People and Capital

How do coastal cities rank among the best cities to live in and do business? What do citizens and enterprises want from territorial authorities and how can government best respond? Since the last decades of the twentieth century, a concert of surveys and ranking systems conclude with an array of top cities, according to definitions, criteria and assessments. They provide interesting insights into specific essentials of coastal cities and their attraction for people and businesses. Attractiveness of cities may well be in the eyes of the beholder. The most recent surveys include personalised elements in order that citizens and enterprises are enabled to do their own particular rankings according to specific weight given to particular criteria. The "Your Better Life Index" introduced by the OECD is a most outstanding example of such approaches (OECD 2011b).

Coastal cities like Copenhagen, Tokyo, Melbourne, Vancouver, Stockholm, and Helsinki rank high in many traditional rankings according to widely accepted criteria that form the complex term of "quality of life". These criteria include identity and engagement, health and well-being, creativity and recreation, heritage and projection into the future, ideas and perceptions, gender and generations.

Urban liveability is cardinal for all citizens. The Economist Intelligence Unit's 2014 global liveability report ranking cities according to the living conditions that they offer to citizens, mainly safety, healthcare, education, infrastructure and environment, without taking into account the cost of living, suggests that cities in Canada, the EU, Australia, and New Zealand are the best to live in, thanks to widespread availability and accessibility of goods and services. According to this liveability ranking, Melbourne remains the most liveable among 140 cities, followed by Vienna. Vancouver, which ranked as the most liveable city until 2011, comes in third place. It is interesting to note that the top cities have not changed much over time. Over the last year, only 20 out of the 140 surveyed cities have experienced changes in scores and the majority of these changes have been driven by deteriorating scores in fragile regions such as Ukraine. Helsinki is the only EU coastal city to rank among the ten most liveable cities. Furthermore, the two European cities are the only non-English speaking top cities in a ranking often criticised to be too English-centric (EIU 2014).

In 2012, the Economist Intelligence Unit teamed up with data sharing company BuzzData to host a competition offering users the opportunity to make their own ranking by combining data from the Cost of Living and Liveability surveys with other sources. Opening up the concept of best cities to a diverse group generated plenty of innovative approaches. Interactivity allowed respondents to propose elements like connectivity and suggest priorities or weights towards a collective choice of best city. This approach had the dual benefit of allowing citizens to make their own choices and effectively crowdsourcing responses to inform future methodologies (EIU 2012).

The results of this survey included adjusted indexes to define the best city ranking. The report suggests that the top ten best cities include a majority of coastal cities, including Hong Kong, Amsterdam, Osaka, Sydney, Stockholm and Tokyo

(EIU 2012). Notable was the introduction of a connectivity indicator, which was a key factor in moving Hong Kong to the top of the ranking, above cities like Amsterdam, Tokyo and Sydney. Although Hong Kong scored relatively poorly for pollution and cultural assets, the city benefited from strong scores in the natural assets and sprawl categories. Stockholm in the sixth position and Tokyo in tenth place complete the league of the best coastal cities to live in.

The lists of the worst cities to live and invest in can be as impactful as the lists of the world's best cities. It also indicates the cities which present the highest potential for improvement. The Economist Intelligence Unit ranked cities according to the "tolerability" of living in a particular place given its crime levels, threat of conflict, quality of medical care, levels of censorship, temperature, schools and transport services. The list of the cities with the highest potential for improvement was topped by Abidjan, in Côte d'Ivoire, with an overall rating of 45.9 out of 100. Lagos, Karachi, Dhaka and Tripoli were also classified among the cities with the worst living conditions.

The quality and cost of life surveys and the ranking of world cities, conducted by Mercer, international consultancy in human resources, also offer interesting insights and information. These surveys are very useful for governments and companies which have executives working in various cities of the world and have to take into account a series of factors when structuring their remuneration packages. Each survey is based on an array of assets and services reflecting the typical expenditure of the expatriated population. Mercer's Quality of Living list includes more than 200 cities across five continents, ranked against New York as the base city.

Coastal and European cities dominate the Mercer rankings according to the quality of life, with some illustrious exceptions like Vienna, which, followed by Zurich, reconfirmed in 2014 its leading position in relation to the overall quality of life. The bright examples of non-coastal cities include various German and Swiss cities, such as Düsseldorf, Frankfurt, Munich, Geneva and Bern. Vancouver, Auckland, Sydney and Copenhagen complete the league of cities with the best quality of life (Mercer 2014a).

Next to the quality of living surveys, Mercer's cost of living surveys focus on comparative cost of over 200 items in each location, including transport, food, clothing, household goods and entertainment. The cost of housing is also included and, as it is often the most important expense for expatriates, it plays an important part in determining the ranking of cities. New York is used as the base city and currency movements are measured against the US dollar. Recent world events, including economic and political upheavals, have affected the rankings for many regions through currency fluctuations, inflation, and volatility in real estate prices.

According to Mercer's cost of living 2014 survey, two African cities top the list of most expensive cities for expatriates. Coastal Luanda is the world's most expensive city for expatriates, followed by N'Djamena in landlocked Chad. European and Asian cities also continue to dominate as the costliest cities, with Hong Kong in third place, followed by Singapore. Tokyo dropped four spots to rank seventh. Karachi, ranked 211th, is the world's least expensive city, three times less costly than Luanda (Mercer 2014b).

Another survey by the global property service provider Cushman and Wakefield sheds light into European business environments since 1990. Each annual survey is based on the opinions of the senior management of 500 European companies and the criteria considered important in location decisions. The key factors taken into account by the companies include accessibility to markets, availability of qualified personnel, transport and communications and cost of living and personnel. London was the top rated city, including in regard to transport links with other cities and internationally, ease of travelling within the city, easy access to markets, customers or clients and availability of quality staff and quality of telecommunications and languages spoken. Dublin once more comes out top for the climate created by government, and Warsaw has taken over from Leeds as the top location for value for money in office space. Barcelona and Stockholm retain the top positions for quality of life and the environment (Cushman and Wakefield 2012).

The state of a city and its environment has a strong impact on the state of the citizens' lives. The Veolia observatories of urban lifestyles revealed that cities are sources of intense feelings, much more positive than negative. City residents appreciate urban advantages, but also recognise that their price is often rather high. The most important factors for quality of life are considered to be safety and the cost of acceptable living conditions, followed by the environment, infrastructures, public transport and leisure (Veolia 2008, 2010).

Those best to live and invest in or most expensive cities are not necessarily the most powerful global cities, which are not necessarily the most competitive. Various indexes try to capture the power of global cities, including the Global Economic Power Index by the Martin Prosperity Institute, and the Global Cities Index by A.T. Kearney. It is interesting to highlight that New York, London and Tokyo rank first in all major rankings.

Coastal cities top the ranking of the world's most economically powerful cities by R. Florida after the rankings according to the above-mentioned major indexes on the relative economic strengths of global cities and metropolitan areas. New York stands on top of the global urban hierarchy and its position is proposed as secure, at least for the medium term. It is the world's most open and diverse large city, the only one to come first or second according to all criteria. London is overall second and Tokyo third. Hong Kong is fifth and Singapore, Shanghai, and Los Angeles follow in the race to the top.

London and New York also rank at the top of the 2014 edition of the Cities of Opportunity list established for the sixth time by PricewaterhouseCoopers. These exercises have been instrumental in benchmarking cities from various perspectives, including unmet demand and potential for future growth. The sixth edition analyses the trajectory of 30 cities, all capitals of finance, commerce, and culture, and identifies strengths and weaknesses. The survey is based on the assessment of ten indicators composed out of a number of variables organised in three groups, focusing on quality of life, economics and instruments for a changing world. In the latest edition, six out of the ten top cities of opportunity are coastal cities. London comes first with a good margin and marks the highest scores in technology readiness, economic influence and city gateway, three indicators which reflect its leading position in the global economy. New York displays a balanced performance across domains, while Singapore climbs to the third place as world leader in transport and infrastructure, as well as in enabling business environment.

Australia's largest city scores in the top ten in four indicators and in more than half of the variables. As in 2012, the comparative advantages of Sydney shine in the three indicators that measure living standards. Sydney ranks first in demographics and liveability, shares the first place with Stockholm in sustainability, and comes second in health, safety and security. The city ranks second for quality of living and lack of traffic congestion and third for relocation attractiveness. This was a new variable of the last survey, in which responses were also provided by 15,000 global staff members of the multinational company (PricewaterhouseCoopers 2014).

This last survey put special emphasis on the connectivity of cities and their role as gateways to territorial economies. Dubai ranked 16th out of 30 Cities of Opportunity in 2014. It ranked first for airport connections to business districts, fifth in most affordable city and eighth in world city as an urban gateway. The city performs remarkably well across several domains, including transportation and infrastructure, potential for future investment growth and safety. It is also very interesting that the survey suggests that the sustainability and intellectual capital of the city hold a high potential for improvement (PricewaterhouseCoopers 2011, 2014).

Is there a Perfect Global City? At A.T. Kearney, the Global Cities Index, which, since 2008, has examined a comprehensive list of 84 cities on all continents, according to their global engagement in business activity, human capital, information exchange, cultural experience, and political engagement, provides an interesting definition. The Perfect Global City could be a composite result out of 16 cities, with New York, Tokyo and London representing 50 % and the other 13 cities the other 50 % of the composed entity (Kearney 2014).

Next to the race for Global Cities, the Emerging Cities Outlook examines 34 cities located in countries classified as low or medium income. The outlook evaluates a city's potential according to its ability to attract, retain, and generate flows of ideas, capital, and people, and the time needed to catch up with the global leaders. The proposed indicators can be grouped into three categories. A first range of indicators includes business activity reflecting the evolution of a city's GDP, changes in its infrastructure, ease of doing business, and perceptions regarding public transparency. A second group focuses on human capital, looking at trends in stability and security, healthcare availability and quality, income equality, and environmental sustainability. Finally, the generation of patent filings per capita, new businesses, volume of venture capital deals, gross expenditure in research and demonstration activities, and access to finance compose the innovation profile of a city as a catalyst to attract business and talent.

Two vulnerable coastal Southeast Asian cities, Jakarta and Manila, head up the list of the Emerging Cities Outlook, including the cities with the highest potential to progress. Jakarta, the beating heart of Indonesia, seems to move up quickly in relation to human capital, but also in addressing income inequality and environmental concerns, as well as across several important innovation indicators. Manila also seems favoured by a relatively sharp increase in human capital, with a particularly notable improvement in healthcare. The Philippine capital, located in the coral triangle, enjoys an extraordinary biodiversity reflected on its seal, a pearl embedded in a shell. Both cities bear witness to ceaseless interactions among people and cultures. They can only progress in harmony with their marine environments and in investing in resilience and sustainable development (Kearney 2014).

References

Boston Foundation (2009) A great reckoning: healing a growing divide. Boston Foundation, Boston

Brookings Institution (2012) Global metro monitor 2011. Volatility, growth and recovery. The Brookings Institution, Washington, DC

Carbon Disclosure Project (CDP) (2014a) Protecting our capital. How climate adaptation in cities creates a resilient place for business. CDP, London

CDP (2014b) Global corporate use of carbon pricing. Disclosures to investors. CDP, London

Coalition of Environmentally Responsible Economies and Societies (CERES) (2010) The 21st century corporation: the Ceres roadmap for sustainability. CERES, Boston

CERES (2014) Investing in the clean trillion: closing the clean energy investment gap. CERES, Boston

Cushman & Wakefield (2012) European cities monitor 2011. Cushman & Wakefield, London

Economist Intelligence Unit (The) (EIU) (2012) Best cities ranking and report. EIU, London

EIU (2014) A summary of the liveability ranking and overview. EIU, London

Ecorys et al (2014) Support activities for the development of maritime clusters in the Mediterranean and Black Sea areas. Brussels

Eurocities (2010) Cities accommodating diversity. Eurocities, Brussels

European Commission (EC) (2012a) Smart cities and communities. European innovation partnership. EC, Brussels

EC (2012b) Blue growth. EC, Brussels

EC (2013) 7th Environment action programme. EC, Brussels

EC (2014a) Blue energy. Action needed to deliver on the potential of ocean energy in European seas and oceans by 2020 and beyond. EC, Brussels

EC (2014b) Micro-algae-based products for the food and feed sector: an outlook on Europe. JRC Science and policy papers. IPTS, Seville

EC (2014c) The future of Europe is science. EC, Brussels

EC (2014d) A European strategy for more growth and jobs in coastal and maritime tourism. EC, Brussels

EC (2014e) Directive of the European Parliament and of the Council establishing a framework for maritime spatial planning and integrated coastal management SWD (2013) 64 final. EC, Brussels

European Environment Agency (EEA) (2012) Urban adaptation to climate change in Europe. EEA, Copenhagen

Food and Agriculture Organisation (FAO) (2014) The state of world fisheries and aquaculture. FAO, Rome

Global Ocean Commission (GOC) (2014) From decline to recovery. A rescue package for the global ocean. GOC, Oxford

Global Partnership for Oceans (GPO) (2013) Fact sheet. GPO, Washington, DC

Gore A (2013) The future: six drivers of global change. Random House, London

International Energy Agency (IEA) (2014) World energy outlook. IEA, Paris

Katz B, Bradley J (2013) The metropolitan revolution. How cities and metros are fixing our broken politics and fragile economy. Brookings Institution Press, Washington, DC

Kearney AT (2014) Global cities index and emerging cities outlook. Global Cities, Present and Future

MGI (2011) Urban world: mapping the economic power of cities

MGI (2012) Urban America: US cities in the global economy

Mega V (2010) Sustainable cities for the third millennium. The odyssey of urban excellence. Springer, New York

Mega V (2013) Quintessential cities, accountable to the future. Innovation, sustainability and leadership. Springer, New York

Mercer (2014a) Quality of living survey. Mercer, New York

Mercer (2014b) Cost of living survey. Mercer, New York

Organisation for Economic Co-operation and Development (OECD) (2006) OECD territorial reviews: competitive cities in the global economy. OECD, Paris
OECD (2011a) Towards green growth. A summary for policy makers. OECD, Paris
OECD (2011b) OECD forum. Better policies for better lives. OECD, Paris
OECD (2013) The competitiveness of global port cities. Synthesis report. OECD, Paris
OECD–FAO (2014) Agriculture outlook 2014–2023. OECD–FAO, Paris
PricewaterhouseCoopers (2011) Cities of opportunity. PricewaterhouseCoopers, New York
PricewaterhouseCoopers (2014) Cities of opportunity. PricewaterhouseCoopers, New York
SEA EUROPE Shipbuilding (2014) Market monitoring. Report 35. June 2014
United Nations World Tourism Organisation (UNWTO) (2014) Tourism highlights. UNWTO, Madrid
Veolia Environnement (2008) Observatoire Veolia des modes de vie urbains 2008. L'état de la vie dans les villes. Étude réalisée par Ipsos. Veolia Environnement, Paris
Veolia Environnement (2010) Cities for living. Veolia Environnement, Paris
World Bank (2012a) Inclusive green growth. The pathway to sustainable development. World Bank, Washington, DC
World Bank (2012b) Toward a green, clean, and resilient world for all. A world bank group environment strategy 2012–2022. World Bank, Washington, DC
World Business Council on Sustainable Development (WBCSD) (2010a) Vision 2050. WBCSD, Geneva
WBCSD (2010b) A world of sustainable cities. WBCSD, Geneva
WBCSD (2014) Action 2020. WBCSD, Geneva
World Economic Forum (WEF) (2014) The competitiveness of cities. WEF, Geneva
WEF (2015) The global risks report 2015, 10th edn. WEF, Geneva

Selected Internet Links

www.weforum.org
www.kobenhavn.dk
www.mind-lab.dk/en
www.rotterdam.nl
www.stockholm.se
www.22barcelona.com
www.aivp.org
www.cici.org
www.mercerhr.com
www.cushwake.com
www.mori-m-foundation.or.jp/english/research/project/6/index
www.arup.com/Publications/SlimCity.aspx
www.CoFoundersLab.com
www.barcelonaactive.cat
www.mckinsey.com
www.theatlanticcities.com
www.coastalsocioeconomics.noaa.gov
http://amsterdamsmartcity.com/knowledgecentre
www.unwto.org
http://ec.europa.eu/maritimeaffairs/policy/coastal_tourism/index_en.htm

Watercolour 6
Shanghai, the Emerging World Port City

Chapter 6
Diversity, Solidarity and Engagement in Coastal Cities

Abstract Responsible, fair coastal cities care much about the distribution of wealth, especially when the number of citizens not engaged in employment, education or training reaches unprecedented highs. Concentration of very diverse people and activities are invaluable assets for cities, acting as schools for respecting difference and living harmoniously together. This chapter sheds light on the evolving social capital of coastal cities and their capacity for intergenerational and intercultural equity, social justice, public health, safety and solidarity. Coastal cities themselves are important direct and indirect sources of new employment, the first factor of social integration, followed by housing and education. The enhancement of urban social capital is of increasing importance in cities that face new forms of unemployment, poverty and exclusion, where more than three generations co-exist, and immigrants also come from the sea in search of better opportunities on a blue horizon. They should all be offered the conditions to enjoy a healthy and meaningful life.

6.1 Coastal Cities as Incubators of Decent Blue Green Jobs

Sustainability is "equity extended into the future", a responsible resource-conscious journey for the well-being of all. Social justice, employment and participation are key sustainability dimensions. High unemployment, endemic violence and exclusion, and food and water shortages can seriously undermine the sustainability prospects of a city. The blue green economy has high potential for decent jobs, food security and public health. Three billion citizens rely on the ocean for their livelihoods (GOC 2014).

Responsible fair coastal cities must invest in fairness and the sustainable distribution of wealth, including from the exploitation of their marine assets, especially at times of protracted crisis. The blue economy has an important estimated potential to attract a more diverse workforce in coastal cities. The development of skills and activities, the integration into the labour and the housing market and the welcome of mobile offshore workers such as seafarers make the role of coastal cities crucial as central institutions organising the dialogue and engaging all stakeholders.

V.P. Mega, *Conscious Coastal Cities*, DOI 10.1007/978-3-319-20218-1_6

Fairness constitutes the ultimate ethos of the social architecture of sustainable coastal cities that strive to offer citizens a fulfilling life worth living. Skills taking care of the marine environment provide a valuable community service and can be one of the alternatives in addressing the risk of a jobless recovery. The weak global economic recovery has failed to lead to a substantial improvement in global labour markets, and the number of global unemployed citizens reached 200 million in 2013. The epicentre of the crisis has been the industrialised world, accounting for half of the total increase in unemployment, but, in a highly interconnected world, this has had significant spillovers into the labour markets of emerging economies. A quarter of the increase of four million in global unemployment in 2012 has been in advanced economies, while three quarters has been in other regions, with marked effects in East Asia, South Asia and sub-Saharan Africa (ILO 2014).

Unemployment is a major waste of resources and talent which undermines urban dynamism. The unemployment rate in the USA reached 5.8 % in 2014. In the European Union, the world region most affected by the crisis, the unemployment rate in EU28 rose to 10.0 % in November 2014. Among the Member States, the lowest unemployment rates were recorded in Austria (4.9 %) and Germany (5.0 %), and the highest in Greece (25.7 %) and Spain (23.9 %). The surge of citizen movements is indicative of the political repercussions which find their privileged spaces for protest in cities (Harvey 2012).

Youth unemployment is particularly alarming, as young citizens are usually the main victims of crises. In November 2014, more than five million citizens under 25 were unemployed in the EU28, with a youth unemployment rate of 21.9 %, compared with 23.2 % in November 2013. Structural trends underline the need for investment in human capital. A protracted recovery which does not generate employment has not only important economic and social costs but signals structural deficiencies in the labour market that weaken the seeds of growth. The concurrent rise in job vacancies and unemployment indicate labour market mismatches among required skills and education and training. Education and life-long training policies are essential to equip the unemployed with the skills required for promising sectors which contribute to sustainable development. Training portfolios related to the exploration and care of the ocean can initiate a new blue green employment dynamic (EC 2015).

There is no evidence that coastal cities are in better condition in terms of employment, but some coastal and maritime sectors have a great deal of potential. Fisheries still employ 4.4 % of the estimated total economically active people, of which 84 % are in Asia and 10 % in Africa. Women are estimated to account for more than 15 % of these. Employment in aquaculture seems to have much potential. Cities can create a wide array of direct and indirect jobs to stimulate, design, regulate, and support the marine and maritime economies. Developing synergies between aquaculture and other activities can, for example, require a range of skills. From the conception of ocean gardens to offshore renewable energy and sustainable tourism, the great diversity of coastal and maritime activities is promising. In the EU, the coastal and maritime tourism sector has been identified as an area with special potential to foster a smart, sustainable and inclusive Europe and create new employment in the EU regional sea basins.

A rising blue green economy could provide SMEs with the momentum needed after the crisis. The creation of clean SMEs in innovative maritime clusters is assessed to be a quality job generator. The results of a Survey by the Committee of the Regions on SME-friendly Regions and Cities of the European Union suggest that the financial and economic crisis has generally been seen by Local and Regional Authorities as having a negative impact on SMEs. The nature of the impact is perceived in very diverse forms but more painfully in regard to employment losses. A decrease in investment or liquidity, falling consumer demand and orders, and declining performance, such as a decrease in production, research and demonstration, were some of the main effects reported by cities and regions (CoR 2012).

Port cities also have to develop the necessary infrastructure and skills to receive the population of passengers and visitors they desire. The global economy also accounts for over one-and-a-quarter million seafarers, predominantly from OECD countries, Eastern Europe and Eastern and Southern Asia. Africa and Latin America are noticeably under-represented. Only 2 % of seafarers are women, mainly in the ferry and cruise ship sectors. Over the past decades, piracy and armed robbery have re-emerged as a serious risk to seafarers. Much attention has been focused on action against such attacks on shipping in waters off eastern Africa, and in the South China Sea, the location of over half the international incidents reported in 2013.

Coastal cities are gateways to regional and maritime tourism, which contributes significantly to direct and indirect employment. Globally, it is estimated that, in 2013, tourism provided 3.3 % of direct employment and 8.9 % when the multiplier effects are taken into account. The capacity in tourism infrastructures in most regions is concentrated in coastal areas. Various activities, such as yachting and cruising, are specific to coastal and maritime tourism. Demand for cruise tourism roughly doubled worldwide during the first decade of the century. The development of high-quality coastal and maritime tourism and jobs can help address diversification of supply, fluctuating demand and seasonality (EC 2014b).

Many green economy initiatives demonstrate their potential for job creation. A prime example is the US Better Buildings Initiative, ensuring that America invests in innovative clean energy technologies and doubles the share of electricity from clean energy sources by 2035. The initiative targets commercial and industrial buildings which have to become 20 % more energy efficient by 2020 and accelerates private sector investment in energy efficiency. The Political Economy Research Institute suggested that the entire Better Buildings Initiative could create up to 114,000 jobs in sectors as diverse as lighting, insulation, heating and cooling, information and access to financing.

Decent green jobs are already part of urban portfolios focused on post-carbon societies. New York's Greener Greater Buildings Plan is expected to create some 17,800 green construction-related jobs by 2030. The Plan consists of four regulatory parts supported by extensive jobs training. It asks for large buildings to annually benchmark their energy performance, that a local energy code be adopted, that every 10 years these buildings conduct an energy audit and a retro-commissioning and that, by 2025, the lighting in the non-residential spaces be upgraded. The first benchmarking report on private buildings suggests that all these activities are goldmines for decent jobs (New York City Mayors' Office of Long-term Planning and Sustainability 2012).

The experience of Boston suggests that climate action can generate many high quality jobs and bring significant economic benefits. Demand for energy and climate-related services have the potential to create thousands of jobs. Energy efficiency and behaviour change are promising multi-win domains with multiplier effects, and are expected to produce a net savings of over $2 billion by 2020 through lower energy bills (City of Boston 2011).

The creation of green jobs, central to the transition to a zero/low carbon economy and society, is considered by the World Bank to be the most human face of green solidarity. Blue green jobs signal the transformation of economies, enterprises, workplaces and labour markets into a sustainable, low-carbon economy providing decent work in balance with the sea. But innovative strategies to promote blue green jobs can only succeed with the full involvement of local authorities and participation of workers and enterprises (ILO 2012; World Bank 2012a, b, c).

The emerging blue green economy may offer multi-sector and cross-sector jobs, and coastal cities could play an important role in human capital creation. From a citizen perspective, choosing the right skills and pathways is vital for a successful professional and social life. From the perspective of the sustainable urban economy, it is crucial to improve skills prospects, labour market adjustments, adaptability of enterprises and workers to change, and to develop new sectors. The concept of flexicurity, "flexibility within security", can also be extended to blue-green sectors.

Blue green growth can reshape labour markets, while the adequate labour and skill policies can help in maximising and sharing the potential benefits. The impact of a transition towards a greener economy on labour markets could extend far beyond the generation of new green jobs, such as those related to the bioeconomy and renewable energy sources. New opportunities, but also new risks, could emerge and alter labour market dynamics. The challenge for skill agendas and labour markets is to maximise the sustainable value and contribute to a fair sharing of benefits and costs. Suggested policy priorities include the smooth reallocation of workers to growing eco-sectors, such as the blue economy, eco-innovation and the diffusion of blue green technologies by strengthening initial education and professional training, and reform of tax and benefit systems to ensure that pressures do not become a barrier to employment (OECD 2012).

Offshore energy developments offer ample employment opportunities. Many direct and indirect jobs have already been generated in research, technology development and demonstration, component manufacturing and the operation of infrastructures and services. In the EU, a world leader in wind technologies, offshore wind activities employed, directly and indirectly, 58,000 citizens. Ocean energy also has the potential to create new, high-quality jobs in project development, component manufacturing and operations. Indicative job estimates suggest that 10,500–26,500 permanent jobs and up to 14,000 temporary jobs could be created in the EU by 2035 (EC 2014a).

Offshore wind developments are very important for employment creation. In Belgium, the offshore wind activities, centred in Oostende, directly employed

2615 professionals in 2012, while the indirect impact on employment accounted for 3522 additional jobs. Employment expectations from the North Sea offshore wind are high and the European Wind Energy Association warned about the possible forthcoming lack of adequate skills and the need for new training schemes (EWEA 2014).

Urban blue economy projects could target young people out of the traditional education and training schemes and offer them a fulfilling first employment experience, for example, in a renewable energy or green tourism recycling project. Shipping and maritime activities which respect the environment can offer rewarding onshore, offshore and fleet careers. Maritime jobs can involve working with state-of-the-art technology, for example, in exploratory ships or a multipurpose platform, maintaining high quality and safety standards, supporting marine technical know-how and preventing harm to marine ecosystems. Coastal cities can bring together all stakeholders and organise training sessions and support schemes.

The European ships and maritime equipment industry employs more than 500,000 people and have an average annual turnover of around €72 billion. The European shipbuilding and ship repairing industry is made up of around 300 yards, of which more than 80 % can be considered to be dedicated to small to medium ships of 60–150 m. The remaining yards can be defined as large. Around 90 % of the demand is for export markets.

Shipbuilding has virtually vanished in Belgian ports but some small companies for ship repair and maintenance are left in Antwerp and Oostende. Maritime transport is an important activity centred in the highly complementary ports of Antwerp, Oostende and Zeebrugge and Ghent. The total direct employment in the ports amounted to 120,528 full time equivalents in 2009. The direct maritime employment of the four ports has shrunk and most of it concentrates in the port of Antwerp, which hosts the largest Belgian sea and coastal freight water transport companies. Short-sea shipping is of higher importance than deep-sea shipping but passenger ferry activities, traditionally concentrated in Oostende, have declined significantly following the construction of the Channel tunnel.

European marine equipment manufacturing and industry, including propulsion, cargo handling, communication, automation, and integrated systems, is made up of around 7500 companies, the vast majority of which can be considered to be small to medium. Around 70 % of production is for export markets.

Coastal cities have to care about the quality of jobs but also the quality of life of those not engaged in employment, education or training or at risk of poverty. Although employment is considered to be the first factor of social integration, over 8 % of the EU population in employment is estimated to be at risk of poverty and the trend in income inequality remains a generally upward one. Cities have therefore to ensure that the risk of poverty is contained and exclusion is prevented, while investing in the interconnected spheres of education and training, labour and housing markets (Mega 2013).

Employment still remains the best defence against poverty and social inclusion but it does not eliminate the risk of social exclusion. Six in ten working-age Europeans at risk of poverty or social exclusion have a job and may be qualified as working poor. This exemplifies the need for an integrated approach addressing both social and employment situations. Working on a temporary contract, which is the case for 40 % of youths, is another crucial feature. Temporary contracts often carry a wage penalty, and this is a particular concern in countries where uncertainty and temporary work are high. Emerging and potential working poor must be understood from a household perspective, notably as regards the composition of the nucleus, the presence of children and the involvement of all adult members in employment.

The legacy of the protracted crisis continues to seriously affect the lives of many citizens. Labour underutilisation, in particular of vulnerable groups, such as single parents, older and disabled citizens and migrants, is a major cause of the risk of poverty or social exclusion concerning four in ten working-age Europeans. Labour market flexibility in the sustainable blue economy may offer an array of quality jobs and help the integration of more workers. In times of uncertainty, the multiplicity and complexity of social inclusion require structural reforms combining employment and social policies and addressing both competitiveness and fairness challenges (EC 2015).

6.2 Urban Coastal Communities and Intercultural Dialogue

Cities are the mirrors of societies. Social cohesion has an utmost value for sustainable cities, places of harmonious economic, gender, social, ethnic, racial, generational and cultural coexistence and cross-fertilisation. Social justice is a precondition for the creation and sharing of sustainable wealth. Diversity is not an obvious asset, since otherness and difference often raise feelings of suspicion, which may generate incomprehension and engender violence. Openness, understanding and mutual recognition and acceptance have made many thriving cities and are considered to be critical factors for sustainable prosperity.

A strong human and social capital is an appreciated high feature of all cities striving for sustainability, and it increases the urban capacity to withstand crises. Given the powerful relationship between urbanisation and capital accumulation, the city can easily become a social arena within which conflicts are often unavoidable. The role of equity in shaping sustainability processes is unparalleled. Unequal distribution of wealth may result in revolt, unsustainable lifestyles and obstacles to cultural change. The breadth, scale and impact of these tensions are very diverse. The escalation of conflicts, especially in times of crisis, may transform movements for the right to the city into urban uprisings and revolutionary actions (Harvey 2012).

The main social challenges for cities include the fair distribution of wealth and solidarity, the harmonious co-existence of more than three generations, high-quality employment and home environments, the creation of a welcoming and friendly environment for newcomers and the most vulnerable and disadvantaged citizens, public well-being and security, and the enhancement of opportunities linked to urban diversity. Its people living longer healthy lives can be a great chance for a city wishing to address the expectations of an older population, with the added benefit of the greater experience and contribution aiding in the quest to care for future generations.

Urban poverty is a very serious problem. According to the United Nations Development Programme Multidimensional Poverty Index, almost 1.5 billion people in 91 developing countries are living in poverty, with overlapping deprivations in health, education and living standards. Although poverty is declining overall, almost 800 million people are at risk of falling back into poverty traps. Environmental disasters and extreme phenomena linked to climate change may further aggravate their condition. Many people face either structural or life-cycle and ad hoc vulnerabilities that sap their potential for fulfilment (UNDP 2014).

Universal access to fundamental services, such as health and education, is crucial for strengthening human capital and preventing social exclusion and multiple inequalities. Those working for local services in everyday contact with people are best placed to see the persons behind the statistics. In the framework of the Europe 2020 strategy, a three-pronged approach addresses the three main dimensions of poverty and exclusion, including income poverty, material deprivation, and exclusion from the labour market.

The reduction and eradication of poverty is a strong priority for all sustainable development policies, especially in ageing societies. Poverty-trap conditions of low income and welfare-dependent structures can exacerbate exclusion and seriously compromise chances for advancing towards sustainable development. Increased financial pressures, in a complex and fragmented institutional environment, have to be addressed through the horizontal and vertical integration of decision-making systems and the optimisation of the contribution and commitment of the public, private and social economy sectors (Parkinson 1998).

Sustainable enterprises can play a great role for local development and social cohesion. Cities can increase awareness of the role of social economy and develop a favourable environment for enterprises which incorporate sustainability values and offer opportunities for the empowerment of minorities. The social enterprises, rooted in local realities, demonstrated greater resilience in the face of the financial crisis compared to other enterprises. This resilience is a good indicator of their capacity to sustain their economic activities or to engage in markets with growth potential.

Social cohesion reinforces the resilience of places and communities. A quartered city mirrors multiple divisions due to socio-economic developments, and new waves of migration. The combination of these factors often results in multi-fragmented spaces and functions. The decline of many urban port areas has led to distressed areas suffering functional impoverishment, with destitute housing, failing services and insufficient equipment, delinquency and crime, high unemployment, low mobility, little access to information, education and training, and high levels of substance abuse. Very often, transport infrastructures, whether operating or disused, further fragment the urban web and further isolate distressed spaces from vibrant urban areas. Comprehensive urban policies have to simultaneously address all these aspects and create a new momentum for attracting and integrating a diverse population.

As with all cities, coastal cities have to ensure the regeneration of their spatial and social fabric if they are committed to preserving their social capital. They must offer all citizens access to information, education and training, adequate housing and noble public, green and recreational spaces, social services and the possibility to participate in co-designing the future of the city in harmony with the sea. Vulnerable coastal cities must in particular inform citizens about the risks that the proximity of the sea involves and engage them in collaborative prevention design processes. Building social cohesion contributes to the creation of a resilient society, able to endure threats (Galbraith 1996).

Cultural diversity, a resource for the dynamism and energy of a city, can also increase social cohesion. Developing a sense of belonging and identity is crucial for cities wishing to adapt to change and enhance their diverse resources to attract investment. If cities manage diversity properly, they can benefit hugely from the potential of all minorities for entrepreneurship and innovation. For this, they should review the array of policies, services and instruments to create the appropriate conditions and governance structures.

The human face of cities is composed out of very diverse singular social attributes. Its richness springs out of the plurality of otherness. In coastal cities, many diverse races and cultures coexist and interact. It is crucial to promote intercultural dialogue in order to avoid "ghettoisation", a term invented in a coastal city, and prevent spirals of identity violence. The blue economy can generate many precious synergies which can be captured and harvested for the benefit of all citizens. Working together with civil society can create a climate which values diversity, openness and freedom by offering everyone the opportunity to share knowledge, exchange ideas and strengthen the elements that compose the marine identity of a city.

Intergenerational solidarity is an essential dimension of sustainable cities. The noble cause of sustainable development can offer a shared value to adhere and make a trans-generational contribution. The 2012 European year for active ageing and solidarity between generations suggested that, instead of creating conflict over scarce resources, ageing could be turned into an opportunity to enrich citizens with the possibility of contributing to community life and keeping contact across generations conscious and alive. Otherness of cultures can also enrich the dynamics of difference. Older scientists from other places have precious capital to transfer to the younger generations, as have older leaders and businessmen. Sea navigators and explorers are particularly appreciated as role models for younger generations.

The generation project, in Portugal, emerged from the need to improve the prospects of youths in the periphery of Lisbon. Older adults can invest time and experience in activities ranging from care for the very young, enabling parents to stay in full-time employment, to a number of programmes with children that encourage them to persevere in education. The "If You Keep Studying" project provides education and training in the fields of martial arts and music (EC 2010a). Ocean projects involving various generations could bring multiple benefits.

Intergenerational cities are those in which every generation brings its best to the service of all. This is a process that the city leaders have to manage, including the enhancing of the talents of the elderly in order to train the next generation of city leaders. The rights of the elderly are enshrined in the Treaty for the Functioning of the European Union. Active ageing policies should include supportive measures such as discouraging early retirement, stimulating learning and training to avoid skills obsolescence, adapting working conditions to the specific characteristics of older adults, valuing volunteering community work, promoting good health of older workers, and providing care for the elderly, while making them most valuable to the city. Universities for the elderly expand their offer for sustainable development to include a great variety of courses of interest to older adults. Coastal cities can sponsor courses that have a special interest for the marine communities and the exploration of the oceans.

Ethnic, religious and cultural diversity are extremely important for sustainable cities, since migration is one of the defining global social phenomena of the twenty-first century, with high potential and impact. A key feature of the increasingly interdependent world, it enables the exchange of talents, services, skills and experiences. An estimated 64 million immigrants live in Europe making up 9 % of the population, while in the USA, the immigrant population reached 13 % (OECD 2014).

The world population of migrants is expected to continue to increase and the majority of them are established in developed countries and especially in cities. Foresight studies emphasise that international migration will increase and counterbalance demographic dynamics. Migration offers a means to alleviate pressures on the social security systems of countries with rapidly ageing populations that face workforce shortages. An OECD study, however, suggests that, although global migration flows are very likely to rise, they would not suffice to offset the effects of population decline in OECD countries (OECD 2009b).

The emphasis on and public funds devoted to integration policies vary substantially across countries, despite a common need to support migrants' labour market integration in order to avoid possible long-lasting effects, notably on young migrants and native-born children of immigrants. Some countries continue to invest significant public resources in integration initiatives.

Most migration journeys choose metropolitan areas, cities and their suburbs, as destination. Consequently, the local environment and experience play a significant role in their settlement. The motivations and incentives, the opportunities and associated risks, and the potential for migration to benefit labour balances and development have been the subject of many analyses. A politically sensitive issue, migration requires public authorities to clear up stereotypes and misunderstandings and create a welcome culture. From refugees and asylum seekers to expats, the spectrum of immigration is increasingly diversified and defies standardised approaches (Cities of migration 2012).

Many migrants choose coastal cities, like New York, Boston and San Francisco, the largest continuous immigrant gateways in the US. London, Rotterdam and Amsterdam are gateways for Europe. Almost half of the population of Rotterdam is non-native residents. The labour and housing market situations are defining attraction factors, but also local openness and acceptance of new citizens. Long-term unemployment of migrants is, however, a serious challenge in many countries. The average unemployment rate of the foreign residents in OECD countries has increased by 5 % between 2008 and 2012, compared with 3 % for the native born (OECD 2013). An expanding blue green economy could provide an integration nexus for newcomers to cities by the sea.

Many coastal cities owe much of their wealth to foreigners. International statistics suggest that Dubai welcomes the largest immigrant population in the world, with more than 70 % of its population being foreign nationals. While this diaspora is predominantly from South Asia, there is also a significant western population. Vancouver is another notoriously multi-cultural city. More than 50 % of the Vancouver population is immigrants, many of whom have claimed Canadian citizenship. Many of these immigrants are of European or Asian origin, bringing with them a wide range of cultures and traditions.

Coastal cities can play a role in changing the perception of migrants, both in the societies of origin and destination, and forge bonds of solidarity among world communities. The religious and spiritual value of the sea can help in bridging differences. Social services and innovative projects involving young emigrants can have significant impact. Interfaith cultural centres and events, especially in disadvantaged areas, help minorities coexist harmoniously and interact peacefully. The organisation of exceptional events, for example, inter-religious open days, can stimulate dialogue among populations that do not usually communicate amongst each other.

After decades of established large cities drawing the majority of immigrants, new opportunities appear for cities like Barcelona, Copenhagen and Malmö in Europe. They can learn from Amsterdam, the European capital of tolerance, one of the best examples of urban openness and willingness to accept and integrate newcomers coming to be considered a source of wealth and progress. Since the city's golden age, political refugees, intellectuals and workers of every race and belief have always lived side by side with the local population and enriched all activities. Plurality and respect to the difference, key, elements in many civilisations, have played an important role for the prosperity of the city.

Openness and tolerance, understanding, solidarity and equal opportunities are key principles for integrating immigrants and enriching cities. The Barcelona Interculturality Plan, published in 2010, is the result of more than a decade of work by the Barcelona City Council. Intended to serve as a roadmap for the city's action to address the challenges of "coexistence in diversity", the plan initiated a different policy that focuses on the relationships and interaction between citizens, the dialogue and the mutual opening of both newcomers and long-term residents. The purpose is to create new foundations for community life which transcends cultural divergences.

If cities wish to attract talent in a competitive world, they should develop a welcoming culture. Boston created a Mayor's Office of New Bostonians to better recognise the resources and skills of immigrants and provide them with the necessary support to enhance them. The settlement and integration of newcomers is a fundamentally local experience and the quality of the welcome has a huge influence on their future success and ultimately on the prosperity of the cities. Bilbao developed a multicultural resource guide offering information on a wide array of city services, including healthcare, education and employment services. With the support of the Maytree Foundation, cities started exchanging their experiences in welcoming migrants and contributed to over 100 "Good ideas in integration" about cities which have been enriched by the energy and opportunity of immigration flows. Each good idea includes success steps and resources to help policy makers (Fundacion Bertelsmann and Cities of Migration 2012).

Many coastal cities have adopted active integration strategies and have mainstreamed them into their policies to strengthen their collaborative gene and aim for leadership which reflects the full diversity of the urban society in the public, non-profit, and private sectors. The number of languages spoken in a given city is becoming an important indicator of cultural diversity. It is estimated that 233 languages are spoken in London, a number indicative of the ethnic communities living in the British capital. Schools are the roots of integration, allowing for the experience of cultural and religious diversity facilitated by the natural open-mindedness of children (Fundacion Bertelsmann and Cities of Migration 2012).

Innovative multi-stakeholder partnerships can best address complex situations and groups in society, and stimulate bottom-up and mutual-trust approaches to help the integration of ethnic minorities. Schemes empowering poor immigrants and especially women, through micro-enterprise development and initiatives promoting sustainable value chains and training in vocational and business skills, are most promising. The Suikeroom project in Amsterdam is a fund for ethnic start-up companies financed by established companies. The fund, created in 2006, helps ethnic minority entrepreneurs, appraised as potential profit makers, successfully start a business. The young entrepreneurs receive guidance for preparing a solid business plan and are introduced to investors (Eurocities 2010).

Antwerp, the dynamic fashion city, integrated immigrants in its web of creativity through Betet Skara, a local weaving house employing Assyrian Christian refugees using traditional weaving techniques. As with most newcomers and political refugees, the Assyrian community arriving in Antwerp had to find sources of employment and create wealth. The professional craft of hand weaving, once a declining industry, is being revitalised by this local social enterprise, through techniques and weaving patterns passed on from generation to generation. They form an intangible asset which gave rise to a full enterprise, offering its services to fashion and interior design companies across Europe.

Education, housing, health, sports, culture and local politics are the sectors with the highest potential for the integration of migrants in the city (EC 2010b). The urban contexts are important and impact the ways in which immigrants interact with institutions such as schools and healthcare systems and contribute to the wealth of the city. Cities should invest in preventing tensions between spatial proximity and cultural distance, also linked to the coexistence of migrants with other very diverse established social groups.

Auckland is a city reputed to be very welcoming to immigrant populations. The city enjoys one of the highest quality environments in the Southern Hemisphere, with well-developed businesses and infrastructure. Many world citizens are attracted to

New Zealand's capital, which organises a great number of events to honour their origins and their cultures. The city enhances the universal languages of arts and sports for promoting communication and intercultural dialogue with the newcomers.

Socio-economic integration can start with music or dance, video or soccer. Such activities forge bonds among populations divided by mother tongue and tradition and help the integration of young migrants or refugees while fostering a sense of identity and belonging. In Auckland, soccer has been used as a strategic tool to reduce the social isolation of young refugees and to help them feel connected with each other and the wider community. The Refugees in Sport Initiative was launched in 2006 by "Refugees as Survivors", a non-profit refugee mental health agency enabling young refugees to achieve better access into mainstream sports and a safe place to meet with others sharing and understanding their experience. The programme has expanded to offer other sports such as cricket and martial arts and also to encourage girls from refugee backgrounds to participate in both team and individual sports (Fundacion Bertelsmann and Cities of Migration 2012).

In Heraklion, the project "I am Roma" tried to promote integration, raise awareness of Roma culture and rights, fight discrimination and improve interactions between Roma and non-Roma populations. The city developed concrete support actions to foster the social inclusion of Roma people, such as literacy, learning and mediation seminars. In 2006, the city set up a socio-medical centre which provides health, education and employment services to Roma citizens. Educational programmes were complemented with vocational and entrepreneurial training schemes. A series of actions were undertaken focusing on the awareness of Roma rights and culture through the media and education system. An antidiscrimination campaign targeted children through school events and intercultural discussions. Approximately 550 people from the Roma camp and 250 more Roma from the wider urban area benefited, while the local society gained a more harmonious social climate (EFUS and ICPC 2014).

6.3 Housing and Active Inclusion in Coastal Cities

Decent and sound housing is considered to be the second factor of social integration, after employment. As with all cities, coastal cities have to provide the conditions for good quality private and social housing, allowing for personal and collective fulfilment. In the path to sustainable development, housing schemes go through radical renewal to enhance ecological performance and become socially uplifting and architecturally compelling.

In the emerging world, global partnerships, such as Cities Alliance, link housing to the fight against poverty and bring together cities striving for urban poverty reduction and the promotion of sustainable development. The partnership features a broad range of stakeholders, including local authorities, national governments, non-governmental organisations, and multilateral organisations. The Cities Alliance's overall strategic objectives are to support cities providing effective local government, an active citizenship and an economy characterised by both public and private investment. The Cities Alliance seeks to achieve its objectives through national policy frameworks to address urban development needs, local inclusive strategies and plans, improved services to the urban poor and mechanisms to engage citizens in urban governance for sustainable development.

Before the turn of the millennium, the "Cities Without Slums Action Plan" was the first endeavour of the Cities Alliance, which proposed a target of improving the lives of 100 million slum dwellers by the year 2020, introducing for the first time such a measurable target in the international development arena. South African President N. Mandela launched the Action Plan, which was subsequently incorporated into the United Nations Millennium Declaration in 2000 as Target 11 of the Millennium Development Goals.

In industrialised cities, homelessness is still a serious challenge for all cities and there is no evidence that coastal cities suffer more than other cities (Eurocities 2009). In the USA, research by the Martin Prosperity Institute mapped homelessness across the 100 largest US metro areas. The surveys on Hunger and Homelessness by the US Conference of Mayors highlight that most of the surveyed cities reported that requests for emergency food assistance increased over the years and that more than 10 % among those asking for assistance were homeless. Unemployment led the chorus of causes of hunger as reported by the cities, followed by poverty, low wages, and high housing costs. A great majority of the cities reported that their total budget for emergency food purchases increased over the last year. And still, across the surveyed cities, 27 % of the people needing emergency food assistance did not receive it (US Conference of Mayors 2013).

Los Angeles and New York alone accounted for almost one in five homeless people in United States, with more than 12 % in Los Angeles and more than 6 % in New York. The 2014 US annual homelessness assessment report to Congress stated that 610,042 citizens were homeless on a given night in January 2013, and more than 100,000 suffered from chronic homelessness. Almost one quarter of all homeless were children and most were living in emergency shelters or transitional housing. The situation has much improved since the implementation of the "Opening doors: Federal strategic plan to prevent and end homelessness". Total homelessness declined more than 6 %, and this was driven by the reduction of the unsheltered chronic homeless (Department of Housing and Urban Development 2014).

Some pioneer cities have given the homeless the right to use the details of the municipality as their postal address. Other cities allowed solidarity action to extend the possibilities of housing. At the other end of the spectrum, many cities use the criminal justice system to punish those trying to survive on the street. In response to the homelessness crisis, the US National Law Centre on Homelessness and Poverty and the National Coalition for the Homeless released the report "Homes Not Handcuffs", identifying and pinpointing the criminalisation of homelessness. A list of the meanest cities has been proposed to name and shame cities which have even enacted food sharing restrictions that punish groups and individuals for serving homeless people. Many of these measures appear to have the purpose of moving the homeless out of sight, if not out of the city.

In many developed coastal cities, waterfront spaces are usually considered to offer privileged housing spaces and have a higher location value if surrounded by clean sea environments. Mixed waterfront residential neighbourhoods expand on often artificial land. Artificial islands are already common extensions of overcrowded frontline cities and harbouring airports, but dredging activities can bring considerable pressure on marine ecosystems. Innovative sustainable housing on artificial islands can be a response to severe housing shortages in coastal cities. This was the case of Ijburg, one of the latest neighbourhoods in Amsterdam. It is a unique urban planning experiment that consisted of creating a neighbourhood from scratch. Ijburg is composed of an archipelago of seven artificial islands responding to the Dutch tradition and culture of living with and on the water.

The vision for Ijburg was of a diversified urban neighbourhood in symbiosis with the surrounding nature and the sea. The initiation of the project faced strong opposition by environmental activists on the grounds that it would be a threat to the ecological balance of the area. The design of the project adopted an ecologically sensitive approach through the involvement of environmental groups in the conception and planning process and ensured an engaging outcome. Ijburg was granted final approval in 1997. The district was developed using an innovative public-private partnership. The private partners committed to purchase land in advance to guarantee the future success of the neighbourhood, developed with a relatively high number of houses that are suitable for families and children. The diversity of building and architecture styles has been achieved with two islands reserved for plot-based housing development. Parcels were acquired by individual citizens developing their properties with the architects of their choice.

The creation of land on sea and the endowment with infrastructure, new roads, bridges, and services was facilitated through public funding. A new tram line was also extended to connect Ijburg to Amsterdam's city centre. Ijburg residents have access to the fastest public fibre optic broadband Internet. Amsterdam Smart City is working with them to develop new products and services, such as education, healthcare, transport, and energy, and improve the local quality of life.

New neighbourhoods offer ample opportunities for improved environmental management. The Beddington Zero Energy Development in the London Borough of Sutton, or BedZED, has been the UK's ultimate eco-district, the first housing development designed to be environmentally friendly. A multi-award winning project, created in 2000–2002 by BioRegional, BedZED is owned and managed by Peabody, one of London's largest housing associations. The development encourages energy efficiency and use of renewables, public transport, cycling, and walking, and asks for limits to private cars and parking space. The residents' car mileage is 65 % less than the UK average (Bioregional Solutions for Sustainability 2009).

Bioclimatic architecture was well enhanced in BedZED. All houses face south to take advantage of solar gain, are triple glazed, and have high thermal insulation. Low-impact materials were used for the building, selected from renewable or recycled sources within 35 miles of the site, to minimise the energy required for transportation. The project is designed to use only decentralised renewable energy. Tree waste fuels the development's cogeneration plant to provide district heating and electricity. Most rain water is collected and reused. Appliances are chosen to be water-efficient and to use recycled water when possible. A review of the BedZed development in 2010 drew mainly positive comments from the residents and neighbours. It has also highlighted some negative facts, including the reliability problems of the combined heat and power system due to technical failures and the intermittent schedule of operation imposed by the local authority, the insufficient cleaning of the water by the "living machine" water recycling facility and the still high, ecological footprint of the residents, which is, however, about to become half of the UK average ecological footprint.

6.4 Fit Marine Cities, Public Health and Safety

Healthy and fit cities provide the living cells for a healthy planet. In harmonious symbiosis with healthy seas, coastal cities can generate virtuous circles and reverse cycles of environmental degradation, physical isolation, obsolete infrastructures and neglected public spaces, which go hand in hand with poor education, unemployment, weak health and public participation, crime and violence. Fulfilling homes and welcoming public spaces generate dynamics for good education to work patterns, public health and participation.

Coastal cities could be inspired by the US place-based programmes of choice and promise neighbourhoods. People living in fragile communities stamped by poverty have a greater risk of being excluded. Policy interest in comprehensive community well-being initiatives has surged with two federal programmes, the Department of Education's Promise Neighbourhoods, modelled on the acclaimed Harlem Children's Zone, and Choice Neighbourhoods by the Department of Housing and Urban Development. They both intend to address vital local needs with the active participation of the targeted population of all generations.

Promise neighbourhoods tried to foster a culture that creates a "cradle-through-college-to-career" continuum and promotes multiple positive effects across sectors. Neighbourhood students have to be safe, healthy and improve their life and career prospects, through coordinated community efforts. Choice neighbourhoods focus on communities dependent on social housing to be transformed into communities of opportunity with good-quality affordable housing and high-performing schools and services. The contextual dynamics of the local communities are crucial in shaping the transformation effort and capturing the longer-term benefits (Urban Institute 2011).

Coastal cities can play a major role for the well-being of their citizens, through preventing and combatting diseases and forging healthy lifestyles. Contemplating the sea is already a source of mental relaxation and brings peace and well-being. Good quality seafood could help change eating patterns and address, for example, the epidemic of overweight and obesity, a critical public health issue. Obesity can lead to serious and chronic diseases and is a major contributor to increasingly unaffordable public health care costs. Studies concluded that obesity, diabetes and hypertension seem to increase in tandem with rapid urbanisation and pointed to "obesogenic", predominantly urban environments as a risk factor. Many urban activities often promoted convenience and minimal physical effort. The dense urban transport network, the easy-to-access services, and the absence of adequate recreational spaces, parks and athletic facilities do not invite citizens to exercise (Institute of Medicine 2012).

Coastal cities could become pioneers in initiating programmes promoting sea nutrition and physical activity through nautical sports. Public beaches accessible to all could help cities in making physical activity and swimming an integral part of everyday life. They could also make creative partnerships with the private sector and local businesses to create healthy and, if possible, marine food environments that ensure that healthy options are the routine, easy choice in public places. Last but not least, they should partner with schools in which so many health and environmental projects begin, but also with local chefs and athletes.

Schools could be perfect focal points for obesity prevention, ensuring that all students have adequate opportunities to engage in physical activity and access to nutrition literacy, as well as enhancing access to the sea as being access to health and well-being. Marine communities have to be empowered to work for change and cities can support their efforts in many ways to achieve and maintain a healthy weight. Safe public places to walk or play, outlets offering affordable healthy food, and awareness campaigns for healthy foods and beverages can encourage citizens to engage in healthier lifestyles and achieve and maintain a healthy weight (Institute of Medicine 2012).

Coastal and port cities are places of large populations in transit. This could make them more susceptible to the spread of infectious diseases. During the last two decades, Singapore suffered from two major epidemics, Severe Acute Respiratory Syndrome and H1N1, both imported. Three months following its identification in Mexico, the 2009 H1N1 influenza virus became a global pandemic affecting practically every country. In Singapore, researchers tracked the transmission of the virus by looking for antibodies, a type of immune response to infection.

Well-being for the ageing is very important in cities which must prepare for increases in their older populations. In Bangkok, the Health Department set up operation plans and projects to care for the ageing population. The projects are steered by 68 Public Health Centres located in 50 districts in Bangkok. The Public Health Centre in the Nong Khaem District, bringing together medical doctors, dentists, medical scientists, pharmacists, nurses, nutritionists, psychologists, general staff, and elderly care volunteers, is emerging as the first Public Health Centre. Its official operation was initiated in 2011 with the concept of long-term care treatment for the elderly in three phases, including the primary phase (for ages 60–69), secondary phase (for ages 70–79), and tertiary phase (for ages over 80). Elderly citizens in the surrounding communities approved the scheme, which also welcomes experts from other cities wishing to shape projects for older adults (Asian-Pacific City Summit 2014).

A healthy city is a city that puts promoting public health high on its political agenda for the well-being of its citizens. Public health is not merely the absence of epidemics or disease, but is a state of complete public physical, mental and social well-being. The Healthy Cities movement, launched by the World Health Organisation in 1987, brings together cities which have developed and implemented a wide range of policies, including health profiles and urban health strategies through community initiatives and programmes that address the needs of citizens, cultural lifestyles, environmental health, and sustainable development.

The Healthy Cities movement aims at improving health for all and reducing health inequalities, and also improving leadership and participatory governance for health. The Zagreb Declaration expressed a strong commitment by the political leaders of cities to bolster and champion action on health, equity, sustainable development and social justice. It highlighted continuing values and priorities and identified new challenges and approaches for cities to address and adopt as they work to protect and enhance the health and well-being of all their citizens (WHO 2009).

Interconnected healthy cities offer a pragmatic framework for concrete cooperation and progress at the local level. Healthy Cities facilitate urban health promotion through projects, training and seminars, many of which link to the sea. Benchmarking between the network cities in relation to many different public health themes is promoted. The "Cities for a Healthier Sea", conceived in 2010, in close connection with the Baltic Sea Challenge campaign, promoted voluntary water protection measures in the Baltic cities.

Urban safety is a critical dimension of well-being in cities. The European Forum of Urban Safety (EFUS) shed light on the interconnected issues of Cities, democracy and safety. The "Cities' manifesto for Safety and Democracy", adopted in Naples in 2000, expressed the desire for quality cities, defined as safe, vital places of harmonious development and immune to insecurity, violence and extremism. Recurrent issues include violence against vulnerable social groups and sensitive places like schools and streets and innovative prevention measures. The exchange of experience and cooperation are judged essential for guaranteeing the legitimate right to safety (EFUS 2003, 2006).

In Tanzania, the Dar es Salaam Safer Cities project tried to raise awareness and strengthen the capacity of local authorities to manage urban safety and reduce crime, including violence against women, children, and other vulnerable groups, and to create a culture of crime prevention and safety at the neighbourhood level, initiated by the local communities affected by insecurity. All stakeholders worked together to develop local solutions to prevent crime and build awareness among communities. The first phase of the project identified youth crime and violence against women as priority concerns. Victimisation surveys and a young offender survey provided insights for the conception of an integrated crime prevention strategy. The strategy encompasses law enforcement, including the creation of the City Auxiliary Police, the development of neighbourhood watch groups and the Ward Tribunals to expedite minor offences, social and economic issues including job creation and skills training, cultural and recreational activities for youth at risk, and improvements of the local environment (EFUS and ICPC 2014).

The project strengthened and embedded institutional capacity and skills in crime prevention. The neighbourhood watch groups were engaged by the community for neighbourhood watch, after some basic militia training from retired officers. The project enhanced security in the local society and promoted women's safety and empowerment by training community members to conduct the safety audits and implement recommendations. Continuous public awareness campaigns also helped to create a culture of adherence to the law and partnerships in local innovations for crime prevention.

In Rotterdam, the City Marines scheme achieved impressive results in improving conditions in the most unsafe neighbourhoods. It is a local approach that combines authority and law enforcement, mobilisation of stakeholders, and mediation. A City Marine is an experienced respected senior person acting as a special delegate of the mayor with the responsibility for tackling problems in the most unsafe neighbourhoods. Appointed directly by the Mayor of Rotterdam, each city marine reports to the "Safety steering group" comprised of the mayor, attorney general and chief of police. The city marines have the authority to introduce change and access funds, and also work on mobilising the community, making partnerships between local stakeholders such as the employment service, housing agencies or businesses, to bring immediate solutions and produce quick results (EFUS and ICPC 2014).

The well-being of children is a litmus test for the future of urban societies. Each place can make a difference. Children's use of space and time has to be valued and enhanced. Environmental quality is important to children's socio-emotional and mental health, cognitive development and respiratory health. Children with attention deficit disorders can benefit from activities in green playground areas or at the seashore. The sea experience seems to have a well-being effect both as a permanent residential horizon and as a holiday destination (Children's Society 2014; OECD 2009a).

Sea and coastal sports can make a major contribution to the well-being of urban populations. The Ellen MacArthur Cancer Trust takes young people, aged between 8 and 24 years old, sailing to help them regain their confidence, on their way to recovery from cancer. The idea came in 2000, when, some months before the start of the Vendee Globe Round the World race, Ellen MacArthur went sailing in France with a group of children suffering from cancer. Supported by a number of dedicated volunteers, the trust celebrated 10 years in 2013. The sailing trips formed the platform for the start of a 3-year plan to double the amount of young people in recovery from cancer.

Travelling far beyond the school is a dream shared by most children and the navigators often allow them to be part of that dream. The shared maritime heritage can weave a strong linkage among citizens. Marine heritage routes, like the Route of the Rum or the Silk Road, bring together far more than the citizens of two coastal cities united by an ocean. The Route of the Rum, a transatlantic single-handed yacht race on more than 6500 km between Saint Malo, in Brittany, and Pointe-à-Pitre, in Guadeloupe, taking place every 4 years in November, is a much-anticipated event, followed by a great number of citizens, sponsors and the media. The route completed ten events in 2014, with a record number of 91 competing yachts, and has attracted much attention to the splendours and dangers of the amazing seas.

The Maritime Silk Road is another cross-cultural concept which can contribute to the future blue economy. The 21st Century Maritime Silk Road International Expo in China in 2014 was marked by high expectations. "From Macao we begin" was the emblematic title of the 2014 Global Tourism Economic Forum, building on the unique role of Macao in the ancient maritime Silk Road, linking China to the Middle East and East Africa more than 600 years ago. It is part of an ambitious project to reinvent the Maritime Silk Road and attract a top-level cruise tourism industry. The aim is to invigorate the maritime business and generate jobs in sustainable tourism, which can lead growth, particularly in the 25,000 islands of the Southeast Asian countries. Although Asia Pacific is leading the world tourism growth, its great potential for cruise tourism has yet to be realised. The concept was rebranded with a vision for improving maritime connectivity, enhancing regional cooperation and reinforcing the economic belt along the 21st Century Maritime Silk Road.

References

Asian-Pacific City Summit (2014) Newsletter No. 27, September 2014

Bioregional Solutions for Sustainability (2009) BedZED seven years on. The impact of the UK's best known eco-village and its residents. Bioregional Solutions for Sustainability, London

Children's Society (The) (2014) Ways to well-being: exploring the links between children's activities and their subjective well-being. Children's Society (The), London

Cities of Migration (2012) From practice to policy: lessons from local leadership on immigrant integration. Maytree Foundation, Toronto

City of Boston (2011) A climate of progress. City of Boston climate action plan update 2011. City of Boston, Boston

Committee of the Regions (CoR) (2012) SME friendly regions and cities. Survey of EU local and regional authorities. CoR, Brussels

Department of Housing and Urban Development (2014) Annual homeless assessment report to congress. Department of Housing and Urban Development, Washington, DC

Eurocities (2009) Intercultural cities—a journey through 23 European cities. Eurocities, Brussels

Eurocities (2010) Cities accommodating diversity. Eurocities, Brussels

European Commission (EC) (2010a) Empowering people, driving change: social innovation in the European Union. EC, Brussels

EC (2010b) This is European social innovation. EC, Brussels

EC (2014a) Blue energy. Action needed to deliver on the potential of ocean energy in European seas and oceans by 2020 and beyond. EC, Brussels

EC (2014b) A European strategy for more growth and jobs in coastal and maritime tourism. EC, Brussels

EC (2015) Employment and social developments in Europe 2014. EC, Brussels

European Forum for Urban Safety (EFUS) (2003) Securities, schools and cities. EFUS, Paris

EFUS (2006) The cities' manifesto for safety and democracy. EFUS, Paris

European Forum for Urban Safety (EFUS), International Crime Prevention Centre (ICPC) (2014) 100 promising practices on safer cities. EFUS/ICPC, Paris

European Wind Energy Association (EWEA) (2014) The European offshore wind industry—key trends and statistics 2013. EWEA, Brussels

Fundacion Bertelsmann and Cities of Migration (2012) Fifteen good ideas. Diversity and inclusion in cities. Fundacion Bertelsmann and Cities of Migration, Toronto

Galbraith JK (1996) The good society: the human agenda. Houghton Mufflin, Boston

Global Ocean Commission (GOC) (2014) From decline to recovery. A rescue package for the global ocean. GOC, Oxford

Harvey D (2012) Rebel cities: from the right to the city to the urban revolution. Verso, New York

International Labour Organization (ILO) (2012) Working towards sustainable development. Opportunities for decent work and social inclusion in a green economy. ILO, Geneva

ILO (2014) Global employment trends 2014. ILO, Geneva

Institute of Medicine (The) (2012) Accelerating progress in obesity prevention: solving the weight of the nation. IOM, Washington, DC
Mega V (2013) Quintessential cities, accountable to the future. Innovation, sustainability and leadership. Springer, New York
New York City Mayor's Office of Long-Term Planning and Sustainability (2012) NYC local law 84 benchmarking report. PlaNYC, New York
Organisation for Economic Co-operation and Development (OECD) (2009a) Doing better for children. OECD, Paris
OECD (2009b) The future of international migration to OECD countries by 2030. OECD, Paris
OECD (2012) The jobs potential of a shift towards a low-carbon economy. OECD, Paris
OECD (2013) International migration outlook. OECD, Paris
OECD (2014) International migration outlook. OECD, Paris
Parkinson M (1998) Combating social exclusion: lessons from area-based programmes in Europe. Policy Press, London
UN Development Programme (UNDP) (2014) Human development report 2014. Sustaining human progress: reducing vulnerabilities and building resilience. UNDP, New York
(The) Urban Institute (2011) How to evaluate choice and promise neighbourhoods. Perspectives brief 19. (The) Urban Institute, Washington, DC
US Conference of Mayors (The) (2013) Status report on hunger and homelessness. US Conference of Mayors (The), Washington, DC
World Bank (2012a) Inclusive green growth. The pathway to sustainable development. World Bank, Washington, DC
World Bank (2012b) Toward a green, clean, and resilient world for all. A world bank group environment strategy 2012–2022. World Bank, Washington, DC
World Bank (2012c) Cities at the crossroads: unlocking the potential for green urban growth. World Bank, Washington, DC
World Health Organisation (WHO) (2009) Zagreb declaration for healthy cities. WHO, Copenhagen

Selected Internet Links

www.citiesofmigration.ca
www.fas.ie
www.socialexclusionunit.gov.uk
www.ilo.org/greenjobs
www.globalcompactfoundation.org
www.who.dk/healthy-cities
www.urbansecurity.org
www.efus.org
www.union-habitat.org
www.socialedge.org
www.ville.gouv.fr
www.youngfoundation.org
www.interculturaldialogue.eu
www.intercultural-europe.org/
www.coe.org
www.healthycommunitiesinstitute.com/
www.theschmidt.org
www.ellenmacarthurcancertrust.org/

Watercolour 7
Istanbul: Multiple Tales on the Coasts of Two Continents

Chapter 7
Cities by the Sea: Ideas and Innovation, Science and the Arts

Abstract Cities are the undisputed intellectual and cultural epicentres of countries, continents and the world. Coastal cities benefit from direct access to the extraordinary beauty and unfathomable diversity of marine life. The bond between the city and the sea is often indissoluble and has a high impact on urban cultures and identities. The urban marine heritage, also including underwater heritage, is as important for humanity as the exceptional natural sites. The proximity of the sea has often inspired citizens and cities to invest time and energy into new discoveries and artistic creation, transforming everyday environments into unique experiences. A sustainable city has to cultivate the seeds of freedom that science and the arts need and offer a space to all for expressing their creativity and enhancing their abilities and skills.

Coastal cities are gateways to both land and the seas. This chapter examines the role of intellectual and cultural resources as key assets for sustainability. Marine cities, with their coastal and offshore dimensions, represent very fertile soil for the incubation of innovation and prosperity. Knowledgeable coastal cities invest in education and ingenuity, observation, surveillance and exploration of the seas, ocean literacy and international cooperation for excellence. The chapter presents a spectrum of inspiring actions to reinforce urban identity and make citizens proud of their heritage on land and sea.

7.1 Coastal Cities of Education, Science and Knowledge

Coastal cities with a propensity for new ideas and actions are particularly attractive for free thinkers, scientists, innovators and artists. Their outward-looking geography exposes ports and coastal communities to the world. This often results in a particular openness to the unknown, which often goes hand in hand with freedom of ideas and lifestyles. The access to very diverse resources, infrastructures and institutions, offers possibilities for very diverse experiences and happenings. R. Florida lists scientists and innovators among the "creative class" of mobile and talented thinkers that a city must attract and who create the cultural urban lifestyles (Florida 2005).

V.P. Mega, *Conscious Coastal Cities*, DOI 10.1007/978-3-319-20218-1_7

The clustering of marine institutions, universities and scientists in coastal cities can spur sustainable development upward. For cities seeking wealth through science, technology and industry, the anchoring presence of private research and development laboratories and an ecosystem of SMEs can bring huge benefits. Moreover, ports are associated with innovation in maritime sectors. Nine out of the ten world regions with the largest amount of patent applications in shipping are home to one or more large global ports, including Los Angeles/Long Beach, Tokyo, San Francisco/Oakland and Rotterdam (OECD 2013).

Marine and maritime education and training are mainly orchestrated by coastal cities. Coastal cities often develop their marine research and innovation capabilities, investing in infrastructures of marine labs and associated facilities and building networks of world excellence. In 1983, the IMO established the World Maritime University in Malmö, as a centre of excellence for maritime post-graduate education and research, promoting the highest standards in disciplines including maritime law and policy, maritime safety and environmental management, education and training, shipping and port management, marine and ocean management, and international maritime transport and logistics.

Most research activities are concentrated around major metropolitan areas. Silicon Valley and European high-tech hubs like Cambridge and Catalonia demonstrate that places play a key catalytic role. They form the spatial horizon bringing together innovation actors to produce higher sustainable value. Knowledge-based blue growth poles, innovation clusters and science parks can attract academic and research institutions, innovative enterprises and business incubators, local organisations and support agencies which cooperate and mutually reinforce regional creativity through strong synergetic links.

Coastal urban conditions can be instrumental for the incubation of innovations and the generation of wealth from science and technology. The size and diversity of a city are very important for the breeding and capturing of synergies, as is leadership and governance. It seems that new patents are granted disproportionately to larger urban centres. Great metropolises such as London, Tokyo and New York are science strongholds, even though their economic strength comes from other areas, such as financial assets. Many experts suggest that being part of the urban mosaic benefits from the presence of enterprises and intense education-research-innovation interactions. Planned science cities have also been conceived to attract a great number of eminent scientists and create multi-beneficial synergies.

In Japan, Tsukuba Science City represents one of the world's most ambitious planned cities. About 50 km northeast of Tokyo and easily accessible from the Japanese capital, the science city encapsulates Japan's most advanced technology. Tsukuba Science City is segmented into the Research and Education District and the Suburban District. These two districts host approximately 220,000 people and more than 300 public and private institutes, universities and firms. The unique features of Tsukuba Science City include the world's largest planetarium, the Science Museum of Map and Survey, whose gigantic parabolic antenna is the landmark of Tsukuba, the Space Centre, the Science Square, and Tsukuba Botanical Garden. Over the past decades, nearly half of Japan's public research and development budget has been attracted by Tsukuba. Important scientific breakthroughs by its researchers include the identification and specification of the molecular structure of superconducting materials, the development of organic optical films that alter their electrical conductivity in response to changing light, and the creation of extreme high-pressure vacuum chambers. Tsukuba has become one of the world's key sites for public-private partnerships in fundamental research, in a broad array of fields, from earthquake safety to microbiology and plant genetics.

Science institutions and port cities often cooperate closely to increase the profile of the diptych city-port. Erasmus University has supported its maritime and port-related research and education in Erasmus Smart Port Rotterdam, an interfaculty centre of excellence. The centre is active in port-related issues, such as port sustainability trajectories, synchromodal transport networks, a flexible and integral deployment of different modes of transport on a network to better satisfy demand, sustainable global supply chains, and inter-organisational systems in logistics. The centre is also a resource for the Rotterdam seaport cluster, including enterprises active in many sectors, such as, transport, logistics, offshore and shipbuilding. The competitiveness of the port and the city of Rotterdam depend on their performance in the framework of the education-research-innovation triangle.

A great city is often praised for its ability to create, retain and attract new talent. Successful cities are considered the ones which provide business and entrepreneurial dynamism, but also cultural and educational opportunity and quality of life. Barcelona ranks high on these fronts. Through extensive public investment and strong strategic planning, Barcelona has reinvented itself since it reconciled with the sea in preparation for the Olympic Games of 1992. The city's Entrepreneurship Centre, created in 2004, serves as a reference for existing and potential entrepreneurs, as well as a hub to promote entrepreneurship and innovation. The centre provides advice and services, and a "pre-incubation space" allowing potential entrepreneurs to design their projects, from their business ideas to the creation of their companies. Barcelona Activa, the local development agency of the municipality for employment, business and the economy, is supporting enterprises in job creation.

In 2014, Barcelona was the first city to be recognised as a European Capital of Innovation ("iCapital"). The European Commission awarded this prize to Barcelona for "introducing the use of new technologies to bring the city closer to citizens". In 2011, Barcelona's city council launched the "Barcelona as a people city" project by introducing the use of new technologies to foster economic growth and citizen welfare through open data initiatives, offering valuable information to individuals and private companies, sustainable initiatives on energy and mobility, social innovation, alliances between research centres, universities, private and public partners, and better "smart services" exploiting the frontiers of information and communication technologies.

The transformation of the physical space was paired with efforts to create a new knowledge and innovation hub. After studying the city's economic and geographic advantages, leaders of 22@Barcelona decide to pursue five industry clusters, including media, medical technologies, information and communications, energy, and design. City authorities deployed many efforts to attract universities, institutions, companies and organisations to the district, and also created meeting and residential spaces for specific industries, a technology centre, incubators, and spaces for students and workers.

The Barcelona urban lab was created to foster the use of the city as an urban laboratory. Through this instrument, the city welcomes companies with innovative projects at the pre-commercial stage wishing to test their services for the future in a real city environment. The Urban Lab can facilitate the use of public spaces in the city of Barcelona to carry out pilot programmes on products and services promising a high potential impact. The Urban Lab acts as a gateway to the City Council and stimulates internal coordination among the different areas. It allows for the creation and testing of pilot programmes, funded by the companies which conceive them, and demonstrates an express need to use public spaces and infrastructures to meet the Barcelona City Council's expectations. The candidate programmes must prove a direct impact on the functioning or on the services provided by the city, and ultimately the benefit to residents. SynergyS22@Barcelona connects the global knowledge community with industry, in order to generate new business opportunities and provide companies with free access to forefront research and new emerging technologies.

In Copenhagen, the MindLab is a cross-governmental innovation bringing together citizens and businesses to develop public innovations and improve public services. The design of any public service has a fundamental component: it puts the citizenry at the centre and tries to understand and anticipate its needs. Citizens' involvement at the earliest stage is a most common denominator in the design of public innovation services. In order to truly enable the public sector to change, it is necessary to reinvent the goals and character of overarching governance models. Moving from the traditional delivery of expected services to the mobilisation of citizens' resources is essential to advancing from the transactional understanding of public welfare to a civic relational experience.

Physical links and their interrelated intangible dimensions can be decisive for the future. The construction of the Øresund Bridge connecting Copenhagen with Malmö, in 2000, brought multiple benefits to both cities and countries. Sweden gained a physical connection to mainland Europe, which impacted socio-economic and cultural cooperation and scientific synergies. The bridge encouraged the establishment of the "Øresund region", a constellation of nine universities, in both countries from both lands. Co-authorship between Copenhagen and the Southernmost province of Sweden has more than doubled in the ten first years of the millennium.

Science and innovation cities invest in the power of ideas and the capacity of their citizens to envision a better future and bring it into being. They try to incorporate key knowledge in all their functions and promote public engagement in science and a culture of change. The Galway Science and Technology Festival, founded in 1998, tries to bridge the gap between education and industry and encourage students to study Science, Technology, Engineering and Mathematics (STEM) subjects at third level. The Festival runs annually for 2 weeks and culminates with a demonstration of best practice from primary to post graduate education, multi-national and local companies. The Marine Institute joins the event and holds open days, inviting secondary school students to discuss the exhibits on display and review opportunities regarding careers in marine engineering.

In 2003, Genoa initiated a Science Festival which offered a stage and a voice to all citizens. Making science accessible is the essence of the festival, inviting all and everyone to a world of excitement and wonder. In Boston, Harvard University, the Massachusetts Institute of Technology (MIT), the Boston Globe, and Massachusetts General Hospital have pulled together their convening powers to bring thought leaders, innovators, and creative disrupters together for a week of seminars, "hackathons", and other events to address global challenges in autumn 2015. Academic and medical institutions are sites of change, places devoted to curiosity, knowledge and progress.

Boston has had to reinvent itself many times in the course of its history. Scientific distinction is the latest in a series of economic revivals, from being the largest city in early colonial America, to a port for global shipping and sailing in the nineteenth century, before attaining its current position as a biotech and innovation hub. The prime world universities, Harvard and MIT, are vital resources in the region, along other universities that have developed specific technology expertise. The Harvard innovation lab tries to foster entrepreneurial activities and deepens interaction among Harvard students, faculty, business and professionals of the Greater Boston community. The Harvard ecosystem has brought a stream of innovations, including the pacemaker, surgical anaesthesia, Facebook, and even breathable chocolate. MIT is a leading institution for technology business creation.

The democratisation of access to knowledge is considered very important for fostering informed public discourse and the capacity for shared civic projects. Education measures to enhance paths from cradle to career are given full attention to make the best possible use of endogenous resources and project them to the world. "Innovate locally, exchange regionally, export globally" became a worthy paradigm for the city's world class human capital and its open and dynamic civic culture (Boston Foundation and the Citistates Group 2004).

The vision for a world-leading city-state has been proposed for Boston after interviews with opinion leaders and observers from government, academia, business and advocacy groups. The vision highlights that it is hard to think of a twenty-first century city-state so ideally positioned as Boston in the "century of the intellect". The suggested decisive factors are smart ingenuity and collaborative leadership, necessary for Boston to address critical challenges of global warming, public health, energy and water supply, shortfalls of knowledge workers, lack of affordable housing and weakened civic ties. Recognising waterfront heritage as a diminishing asset is crucial for encouraging selective preservation along with quality urban design.

The emerging countries also provide outstanding examples of bridges between education and the future of places, often orchestrated by civil society. In Bangladesh, extreme heavy rain is a frequent problem during the monsoon season and hundreds of schools have to occasionally shut down. To address this, the non-profit organisation Shidhulai Swanirvar Sangstha has developed 54 solar-powered boat schools that provide uninterrupted education all year round for children in flood-prone areas. On-board solar PV modules generate electricity for lighting, internet-linked computers, and video projectors. Many of the boats use solar PV to charge batteries for portable solar home lighting systems. Each boat first serves as a school bus, making periodic stops to collect children from riverside points; it then docks and serves as a classroom providing basic primary education for 30 students. As part of the scheme, Shidhulai introduced a river-based environmental curriculum which educates children in environmental protection and water conservation. The floating boats also provide complementary community services, such as libraries, health services, solar workshops, training and adult education.

7.2 The Ladder of Excellence and Urban Public Innovations

Open responsible innovations interwoven into the fabric of marine cities can lead to quantum leaps. Cities are untapped reservoirs of ideas, labour and commitment. They grow by processes of gradual diversification and differentiation. Innovations can bring dramatic and thorough changes that force conventional frontiers and open up the spectrum of possibilities for a new concept or idea, product or service, process or order to bring about the desired transformation. Invention is often identified with the research and development phase, while innovation includes all the politics and processes of its implementation until the final transformation. Raising the ladder of progress requires enlightened leadership, strategic foresight, a converging and shared vision, and integrated planning and policy to make the vision come true.

The need for sustainability transition is a very strong driver for coastal cities, which face risks from storms and sea level rise. Innovation is an evolutionary process, often involving an impressive array of alternatives, and should be distinguished from pure evolutionary change within the established rules and procedures, or linear optimisation processes. The search for sustainable development requires innovations and alliances mobilising collective intelligence. Innovation progressively leads to a policy response, among many possible options, to the multiple opportunities offered by invention, research and technology. The sustainability challenge is to best manage change and ensure an overall beneficial outcome for present and future generations (BURA 1997).

Cities can stimulate, generate, test, adapt and integrate all kinds of innovation, including product, process and service innovations, social, public and open innovations. Each innovation constitutes a dynamic, which can be very powerful and open new creative trajectories. Innovation can be a highly political process and governments have a broad range of ways to influence this. They can promote innovation directly by investing in research and development infrastructures and activities, and promoting sustainability standards and labels. Indirectly, governments can influence innovation by purchasing blue green products and services and creating new markets for sustainable development.

Cities which leverage innovations and weave them into effective patterns can change the way that they function and improve their performance. Coastal cities should enhance innovations to explore their endless frontiers on land and sea and optimise the condensed knowledge and information that the proximity of the sea confers upon them. Innovation might also be the result of a struggle for survival, for example, after the decline of traditional maritime industries. Crises have forced many cities and their stakeholders to take a hard look at reality, assess drivers and obstacles to change, engage with complexity, uncertainty and risks, and, finally, incubate pioneer actions for sustainable development generating multiple benefits.

Innovations for the conversion of waterfront areas for activities of the future were prompted by previous waves of crises. City-centre harbours have disappeared, leaving behind the husk of an infrastructure inadequate for the future. Disused docks were offered a new life when turned into art exhibition halls, shops, craft workshops and centres for cultural business activities. Changes in the built fabric prompted the transformation of the social fabric. The redesign of the places and the (re)conception of new activities generated many new jobs after waves of dramatic unemployment caused by the decline of the old activities. Careful planning with citizen consultation can ensure that waterfront developments are not reserved as luxurious office and leisure areas but become organic parts of cities. Innovation combined with respect for past structures can result in emblematic buildings which are both beautiful and functional (EFILWC 1997a, b).

Many awards celebrate the role of cities as crucibles of ingenuity and invention. From cradle to adult life, and from structures to the soul of a city, the human and the built environment bear witness to many outstanding efforts for excellence. Bright urban ideas are usually brought forward by individuals trying to extend the breadth of alternative futures. The Guangzhou International Award for Urban Innovation recognised worthy initiatives enhancing sustainable urban development through sharing of insights and rewarded innovations striving to improve social, economic and environmental sustainability. Co-organised by UCLG, Metropolis, and the Guangzhou Municipal Government, the award encourages and recognises outstanding innovative projects and practices in the public domain.

In 2014, Christchurch in New Zealand was rewarded for its remarkable recovery and citizen outreach after a series of deadly earthquakes and aftershocks in 2010. Dakar, in Senegal, was honoured for successfully accessing mainstream capital markets for a major project, a first in sub-Saharan Africa outside of South Africa. Last but not least, Hangzhou was recognised for introducing the first public bike sharing programme, currently the largest in the world. Most innovative projects are the results of stakeholder coalitions, which make each step of the transformational process possible. They are of political, strategic and tactical importance. They are significant because they contribute to the redirection of the flow of power, open urban chains to highly creative approaches and change the city's political culture.

Christchurch's Transitional City initiative was testament to the intensive work taking place in the city's post-earthquake environment. Community-led transitional initiatives encouraged participation in creating the places that matter for Christchurch citizens and strengthening the city's commitment to supporting local action. Transitional initiatives have made a real difference to the Christchurch community, and have brought the city to this interim stage of recovery. Transitional city approaches can also inspire cities which evolve even if they have not gone through a natural disaster. Together with Cape Town in South Africa, Hobart in Australia, Punta Arenas in Chile, and Ushuaia in Argentina, Christchurch is one of five official Gateway cities to the Antarctic. The city has been a base for Antarctic expeditions since 1900. Antarctic connections permeate the city's economic and cultural life. Activities have been developed to meet the needs of the scientists, artists, tourists, explorers and enthusiasts.

Coalitions that place citizens at the centre of a genuine strategy and mobilise them for its implementation seem to have unparalleled potential. Charismatic leaders, business actors, scientists or simply local citizens and workers are all potential bearers, initiators or adapters of innovation. A common problem or a shared perspective often ferments the common ground for the coalition. The architecture of coalitions is very diverse and challenges general rules. Flexible but strong alliances are often needed to create the space and the conditions for the future. Alliances based on agreement, mediation, political manoeuvring and negotiation can best direct the wave towards the desired vision. Agreement is much more powerful than arbitration. Consensus and persistent commitment are a necessary front for projects of long gestation.

Innovations may bring impressive leaps out of value-creating activities in a continuous flow to improve quality. They may also lead to a point of no return and affect the cultural equilibrium of a city. They may permanently change a city's sense of possible futures. Strong leadership is needed to orchestrate cooperation. Citizen participation can act as a net in communicating vision, in sharing the costs and fruits of change, and in ensuring that innovation does not become excessive, irreverent to traditions or socially unacceptable. Business participation can act as a net to ensure that costs are kept to a minimum.

Flagship renewal projects offer ample ground for innovation, including in the areas of planning and financing. Each successful initiative is probably the result of various purposeful trials and constructive errors and a step on the innovation ladder of excellence. Sometimes, the purposeful and knowledgeable use of capital is impossible unless initial investments have fecundated a multitude of new departures. The success of innovations is an expression of the creativity invested in the chain of experiments (Jacobs 1969).

Place-based and human-centred innovations can be extraordinarily powerful and completely transform the status of a city. Efficient but non-creative use of capital or technology in cities can lead to the systematic imitation of ingenious solutions. Risk taking and trial are limited in the search for the optimal and most efficient conditions for transplanting innovations. Continuous imitation kills the productive seeds and weakens the creative capacity of cities. A permanent environment for the peaceful incubation of genuinely new ideas and unproven methods, goods and services is the hallmark of many innovative cities.

In US in 2012, Portland, Oregon, joined the metropolitan areas stepping up to help achieve the Obama administration's goal of doubling exports. Portland is one of the nation's leading green economies, with an export sector that accounts for 18 % of the area's economic output, third among the 100 largest US metropolises.

In 2011, the Mayor and the Portland Development Commission, in partnership with regional business, innovation and university leaders, had conducted an analysis to better understand the region's economic strengths, global position, and opportunities for growth. The assessment revealed a strong computer and electronics sector, which accounts for more than half of regional exports, and vibrant but under-exporting sectors like clean technologies and software. In light of these conclusions, Greater Portland Inc., a regional economic development organisation, coordinated an export plan aligned with green policies. A diverse coalition of government, business, and non-profit leaders worked together to create a full-fledged Greater Portland Export Initiative Business Plan. The creation of a "We Build Green Cities" brand was promoted to leverage regional green economy exports to the rapidly-urbanising emerging world.

The Helsinki Region Infoshare service provides regional information quickly, easily accessible, and free to all, citizens, businesses, universities, research facilities or municipal administration. The data give a comprehensive and diverse outlook on living conditions, economics, well-being and transport. A unique new service is the public access to Ahjo, the electronic decision making system of the City of Helsinki. Ahjo comprises all agendas and minutes of Helsinki's City Council and other boards. The aim is to improve the compatibility of the data with other city data, including financial and location information, and encourage the development of new innovative services for citizens.

Many cities can get inspiration and insights from the OECD observatory of public sector innovations, collecting and analysing experiences of practical advice for sharing, discussing and co-creating policy options. It also includes social and cultural innovations. In France, the JocondeLab website sprung from an experiment by the Ministry of Culture and Communication highlighting the advantages of the semantic web 3.0 and the ways that the interweaving of cultural data and multilingualism can enhance ergonomics and interactivity. The innovation gives access to over 300,000 detailed records retrieved from the Joconde Catalogue, including the whole collections of the French National Museums, in 14 languages. This shared experience can serve as an example for other heritage websites, including marine and underwater heritages. It also provides opportunities for exploring new ways of browsing online cultural resources.

7.3 Cities and Blue Horizons: The Marine Urban Heritage

Every city being absolutely exceptional, the urban archipelago contains only prototypes. Each one is a unique civilisation involving places, links, visions, facts, notions, perceptions, smells, colours, sounds, emotions, symbols. It also has its subconscious, its interwoven bonds and conflicts, convergences and divergences, myths and legends. Urban culture is made out of tangible and intangible assets, reinvented and enriched every day. Monuments and sites of local, national and international importance, unique and universal, are the most visible and tangible collective assets and it is the responsibility of world citizens to ensure their preservation. But culture is also created through the intangible marine traditions, the gastronomic experiences by the sea and social events which are a source of joy and wonder. Culture constitutes a precious capital constantly reinvigorated by citizens' projections of hopes and desires into the urban reality and marvel (Calvet 1994).

Coastal urban cultural spaces have high existence and bequest values. Cultural heritage bears witness to the ways that citizens and communities have dreamed about and transformed the environment they live in and etched this into the landscape. Coastal landscapes and monuments play the role of catalyst as vectors of common memory and can generate new dynamics for participation and citizenship.

The cultural understanding and enjoyment of urban places is critical for personal and collective well-being. In all cities, exceptional events such as heritage nights or light festivals reveal unexpected jewels and invite citizens and children to rediscover urban environments with new eyes. Exceptional events such as carnival seasons disrupt everyday routines, help discover the ignored assets of neighbourhoods and bring local imagination and life to an apogee.

Culture links with the psyche of a city and makes it a space of belonging. The city can serve as the mirror in which diverse cultures observe each other, discover their convergences and divergences, interpenetrate and fecundate. Strongholds of civilisation, solidly anchored in local traditions but open to the world, cities can promote multicultural cooperation and intercultural understanding (Sansot 1973).

Heritage is invaluable and irreplaceable. Coastal urban biographies are enhanced by the presence of the sea. The seaside spaces are sometimes disrupted by projects disrespectful to the collective culture. The resistance by citizens is often powerful and effective, especially in cities with a strong democratic tradition. International pressure can also be very influential and UNESCO's World Heritage in Danger list has often brought good results.

Culture is a driving force of tourism. High-quality cultural tourism can find in coastal cities blue green landscapes of choice. Sustainable coastal cultural tourism tries to bridge physical and cultural resource conservation and quality of leisure for personal and collective fulfilment. Cultural and marine parks and itineraries are two components of cultural tourism that involve public heritage spaces and merit particular attention for sustainability. Linked to green ethics, the design of cultural parks has to take into account all natural and cultural assets of an area, and support ecosystem-based practices. Cultural itineraries can serve as metronomes of urban beauty, with its asymmetry, its paradoxes, and extraordinary contradictions. Each city is a world and has its place in the global community as it becomes a cultural reference and a source of wonder.

Public heritage sites are often exceptional cultural spaces, closely linked to the identity of a city. They have the potential to stimulate the collective memory. They can be a place of sharing and not just places of passage. They belong, by definition, equally to all and must be accessible to all. R. Koolhaas describes them as fortresses of freedom. They offer great opportunities as islands of civilisation in the archipelago of the city. Streets, squares, parks, and public spaces should be the object of special attention as nerve centres of the city.

The European Urban Charter II "Manifesto for a new urbanity", adopted in 2008, pleads for a new culture of urban life and encourages progress towards the sustainable city. The manifesto released a set of principles and concepts enabling cities and their inhabitants to address urban challenges. It invites local actors, in all their diversity and shared values, to implement public policies upholding the principles of ethical governance and sustainable development solidarity (Council of Europe 2008).

Sustainable cities have to transmit their cultural capital enhanced or at least intact to future generations. The Organisation of the World Heritage Cities (OWHC), hosting monuments and sites included in UNESCO's World Heritage list, involves hundreds of cities committed to actively preserving their unique heritage. The OWHC was created on 8 September 1993, in Fez. In commemoration of this event, September 8 was declared "Solidarity Day of the World Heritage Cities", and all cities are welcomed to celebrate it every year by organising different activities highlighting their uniqueness. More than half of them are located in Europe, where Italy alone hosts more than 5 % of the world's listed monuments and sites. Governments must preserve the characteristics and the qualities that justified the inscription of the cultural sites on the World Heritage list (UNESCO World heritage Centre 2014). The OWHC's "City2City" programme aims at strengthening all types of exchanges and dialogues on preservation among its World Heritage Cities. Many of the 250 World Heritage Cities and the 350 World Heritage Sites in urban areas experience a long-term relationship with the sea.

Monuments and sites are concrete signs of the biography of cities, shaped by historical, socio-economic and cultural waves and tides. The Hanseatic City of Lübeck provides an example. Founded in the twelfth century on the Baltic coast of Germany, it was, for three centuries, one of the principal cities of the Hanseatic League of merchant cities, which came to hold a monopoly over the trade of the Baltic and North Seas. The plan of the Old Town island of Lübeck, with its blade-like outline determined by two parallel routes of traffic running along the crest of the island, dates back to the beginnings of the city and attests to its expansion as a trading centre. To the West are located the richest quarters with expensive homes, and to the East, small shops and artisans. Lübeck suffered much during the Second World War, with almost 20 % of its buildings destroyed, including most of its famous monumental complexes, such as the Cathedral. Selective reconstruction has permitted the replacement of the most important churches and monuments (UNESCO World Heritage Centre 2014).

Venice is an eternal city of cultural intensity, an unbeatable art destination, which tries to advance in symbiosis with the sea. The Grand Canal, considered the most attractive urban avenue in the world, winds across the city, also taking care of its green spaces, not least the Green theatre. Crumbling waterside palaces or exquisite churches in perilous state are being renovated to combat physical decay. Many organisations, such as Save Venice Inc., dedicated to preserving the artistic heritage of the city, celebrated more than 40 years of preservation work in Venice (Save Venice Inc. 2011).

Venice reminds us that the built environment is a testament to the permanent evolution of human ingenuity and appreciation of the world. Architecture involves diverse and often contradictory concepts, but also common values and selective affinities which create a vibrant and fertile dialogue amongst architects, city planners and citizens, but also eras and places, individual and civic life. New high-quality exhibition venues, like the monumental triangle Punta della Dogana, compete in aesthetics with majestic open air public spaces.

The Old City of Dubrovnik, dotted with beautiful Gothic, Renaissance and Baroque buildings had withstood the passage of centuries and survived several earthquakes. In 1991, seriously damaged by artillery fire, the city was included on the list of World Heritage in Danger. With UNESCO's technical advice and financial assistance, the Croatian Government restored the facades of the Franciscan and Dominican cloisters, repaired roofs and rebuilt palaces. As a result, in December 1998, it became possible to remove the city from the list of World Heritage in Danger. The millennium started with new promise for this exceptional city in unique symbiosis with the Adriatic Sea.

Coastal urban sites on the list of World Heritage in Danger deserve priority attention. The Liverpool Maritime Mercantile City, inscribed on the list in 2012, is the only World Heritage in Danger from the European Union. The World Heritage Committee has placed Liverpool on this list due to the proposed construction of Liverpool Waters, a massive redevelopment of the historic docklands North of the city centre. The Committee contended that the development would alter the visual profile and skyline of the site inscribed in the list of World Heritage in 2004.

Liverpool Maritime Mercantile City includes six areas in the historic heart and docklands of Liverpool bearing witness to the development of a major world port and trading centre from the eighteenth to the twentieth century. Liverpool is a supreme example of a pioneer port city at the time of Britain's greatest global influence and the mass movement of people, including slaves and migrants. Liverpool's waterfront, port and management reflect the prominent role of the city in the past global trading system and dock technology. Docks had a right to architecture and art. The best-known, Albert Dock, is a living textbook of industrial harbour architecture. Part of the old dock complex hosts the Maritime Museum, an anchor point of the European Route of Industrial Heritage, the International Slavery Museum and the Tate Liverpool. The area also still includes the world's largest brick-work building and Pier Head's "Three Graces", as well as the Royal Liver Building (UNESCO 2014).

World Heritage sites include vibrant coastal cities with cultural heritage of outstanding universal value. Many exceptional coastal heritage cities enjoy a strategic location and have inherited a majestic protective infrastructure. The Old Town of Corfu, in a strategic position at the entrance of the Adriatic Sea, is emblematic of fortified Mediterranean ports, demonstrating a high level of integrity and authenticity. The three forts of the town, designed by renowned Venetian engineers, were used for four centuries to defend the maritime trading interests of the Republic of Venice against the Ottoman Empire. In the course of time, the forts were repaired and partly rebuilt, more recently under British rule in the nineteenth century.

The mainly neoclassical residential buildings of the Old Town are partly legacies from the Venetian period.

Naples' urban fabric preserves elements of its long and eventful history. Its exceptional setting on the Bay, its street pattern, and its wealth of diverse historic buildings from many periods give it an outstanding value. Little survives above ground of the original Greek town, but important archaeological discoveries have been made during the various excavations. The surviving Roman remains are more substantial, notably the large theatre, cemeteries and catacombs. The street layout in the earliest parts of the city owes much to its classical origins. The period that followed the collapse of the Roman Empire saw the beginning of church building on a substantial scale. The chapel of Santa Restituta in the fourteenth-century cathedral is reputed to be the first Christian basilica in Naples. The Castel dell' Ovo is one of the most substantial survivals from the Norman period. Built as a fortress-monastery, it was subsequently remodelled and given its present form at the end of the seventeenth century. The arrival of the Angevin kings saw the city expanding and absorbing the suburbs and neighbouring villages. French Gothic pervaded both religious and domestic architecture. The Renaissance heritage of Naples is also worth mentioning. Quarters were gradually enriching the urban fabric both inside and outside the walls. The port also grew to meet the increasing significance of the city. With the Treaty of Vienna in 1738, Naples became once again the capital of an autonomous kingdom, and, with the rest of Italy, the city came under French rule in the Napoleonic period and benefited from some ambitious urban projects.

The physical setting also determined much of Genoa's development. Its historic centre dates from the late sixteenth and early seventeenth centuries when the Republic of Genoa was at the height of its financial and seafaring power. The city represents the first example in Europe of an urban development project parcelled out by a public authority. The site includes an ensemble of Renaissance and Baroque palaces along the so-called "new streets". The Palazzi dei Rolli offers an extraordinary variety of patterns, achieving universal value in adapting to the particular site and to the specific requirements of social and economic organisation. They also offer an original example of a public network of private residences.

The science and art of city planning advanced in the coastal city of Tel Aviv, built as an urban ecosystem with physical, economic, social and human dimensions. Founded in 1909 on the Eastern coast of the Mediterranean Sea, it is an emblematic white city. During the era of British rule in Palestine, it developed into a thriving urban centre, becoming Israel's foremost economic and metropolitan nucleus. Tel Aviv can be seen as an outstanding example of the innovative approaches to town planning during the first part of the twentieth century. The architecture became a synthetic representation of some of the most significant trends taking into account local traditions and climatic conditions. The "white city" forms the heart of Tel Aviv, and is based on the urban master plan by Sir Patrick Geddes, a visionary town planner, in considering the urban landscape as an organism in constant evolution.

Most world heritage sites provide remarkable examples of the ways that cities and citizens imprint the landscape. Rio de Janeiro consists of an extraordinary urban setting encompassing the key natural elements that have shaped and inspired the development of the city. From the highest points of the Tijuca National Park's mountains down to the sea, and from the Corcovado Mountain to the designed landscapes along Copacabana Bay, all have contributed to the outdoor living culture of this spectacular city. Rio de Janeiro is also recognised for the inspiration it has provided to artists, landscapers and city planners.

Another remarkable example of urban cultural symbiosis with the sea is the historic Quarter of Valparaíso, an excellent example of late nineteenth-century urban development in Latin America. The city was the first and most important port on the sea routes of the Pacific coast of South America that linked the Atlantic and Pacific oceans. It had a major commercial influence on the region from the 1880s until the opening of the Panama Canal in 1914. Since then, its development has slowed, and the city attracts more attention due to its unique physiognomy in a natural amphitheatrical setting embracing the sea. Valparaíso has well preserved its interesting early industrial infrastructures, such as the numerous elevators on the steep hillsides.

Valparaíso's historic quarter is composed of five interlaced neighbourhoods and includes many squares, viewing points, promenades, stairways and the top stations of some of the city's distinctive funicular elevators. The attractiveness of Valparaíso was strongly marked by the geography of its location dominated by the bay, the narrow, largely artificial coastal plains and the steep hills forming the city's exceptional layout. Adaptation of the built environment to these geographical conditions produced an innovative and urban composition which attracted many world citizens. The result of this challenging topography, creative impulse and intercultural dialogue led to an original Latin American city with a late nineteenth century character.

Panama City, the oldest continuously occupied European settlement in the Pacific coast of the Americas, was founded in 1519, 6 years after the discovery by the Spanish of the South Sea. The archaeological remains of the original settlement include the Pre-Columbian vestiges of the aboriginal occupation and a protected heritage site. This major colonial site was the seat of a Royal Court of Justice during the sixteenth and seventeenth centuries, until Panama consolidated its position as an intercontinental hub. The importance of the city is reflected in the imposing stone architecture of its public and religious buildings. It was laid out on a rectilinear orderly grid according to European ideals of planned towns. Abandoned in the mid-seventeenth century, it was replaced by a new town, known as the historic district, which has also preserved its original street plan and architecture, and an unusual mix of Spanish, French and early American styles.

Colombia's port of Cartagena is another unique urban site of the coastal World Heritage and an eminent example of the military architecture of the sixteenth, seventeenth, and eighteenth centuries. Situated on a protected bay facing the Caribbean Sea, the city of Cartagena de Indias boasts the most extensive and one of the most complete systems of military fortifications in South America. The narrow streets of the colonial walled city host civil, religious and residential monuments of great beauty. Due to its strategic location, the city was also one of the most important ports of the Caribbean. The port of Cartagena was an essential link in the route of the West Indies and thus an important part of the history of world exploration and the great maritime routes.

Cartagena was for several centuries the scene of confrontation among the principal European powers fighting for control of the New World. Defensive fortifications were built by the Spanish in 1586 and were strengthened and extended to their current dimensions in the eighteenth century, taking advantage of the natural defences. The initial fortifications included the urban enclosure wall, the bastioned harbour of San Matías, and the tower of San Felipe Del Boquerón. All of the harbour's natural ways were protected by fortresses. Within the security of the walls are the neighbourhoods which form the historic heart of the city.

Throughout history, the defensive coastal architecture left precious legacies with outstanding remains in many port cities. In Cuba, Old Havana's extensive network of defensive infrastructure, constructed between the sixteenth and nineteenth centuries, includes some of the oldest and largest stone fortifications still standing in the Americas. With the development of the fleet system in the Spanish West Indies, Havana became the largest port in the region, and developed, during the eighteenth century, to be the most complete dockyard in the New World. The port and its dockyard are protected by a complex system of fortifications, including the Castillo de la Real Fuerza, one of the oldest colonial fortresses.

Old Havana, defined by the extent of the former city walls, has maintained the original urban layout which presents a remarkable unity of character. Urban plazas are surrounded by many outstanding buildings, including the Cathedral, and the ancient convent of San Francisco de Assisi. Narrow streets permeate the historic centre of the city, serving a homogeneous ensemble of private houses with arcades, balconies, gates and courtyards. The overall sense of architectural, historical and environmental continuity makes it one the most extraordinary heritage city centres in the Caribbean.

Also part of the World Heritage sites, the Historic Town of Grand-Bassam, first capital and most imperative port, economic and judicial centre of Côte d'Ivoire, is an example of a late nineteenth- and early twentieth-century colonial town planned with quarters dedicated to commerce, administration, and housing for Europeans and Africans. The site includes an African fishing village alongside colonial architecture in residences marked with galleries, verandas and gardens. The city bears witness to the complex social relations between Europeans and Africans, and to the subsequent independence movement. As a vibrant centre of the territory of French trading posts in the Gulf of Guinea, which preceded modern Côte d'Ivoire, it attracted populations from all parts of Africa, Europe and the Mediterranean region.

In Finland, the fortress of Suomenlinna, built in the second half of the eighteenth century by Sweden on a group of islands located at the entrance of Helsinki's harbour, is an especially interesting example of European coastal defensive architecture. In Denmark, on a strategically important site commanding the Sund, the Royal castle of Kornberg at Helsingor has great symbolic value and played a key role in the history of Northern Europe in the sixteenth to eighteenth centuries. The defences of this outstanding Renaissance castle were inspired by models of military architecture from the late seventeenth century. It has remained intact and is world-renowned as Elsinore, the setting of Shakespeare's Hamlet.

A complex maritime trading hub and strategic gateway between China and the western world, Macao played a crucial role in the development of international trade. A Portuguese territory in the mid-sixteenth century, which returned to Chinese sovereignty in 1999, it is a unique testimony to aesthetic, cultural, religious, architectural and technological influences from East and West. Macao's multicultural identity reflected on Western and Chinese architectural heritage coexisting and even influencing each other. European port city characteristics can also be seen in the urban fabric of the settlement, with public squares and meandering streets. The seascape is also reminiscent of Macao's origin as a trading port city. The Inner Harbour, used over centuries and still functioning, is a vital heritage. Intangible influences on local lives are reflected in religion, education, medicine, and cuisine.

In the Islamic world, the Tunis Medina constitutes a prototype among the best conserved. Located near the sea in a fertile plain region, the Medina of Tunis is one of the first Arabo-Muslim towns of the Maghreb. Capital of several influential dynasties, it represents a human settlement that bears witness to the interaction between architecture, urbanism and the effects of socio-cultural and economic changes of earlier cultures. Under the Almohads and the Hafsids, from the twelfth to the sixteenth centuries, Tunis was considered to be one of the greatest and wealthiest Arab cities. Between the sixteenth and nineteenth centuries, new powers endowed the city with palaces, residences and priceless mosques. The historic heart covers an area of approximately 280 ha and comprises all the features of an Arabo-Muslim city. It is composed of the medina dating from the eighth century and suburbs from the thirteenth century.

The influence of marine and maritime activities on cities is often compelling. In Bahrain, the merchant quarters of the city of Muharraq reflect the wealth of the trade of pearls from oyster beds in the Persian Gulf. As the best-known source of pearls since ancient times, the Gulf industry reached the apex of its prosperity at the end of the nineteenth and the beginning of the twentieth centuries. A few distinctive commercial and residential buildings remain as a testimony to this demanding economic activity, which suffered a sudden decline and demise in the 1930s as a result of the development of cultured pearls from freshwater mussels in Japan. The structures are tangible manifestations of the major functions and institutions associated with the pearl economy and society.

The legacies of Muharraq include 17 buildings embedded in the urban fabric, three oyster beds, and a part of the seashore at the southern tip of the Island, from where the boats set off for the oyster beds. Most of the structures have survived relatively unaltered since the collapse of the pearl industry in the early twentieth century and bear witness to distinctive traditions that the industry fostered, and particularly their high standard of craftsmanship in timber and plaster. The buildings evoke memories related to the pearl industry, its supporting social and economic activities, and the cultural identity of the place.

Exchanges among exceptional cities can be even more far-reaching when involving youth. The Youth on the Trail of World Heritage programme consists of pairing schools located in world heritage cities and promoting exchanges among groups of young people aged between 15 and 18 years old. Students of different cities can discover another world heritage city and share ideas, knowledge and work related to exceptional urban heritage. They thus become more aware of the universal value of their heritage and the challenges involved in the management of heritage on the international level.

Many World Heritage Cities are among the most vulnerable places to experience rapidly occurring changes, not least due to advancing climate change. They can serve as laboratories for monitoring impacts and testing adaptation and mitigation measures to enhance their physical and socio-economic resilience, while ensuring the conservation of their precious heritage for humanity. Heritage coastal cities hold significant keys to understanding long-term evolution of a place by the sea from the past and on to the future.

Underwater heritage is another unique cultural issue, which often reveals lost cities and shipwrecks. Traces of human existence underwater which are one hundred years old or older are protected by the UNESCO Convention on the Protection of the Underwater Cultural Heritage, signed in 2001 and entered into force in 2009, which provides an international legal framework for the protection of underwater cultural heritage. Underwater archaeological sites include submerged sites that have been subsequently covered by water due to storm surge and rising sea levels, but also wrecks from ships and aircrafts, port-related structures and even weapons. Human interference with underwater cultural heritage raises the question of ownership rights in sunken vessels, sovereign immunity, the ethics of commercial exploitation, and the question of jurisdictional competence to regulate activities beyond territorial sea limits (UNESCO World Heritage Centre 2013).

Underwater archaeology can make a significant contribution to the cultural understanding of places. It involves very sophisticated equipment and techniques to support diagnosis, analysis and evidence-making. Underwater archaeology evolved towards action to address threats to underwater cultural heritage arising as a result of advances in deep-water technology, destruction or loss of historic information and looting. For deep sites, submarines or remote sensing equipment are needed. For marine sites, specially equipped ships or floating platforms are often used, in coordination with shore-based activities. Underwater marine sites are often subject to tidal flows or storm damage. Structures may be unexpectedly buried or uncovered. The dynamic nature of the environment may make in-situ priority conservation infeasible. In addition, underwater sites can remain chemically active, and iron can be leached from metal structures. Artefacts recovered from underwater sites need special care. Furthermore, it can be difficult to allow access to the results of the archaeological research as underwater sites do not always provide the safety conditions necessary for access to the general public.

Most sites need at least some form of topographical survey and a site plan showing the locations of artefacts and other archaeological material. Environmental assessment of archaeological sites will also require the recording of site conditions as well as the natural organisms. A variety of techniques are available to divers to record findings underwater. Scale drawing is a basic tool of archaeology and can be undertaken underwater. Photography is the mainstay of recording, and the advent of digital photography has multiplied opportunities. Low visibility underwater and distortion of image due to refraction make perspective photographs difficult. A series of photographs at adjacent points and their combination into a single photomosaic image of the whole site can be very useful. Underwater photography can also be conducted from submarines and assist in the recording process.

Maritime conservation and recording are most important, since inventories of sites and shipwrecks can tell a colourful tale of the maritime heritage. It is estimated that up to 15,000 shipwrecks lie in Ireland's designated waters, shipwrecks that tell of cruel wars, political machinations, private lives and sunken treasures. From the sixteenth century to the two world wars, the ships' stories are compelling and often relate to important moments in history. For the past years, Ireland's offshore waters and coastal seas have been subject to one of the largest seabed surveys in the world, in the course of which a database of over 300 shipwrecks has been created, containing detailed information regarding each wreck's condition on the seafloor, its extent, dimensions and water depth, along with a short background history (Brady et al. 2012).

The Greek seas have revealed many of their hidden treasures and the "Antikythera mechanism" has probably been one of the most enigmatic objects ro arise from the depths of time. In 2012, the archaeological museum of Athens presented the this mechanism found in a shipwreck near the sea of Antikythera, alongside the remains of the original vessel and treasure, which have been reunited for the first time since their discovery in 1901. This first astronomical calculator shows with high precision, according to the inscriptions which have been deciphered, the different cycles of the sun and the moon, the planets and the constellations. The instrument remained a mystery for the scientific community. Obscurantist statements claimed that the artefacts from the Antikythera shipwreck were almost extra-terrestrial in nature. Only 82 fragments of this mechanism still exist, all corroded. Their multidisciplinary analysis, at the beginning of the twenty-first century, concluded that this astronomical instrument dates from the second century BC.

The Antikythera mechanism expresses a mathematical vision of the cosmos and constitutes the first mechanical computer known to humanity. Created more than a thousand years ahead of the first astronomical clocks, the mechanism could have been designed in Rhodes, home to a community of astronomers, including Hipparchus, as well as mechanical engineers. The Antikythera mechanism seems a genuine cosmograph, a device to describe the cosmos, and also a selenograph, a device to describe the movements of the Moon. The instrument could show multiple astronomical cycles, including the Metonic cycle, which runs for 19 years, and the Callippic cycle, which runs for 76 years. The astonishing volume of astronomical data needed for such a mathematical model challenges current assumptions about the conceptual abilities of the ancient scholars and engineers.

7.4 Ocean Cities as Theatres and Masterpieces of Art

Art is the aesthetic expression of individual, collective and universal ethics, the ultimate expression of the creative intelligence of a city, a dynamic component of its cultural capital (Jimenez 2002). It enriches and heightens human and social capital but also influences the man-made environment of cities. Each city should be enjoyed as a collective masterpiece of art (Olsen 1987).

Urban dynamics contributed to the genesis of many arts but have also been impacted by the arts. Seas were among the main sources of inspiration. The art of theatre, one of the legacies of Greece to the world, provides a fine example. Initially consisting of a choral group, part of a rural festival in the honour of the god Dionysus, drama gained importance when brought to the city. The spark of the innovation was the injection of the critical word into the lyrical songs. The chorus became the voice of the citizens, a strong expression of public opinion. Tragedy, comedy and the satyr play were the three dramatic genres to emerge. Athens exported theatre to its numerous colonies and allies in order to promote a common cultural identity. The Athenian drama has had a significant and sustained impact on world culture as a majestic anatomy of all human passions.

Coastal cities are permanent theatres of all expressions of human creativity on land and sea. Arts and culture can also be catalysts of urban regeneration and waterfront development. Efforts to revitalise urban areas often focus on cultural hubs or art districts, which contribute to making the urban core friendly to the creative class (Florida 2008). UNESCO's design cities provide many examples of future oriented interventions that will make cities more attractive to residents and visitors.

The 2014 city of design was Bilbao. The city made a wise investment with the creation of the Guggenheim museum, which acted as a catalyst for urban transformation. Bilbao had to overcome a past of heavy industry and wastelands and needed an impetus to be projected into a dynamic trajectory on the world stage. The museum lifted Bilbao into the twenty-first century. It is a temple dedicated to modern art, which gave to the city a sense of permanent wonder. The titanium vessel designed by F. Gehry was instantly hailed as a most significant accomplishment, a radically new design which challenged the museographic approaches and changed assumptions about art, architecture, and collections. The museum revolves around a broad atrium bathed in light. Its structure undulates smoothly and captures every ray of sunshine. F. Gehry stated that light is free of charge and cities should use and abuse it to the maximum.

Just as a religion spreads through places of peace and worship, culture needs places with a true character that promote wonder. Some of them enjoy a unique relationship with the sea. The Sydney Opera House is a great artistic monument and an icon in an extraordinary waterscape. The design represents an outstanding dialogue with the environment in Sydney Harbour. It bears witness to creativity and innovation in both engineering and architecture. A great urban sculpture set at the tip of a peninsula projecting into Sydney Harbour, the building has had an enduring influence on architecture. The Opera House comprises three groups of interlocking vaulted "shells or sails" which house two main performance halls. These structures are set upon a vast platform and are surrounded by terraces that function as pedestrian concourses. The significance of the building springs from its unparalleled design and construction. It is a daring, visionary experiment with an enduring influence on the later architecture. The original design concept and unique approach to building gave impetus to the collective creativity of architects, engineers and supporters who have already made it part of the UNESCO World Heritage Sites list (UNESCO World Heritage Centre 2014).

Landmarks and cultural public spaces expand, often in unexpected surroundings. The development of the High Line in New York is a remarkable example, built on an old railway raised to 7.5 m above the ground, which was used to transport goods from 1934 to 1980. At the heart of the West Side of Manhattan, the cultural passage of the High Line alludes to the industrial past of a city in eternal transformation. The walk is designed as an urban choreography with performance artists, musicians and walkers, suspended in space and time. The High Line hosts 220 species of plants and has become a new reference for a better social life.

Citizen imagination could transform waterfronts into places of civic engagement. Shopping could be just one small part of a rich mix of activities, including working, communicating, sharing, playing and learning new skills or appreciating cultural events. Attractive public spaces can build the frame of social exchange and promote democracy. Instead, deteriorated public places, victims of neglect or standardisation, may become places of confrontation and exclusion that generate violence. Many cities invest in the creation of public spaces that combine aesthetics and functionality, such as the development of quality urban beaches during the summer and fountains at bus stops. The development of public spaces has no other limits apart from the collective imagination.

A demonstration of unlimited imagination can be provided by Singapore's Gardens by the Bay, a park spanning 101 ha of reclaimed land adjacent to the Marina Reservoir. The park has become an outstanding urban recreation space and a national icon. It consists of three waterfront gardens, conceived as an integral part of a strategy to transform Singapore from a Garden City to a City in a Garden, and raise the quality of life by enhancing greenery and flora. The Supertrees Grove, with tree-like structures, dominates the landscape with heights that range from 25 to 50 m. Supertrees are vertical gardens that perform a multitude of functions, which include planting, shading and performing as environmental systems.

The Supertrees are home to enclaves of unique and exotic ferns, vines, orchids and a vast collection of other plants. They are fitted with environmental technologies that mimic the ecological function of trees, especially photovoltaic cells that harness solar energy, which can be used for some of the functions of the Supertrees, such as lighting, and absorbs ions of rainwater for use in irrigation and fountain displays. The Supertrees also serve air intake and exhaust functions as part of the conservatories' cooling systems. A 3-km elevated waterfront promenade, between two of the larger Supertrees, allows for breathtaking aerial views by the Bay. At night, the Supertrees further expand wonder as they come alive with light and music.

The European Capitals of Culture offer cities an outstanding opportunity to affirm their position on the cultural world chessboard and to benefit from considerable returns in economic, cultural, tourist and media terms. Launched on the initiative of the Greek Minister of Culture M. Mercouri in 1985, the European City of Culture is designed to "contribute to bringing together the citizens of Europe". All designated European Capitals of Culture presented initiatives interweaving strong linkages among citizens and the world and combining international artistic content with local tradition and culture. They often celebrated the close proximity of the sea. The title of the European Capital of Culture has proved to be an extraordinary opportunity to move towards sustainable development through culture, boost tourism, create jobs, and share a powerful creative energy.

The array of the European Capitals of Culture has been very rich and diverse and has often acted as a catalyst for urban transformation and renaissance. Glasgow is often cited as the most outstanding example of a city that enhanced its year as cultural capital as an investment in urban regeneration. A shared title among EU cities and with cities from outside the European Union has led to precious exchanges. The Norwegian Stavanger port, European capital in 2008, together with Liverpool, presented an ambitious programme and displayed a willingness to involve the community in more than 100 events spanning from ancient Greek theatre to opera and ballet ski amid rocky fjords.

Istanbul was the largest ever non-EU city to share the title of the European Capital of Culture in 2010. Since the 1990s, Istanbul has reinvented itself with new communications infrastructure, greater environmental awareness and a heightened attention to culture and the arts. The city, former capital of four empires, features architecture from the Roman, Byzantine, Latin and Ottoman times. Although the city has a rich cultural heritage, its motivation for becoming a European Capital of Culture centred on modern and contemporary culture, reflecting the diversity of cultures and ethnic groups, as well as its young, dynamic population. The event generated transformative energy and built capacity to address the urban and cultural dimensions of citizenship, functioning as a bridge connecting Europe to the East.

In 2011, Tallinn honoured Estonian culture in all its diversity and richness, ranging from traditional to contemporary art. Almost half of the events were open to all citizens and visitors. There were many large-scale events, such as the Tallinn Maritime Days, a song and dance festival, the "Song of Freedom" concert, the "Happy End" during the last day of the event and the Tallinn marathon bringing together many thousands of participants. The greatest ambition of Foundation Tallinn 2011 was to give life to smart ideas and fascinating, unique and original events that place culture at the heart of the city. Cultural start-ups, especially by local designers, were encouraged and embraced unexpected ideas. Tallinn 2011 also initiated a sustained movement of 1500 volunteers to create a more bike-friendly city, presented a programme of hospitality, and launched a project to attract young people and children from outside Tallinn to the cultural events.

As the 2013 European Capital of Culture, Marseille wished to encourage the expression of the creativity of the broader region and extend it to the whole Mediterranean. Nestled between sea and hills, Marseille is an amazing city founded 2600 years ago, the oldest city in France, offering the richness of a unique heritage and an intense socio-cultural life in a mild climate. The metamorphosed old harbour has been at the crossroads of all events and the heart of the Capital of Culture. The Silo, a concert arena in a converted silo on the docks, and a huge hangar at the head of the old port were among the most exciting venues. Exceptional events included an exhibition at the crossroads of science, mathematics and experimental art and a fictional journey around the Mediterranean Sea, tracing its real and mythological history through the eyes of residents and artists. La Marseillaise newspaper presented a serialised novel through weekly instalments by 12 authors in residence in symbolic locations. The European Capital of Culture was also marked by the inauguration of the Museum of Civilisations of Europe and the Mediterranean, devoted to the history and culture of civilisations, and the Villa, an international centre for cultural and artistic exchange largely built underwater.

In 2014, two cities with the sea close to their horizon and their heart, Umeå and Riga, shared the title of European Capital of Culture. They both enjoy cultural richness and diversity. In Riga, the festivities started with a tribute to the 500th anniversary of the first printed book. The move of thousands of books by voluntary citizens from hand to hand, from the old National Library to the new šviesos Pils, was an extraordinary event, suggesting that a public library is not just a place to borrow a book but a repository of true masterpieces and a communication node for the local community. Riga also organised the world select games, a kind of world championship for choirs and 20,000 singers. The Festival of Light featured a large number of international multimedia art projects and encouraged participation of the public for five evenings during November 2014. As part of the festival, the largest graffiti in the Baltic countries, "The Sun, the Thunder, the Daugava", created for the Capital of Culture, was also shown in a new light.

The Riga Carnival at the Festival of Light not only embodied transformation through lighting up the city, buildings, facades and windows, but also wished to give people the feeling of a continuous adventure, exploration and discovery. Carnival stands for happiness and change. The public participation campaign "Light Up Your City!" created colourful installations on the facades of buildings. Lattelecom, the main cooperation partner of the Staro Rīga 2014 festival, set up new Wi-Fi hotspots at bus and trolley-bus stops near the Cabinet of Ministers and Riga Regional Court buildings, accessible to the visitors of the festival, in addition to more than 1000 free Wi-Fi hotspots, making Riga the European capital of free Internet.

The precious stone of amber was celebrated as one of the symbols of Latvian culture from the depths of seas and time. An exhibition by the National History Museum displayed selected archaeological and ethnographic samples of amber, informing visitors about the "amber road" and "crossroads" and highlighting the continuity of amber treatment skills and the importance of amber in Latvian folklore. The main motif at the exhibition was the depiction of the depths of the Baltic Sea, the environment in which amber is created. The multimedia part featured magnified images of the most outstanding amber artefacts to showcase their aesthetic beauty and texture.

Co-creation was the key idea in Umeå 2014's Capital of Culture. The city wished to be a window to all of Europe and show that it is a foremost culture-driven cities in Europe. Before, during and after 2014, cultural exchanges between Umeå and the rest of Europe intensified and new cultural bonds were formed. Raising Northern Sweden's European profile was another cultural ambition of the Capital of Culture, which insisted on the role of cultural exchanges for regional development and identity. To promote a long-term perspective and encourage the continuation of Capital of Culture activities beyond 2014, projects were run by the cultural operators themselves.

Cities can be compared to huge theatre stages hosting many diverse plays. The project "Cities on stage", initiated by the National Theatre in Brussels, invited six European theatres to exchange and confront their art on the issue of "living together" in cities like Brussels, Gothenburg, Paris, Sibiu, Naples and Madrid. Since 2011, theatre artists offer singular, critical and poetic views on this urban world in transition, through seven creations inspired by the experiences of the seven cities. Their questions resound beyond urban and national frontiers.

The initiative also wishes to strengthen the link between artistic creativity and citizenship, closely associating the inhabitants of each city with the creative process. "Moving cities" invites groups of citizens, young and old, to question the challenges of contemporary cities in the present and future world. Citizen workshops, accompanied by artists of multiple disciplines, follow the process of theatrical creation and explore themes such as migration, diversity, multiculturalism and cross-fertilised urban identity. In parallel, each theatre invites a group of young actors of partner countries to work for a month with a director and citizens to support young artists in their discovery of the city.

The oceans have been a prime source of inspiration for many scientists and artists, from Homer's Odyssey to the contemporary art of Bill Viola. The Good Planet Foundation, a film directed by Y. Arthus-Bertrand and M. Pitiot, was intended to be a plea for humanity to respect the blue world supporting life. It presents the greatest natural mystery of the planet, and highlights the scientific missions of Tara, a schooner acting as a platform for state-of-the-art research on marine ecosystems and biodiversity. Tara, supported by the UN and many institutions and foundations, has shed light on the role played by plankton in the earth's life support system and has observed the effects of climate change on this critical element of the marine chain.

Artists and journalists have gladly participated in the Tara Oceans expedition, helping to promote public awareness about the serious threats facing the world's oceans. Twenty-one laboratories in ten countries are collaborating on the mission, and their research findings are immediately being published in open access databases. This work is expected to provide the first complete overview of the world's plankton ecosystem.

Well before the modern tsunamis devastated cities, the awesome waves by the Japanese artist Hokusai in his "Thirty-six Views of Mount Fuji" had shaken humanity. Mount Fuji was traditionally seen as the source of the secret of immortality. Hokusai also inspired the Hugo Award-winning short story by science fiction author R. Zelazny, "24 Views of Mount Fuji, by Hokusai", in which the protagonist visits locations surrounding Mt. Fuji, painted by Hokusai.

References

Boston Foundation and The Citistates Group (2004) Boston unbound Tapping greater Boston's assets and talents to create a world leading citistate. Boston Foundation and The Citistates Group, Boston

Brady K et al (2012) Warships, U-boats and liners—a guide to shipwrecks mapped in Irish waters. Department of the Arts, Heritage & the Gaeltacht, Geological Survey of Ireland and Stationery Office Dublin Ireland, Dublin

British Urban Regeneration Association (BURA) (1997) The future of cities. Conference report. BURA, Belfast
Calvet LJ (1994) Les voix de la Ville. Payot, Paris
Council of Europe (2008) La Charte urbaine européenne II—Manifeste pour une nouvelle urbanité. CoE, Strasbourg
European Foundation for the Improvement of Living and Working Conditions (EFILWC) (1997a) European cities in search of sustainability. A panorama of urban innovations in the European union. EFILWC, Dublin
EFILWC (1997b) Innovative and sustainable cities. EFILWC, Dublin
Florida R (2005) Cities and the creative class. Routledge, New York
Florida R (2008) Who's your city. Basic Books, New York
Jacobs J (1969) The economy of cities. Penguin, London
Jimenez J (2002) Teoria de arte. Tecnos, Madrid
Organisation for Economic Co-operation and Development (OECD) (2013) The competitiveness of global port cities. Synthesis report. OECD, Paris
Olsen D (1987) La città come opera d'arte. Sena e Riva, Milano
Sansot P (1973) La poétique de la Ville. Klincksieck, Paris
Save Venice Inc. (2011) Four decades of restoration in Venice. Save Venice Inc., New York
United Nations Educational, Scientific and Cultural Organization (UNESCO) (2014) World heritage and our protected planet. UNESCO, Paris
UNESCO World Heritage Centre (2013) Safeguarding the crown jewels of the ocean. UNESCO World Heritage Centre, Paris
UNESCO World Heritage Centre (2014) The list. Global strategy. UNESCO World Heritage Centre, Paris

Selected Internet Links

www.kg.eurocean.org
www.intelligentcommunity.org/
www.ilab.harvard.edu
www.citiesofscience.co.uk
www.cite-des-sciences.fr
www.ec.europa.eu/culture
www.europanostra.org
www.icomos.org
http://portal.unesco.org/culture/
www.eit.eu
www.cultureactioneurope.org
www.worldcitiescultureport.com/cities
www.culturemontreal.ca
www.thebreakthrough.org
www.koshland-science-museum.org/explore-the-science/earth-lab/responses
https://fondationjacquesrougerie.squarespace.com/
https://www.oecd.org/governance/observatory-public-sector-innovation/
http://www.guangzhouaward.org/789.html
www.galwayscience.ie

Watercolour 8
Dubrovnik: A Cultural Marvel by the Adriatic Sea

Chapter 8
Sustainable Regeneration of Coastal Cities

Abstract Urban renaissance usually tends towards an ideal nurtured with the ideas of all citizens. Better co-designed policies should help citizens to live more fulfilling lives. They should start with a shared vision for the future of the place they live in and the planet. Strategic, holistic, transparent, ecosystem-based planning and management should strive to make the vision come true. Spatial planning in search of the sustainable regeneration of coastal cities, of their physical parts and of their extraordinary diversity, should not stop at the edge of the water but also consider marine resources. Urban sustainability agendas have to address many spatio-temporal patterns, both on land and the sea. Time is a scarce and precious resource and can serve as the litmus test of the well-being of societies. Local time plans can enhance the capacities of cities and improve resource allocation and quality of life for inhabitants and visitors. Urban observatories and sustainability indicators should take the pulse of coastal cities and their citizens and serve as compasses in the journey towards sustainable development.

This chapter sheds light into the dynamics of coastal renaissance for optimisation of urban functions on land and the sea, in the context of multiple pressures such as climate changes, natural hazards, erosion, and increased maritime activity. Citizens can play a decisive role in shaping vital urban spaces and forging bonds out of degraded spaces, including harbour infrastructures and disused seafronts that turn their back to the sea and the world. Distressed peri-harbour areas should be transformed into vibrant inclusive communities which can withstand shocks and attract sustainable activities. Symbolic and public and cultural projects on the waterfronts can promote collective life and local democracy and bring more value to places.

8.1 Urban Renaissance and Healthy Coastal Ecosystems

In the path to sustainable development, coastal cities hold a high potential to reinvent themselves. Each city is a living organism with infinite possibilities of regeneration. The factors of renewal multiply in coastal maritime and deltaic cities where the highest number and variety of land and sea ecosystems interact. Coastal cities are

V.P. Mega, *Conscious Coastal Cities*, DOI 10.1007/978-3-319-20218-1_8

not simply concentrations of people and activities on land and the sea, but hives of intense relationships and synergies, in which the whole can much exceed the sum of the parts and enhance the conditions for renewal. The balance among co-evolving policy objectives, including for maritime development, is very dynamic and impacts cyclical ecological processes.

Urban renaissance provides a framework for the harmonious and synergetic renewal of urban functions, in the context of multiple global and local, natural and human pressures, such as climate change and extreme phenomena, shoreline dynamics, and increased maritime activities. Inadequate coordination can lead to competition for space and pressure on valuable resources. Shipping, fishing and aquaculture, offshore wind energy, submarine cable and pipeline routes, can have severe impacts on coastal economic development and growth, as well as on coastal and marine ecosystems, leading to often irreversible deterioration of environmental status, loss of biodiversity and degradation of ecosystem services. On the other hand, healthy coastal and marine ecosystems are essential to continue deliver substantial services, in terms of food security, recreation, well-being and tourism, climate change mitigation and adaptation, erosion control and disaster prevention (Elmqvist et al. 2013).

The concentration of activities also leads to competition among sectoral interests, related to development, shipping lines and pipelines, wind energy operators, tidal and ocean energy infrastructures, fisheries and aquaculture sites, emerging marine biotechnology and seabed mining activities, and raises multiple environmental concerns. Ensuring the optimal distribution of space among relevant functions and stakeholders is essential to enable concurrent activities not simply to coexist but to achieve their full potential and generate synergies that can be captured for the benefit of all.

The sustainability paradigm helped cities liberate themselves from rigid concepts, models and practices inherited from the functionalist era and Le Corbusier's Charter of Athens adopted in 1933. A fundamentally interdisciplinary perspective and a multi-stakeholder approach are needed to question concepts, methods and policies towards systemic urban change in balance with nature and with active citizen participation. Cities and innovations have to be open to benefits from other world experiences, to create bridges, and to co-invent new concepts or ways of interaction.

The vision of a blue, and inevitably green, city is one of the last concepts in a chain of concepts for the possible and desirable sustainable city of the future (Beatley 2011). Visions had already embraced many interrelated concepts like the green city, the liveable city, the viable city, the affordable city and the ecological city. Blue cities join the eco-cities and eco-societies league, including cities which provide simultaneously ecological, social and economical models. Integrated urban planning and policy are instrumental for reducing the ecological footprint and improving urban performance towards a blue green city that can strike a sustainable balance with the seas.

Already in the 1990s, the OECD project on the "Ecological city" advocated for strengthening and integrating ecological concerns in all urban policies. The process proposed essential bridges between the macro-level concept of sustainable development and the micro-level of local activities (OECD 1996). Special attention has to be given to all urban infrastructures which have been long lasting and have influenced resource consumption for several decades. Public decisions on the future of infrastructures can foster a city's prospects. Urban planning can enhance opportunities for terrestrial and marine resource-efficient lifestyles.

For a vision to come true, discussion and mediation of multi-stakeholder approaches at the earliest possible stage are most important as the range of stakeholders progressively opens to include all of society. The Tokyo government proposed the concept of an eco-society towards a clean, sound and citizen-friendly metropolis. Comprehensive actions focused on resource and water management, transport, consumption patterns and promotion of environmental education and awareness (UN/Tokyo Metropolitan Government 1998).

Urban coastal planning evolved quickly and had to adapt and embrace already existing sustainability ethics and principles. The New Charter of Athens, issued by the European Council of Spatial Planners (ECSP), signalled a clear shift in planning values and objectives in accordance with post-war functionalistic principles and segregation of spaces for work, living, leisure and communication. The charter for the new millennium advocated for sustainable human settlements for all, based on true involvement, and responsible planning, which would promote socio-economic and environmental enhancement and safeguard traditional values. The 2013 Barcelona ECSP General Assembly approved the "Charter of European Planning", which promoted a more pro-active role for planners in shaping public debate. The Charter allowed development of an Action Plan and triggered discussions for an update of the Code of Conduct (ECSP 2013).

Strategic urban coastal planning has to be comprehensive and integrated and must involve the complete spectrum of urban and maritime actors and activities. In order to ensure the sustainability and environmental health of all urban functions, it has to embrace an ecosystem-based approach for the protection of natural resources and habitats that provide the basis for urban activities. It has to include a public and transparent process for analysing, planning, monitoring and reviewing the spatial and temporal distribution of human activities to achieve economic, social and environmental objectives. The ultimate aim is the optimal use of terrestrial and maritime space for different public, residential, industrial, economic and cultural activities. Transport infrastructure is vital and should be addressed in an integrated way for all modes, including maritime transport.

Sustainable development asks for healthy ecosystems on land and the sea. On land, sustainable cities opt for consolidation and renewal rather than expansion and urbanisation of greenfields. Concentrated and intensified use of space in a well-defined urban territory provides multiple advantages for the integration of urban structures that minimise flows of strategic resources and transport, local pollution and greenhouse emissions, which ultimately also benefits the bordering sea. New visions for synergies stemming from the mix of compatible urban uses affected the evolution of many ports and their articulation to the city, like at the port of Barcelona.

Until recently, cities have rarely accepted responsibility for marine resources, often because of the apparent immensity of the seas and the difficulties of visualising or quantifying the offshore effects of urban life. Urban maps stop at the water's edge, even though the activities that support urban systems extend further into the sea. To enhance vision, the Urban Whale, the New England Aquarium, produced a fascinating map of terrestrial watersheds and offshore waters on the US Atlantic Coast, showing areas of urban activity, including high boat traffic, shipping, fishing and dredging (Beatley 2011).

Cities have jurisdiction over near-shore habitats and can extend zones of planning and management to offshore areas. The Cape Cod Commission created an Ocean Management Planning District in order to extend its regional planning powers towards the open ocean. It also evaluated the scale, location and impact of offshore wind turbines. According to the EU maritime spatial planning system, coastal zones include the geomorphologic area on both sides of the seashore area. The seaward edge is the external limit of the national territorial seas and the landward border is the limit defined by the Member States in their integrated coastal management strategies (EC 2014).

Coastal cities can modify port operations and shipping practices to reduce environmental impacts. In the US Great Lakes and St. Lawrence Seaway, the maritime industry and port cities have formed a partnership to tackle problems including aquatic invasive species, greenhouse gas, air pollution and particulate matter emissions from ships and trucks, residues from local disturbances such as solid bulk cargoes, oil discharges, and noise, dust, odour and light pollution. Green maritime labels can help exemplary efforts to reverberate throughout the maritime economy.

Stronger international cooperation and regulations can also be instrumental for coastal cities as their sea waters unite them with many world regions. In 2009, the International Maritime Organization modified North–South shipping channels near Boston Harbour to reduce collisions with endangered North Atlantic whales. The US National Oceanic and Atmospheric Association also imposed mandatory speed reductions for ships over a certain size in "seasonal management areas". Blue cities have to reinvent the art of negotiation and compromise for the brokerage between diverse industrial and organisational interests, including for the benefit of marine populations.

Sustainable development ethics would hold cities responsible for the solid and chemical wastes that they produce and the wastewaters that should never impact coastal water quality. Coastal cities can take the lead in mitigating waste streams and supporting efforts to clean up existing waste. San Francisco, for example, through its Zero Waste policy, has been a pioneer in banning plastic shopping bags at large supermarkets and chain drugstores and has dramatically reduced the use of chemical pesticides in managing parks and urban infrastructures. Many more cities in the developing world followed the example.

Industrial alliances have already introduced many innovations. The California non-profit organisation Project Kaisei, in partnership with the New Jersey company Covanta Energy, produced energy from waste, to test a new catalytic technology for converting Pacific Ocean garbage into diesel fuel. Blue cities could be important partners in recovery projects by committing to power their buses and municipal vehicles with alternative fuels generated from ocean waste (Sesini 2011).

Sustainable coastal cities are responsible for promoting ethical and sustainable fishing practices. The world's rapidly growing and increasingly urbanised population will put severe pressure on fisheries, many of which are already depleted. Cities can shift fish consumption patterns, which concentrate on a small number of species through sustainable seafood education efforts, such as the Monterey Bay Aquarium's Seafood Watch Programme, and by preventing unethical practice. Urban plans can identify specific places for aquaculture and help develop synergies with non-conflicting land uses and actors.

Coastal cities investing in an ecological urban renaissance take special care of their urban and marine environment, their functions and their balance with human activities. Coastal and marine protected areas should be well integrated with urban and regional plans. The renaissance of Copenhagen, described as the largest on-going recycling project in Denmark, is founded on principles of quality and equality and aims at ensuring that sustainable development enhances all natural, marine, and human resources. Already in 1989, the regional plan for Greater Copenhagen tried to promote a "better city instead of a larger city". New principles can be injected into old plans and thus ensure change within continuity. The Copenhagen "Five Finger Plan" has been a prime example of post-war planning directions for future expansion of the city into the countryside surrounding the coast. Sustainability principles can preserve the green wedges and the marine environment, consolidate the fingers and equip them with highly performing public transport. The organic integration of urban structures, natural spaces and social life played a most critical role.

In Amsterdam, urban renaissance has been linked to the vision of a diverse and compact city optimising scarce and fragile land and hydraulic resources and their capacities to host housing and economic activities. Until the Second World War, Amsterdam had developed a series of concentric rings, embracing the central port area. The 1950 General Extension Plan added lobes like the fingers of a hand spread out into the Dutch green heart. During the reign of the private car, concentric canals started being filled in to provide more space for car traffic. Ecological demands led to reopening of canal rings and intensification of the land use and the promotion of public transport. The marine environment has been rediscovered and enhanced. Many Dutch cities created mixed-use and citizen-friendly environments, including pedestrian and bicycle bridges spanning urban spaces and canals (Pistor et al. 1994).

Under the denominator "Port-City", the Port of Amsterdam and the municipal Spatial Planning Department have explored a future perspective for the Western Dockland, a busy port area within the city centre of Amsterdam. Both the city and the port had expressed in the past conflicting needs for more space and, after a period of divergence, they tried to work on synergies. They called for reorientation of the relationship between the city and the port towards a better urban-maritime symbiosis along the river. A key question was whether the existing port activities can coexist with housing, or that relocation or restructuring of port activities would be required for new housing over the longer term (Port of Amsterdam 2009).

Large sustainability waterfront projects demand foresight, long-term planning, flexibility, and ex-ante impact assessment, citizen participation and communication. The success of the projects depends on public support and a constant and affirmed political determination, capable of withstanding changes in elected representation. Flexibility is imperative for adapting high-scale projects to market fluctuations, whereas continuity is linked to a shared vision for the future of the infrastructures. Citizen consultation and partnership with major stakeholders are extremely important. Outstanding examples include the Tokyo Waterfront, part of the Japanese National Policy for introducing private capital to public high-level projects (Metropolis 1996).

Coastal beacons projects can offer significant landmarks to the city, structure urban territories, and enhance magnificent skylines and shorelines. They can act as strong catalysts for the future of cities and regions. They range from ambitious government plans or unique international events, such as Olympic Games, Universal Exhibitions and high-level fairs, to local maritime and vessel museums. Their planning and construction with respect to sustainability principles are fundamental given their emblematic role and manifold effects.

Resource-conscious architecture has made great strides over the last decades and has created outstanding examples of marine urban conceptions from floating sites to amphibious housing anchored to the land as the structures rise with water levels. The 2014 sixth edition of the International Architecture Biennale in Rotterdam focused on "Urban by Nature" included a lively debate on Delta cities. Jakarta, Alexandria, New Orleans or the cities of the Dutch green heart can all look forward to effectively address major challenges.

8.2 From Vision to Strategic Ecosystem-Based Planning and Management

Since Hippodamus and the first grid plans, the collective search for optimal urban forms for the desired future has produced many interesting visions and concepts. Strategic integrated planning is an important instrument towards the preferred future of sustainable cities and impacts the flows of energy towards programming the various actions. It defines the location and distribution of various diverse activities and infrastructures in place and time. It has to be an open, transparent and inclusive process and involve all stakeholders. For sustainable coastal cities, the ecosystem-based approach to planning is a major fundamental principle.

Coastal cities should undertake lifecycle resource assessments, appraise investments and opportunities, risks and threats, respect geophysical and cultural local limits, seek a symbiosis with the bioregion and mobilise all visible and invisible societal hands. Strategic urban planning for sustainability requires a comprehensive interdisciplinary assessment of urban assets, a natural resource information system and an identification and analysis of policy distortions and bottlenecks. Prevention of pollution in marine ecosystems is a major investment.

Vision building is the first mobilising element for creating sustainable cities of the future. It has to be compelling and it is increasingly subject to consensus and participative governance. The discussion of many possible alternative futures, envisioned by all stakeholders, is a key element in this dialogue, in order for the future not to result as an unwelcomed or linear continuation of the past. Multi- and interdisciplinary, scientific, technical and social approaches are crucial, as many challenges are global, multi-faceted and inter-dependent.

Increased macro-complexity and uncertainty create a space for participatory innovation and cooperative processes. Paradigm shifts and quantum leaps are possible if integrated in a coherent vision for the desired future. Strategic foresight, which has always to be accompanied by hindsight, deliberately cuts across the traditional boundaries of sustainability science disciplines and policy areas. It can act as a driver of social interactions that stimulate the generation of common public visions.

Strategic anticipation includes the exploration of future possible prospects through the early identification of emerging challenges, opportunities and threats and horizon scanning for unforeseen turbulences and weak signals and "black swans", low-probability/high-impact events. Setting the strategic questions, identifying driving forces of change, determining main issues and trends, clarifying levels of impact and risks and degree of uncertainty, creating probable scenario narratives, assessing policy options and identifying inflection points are important elements of the process. Ex-ante impact assessment of possible, probable and preferred futures, exploration of alternative scenarios and consensus building can lead to an ambitious, collective and engaging vision.

Some governments and cities have integrated futures thinking and strategic foresight in their decision-making process and developed policy portfolios to achieve the desired visions according to sustainability criteria and financial constraints. Scenario planning is often complemented by risk assessment approaches, especially for cities facing important storm surge and sea level rising. The process, well integrated in the heart of governments, could help strike a balance between bureaucratic effectiveness and creative thinking and lead to forceful goals backed by all stakeholders.

Cities striving to reach their various intended destinations need to be well aware of their starting positions, their strengths and weaknesses. They have to decide together with their citizens where they want to be in the future and understand the significant trends that may influence the direction in which the future unfolds. Striving in the global knowledge economy for a sustainable future requires a dynamic local balance among economic, social and environmental objectives and demands that the appropriate people, skills and capabilities be developed, and that city leaders demonstrate their abilities to appreciate these assets and invest in the ways that they can be enhanced and allowed to prosper.

Integrated approaches to the management of marine and coastal resources increasingly incorporate systems-oriented approaches based on precautionary and ecosystem management principles. Implementation of Integrated Marine and Coastal Area Management (IMCAM) is gaining ground for the sustainable use of marine and coastal ecosystems and habitats.

Community-based and ecosystem management approaches have proven particularly important. IMCAM is a participatory transparent process for decision making to prevent, control, and mitigate adverse impacts from human activities in marine and coastal environments, and to contribute to the restoration of degraded areas. It involves all stakeholders, including policy makers, the business community, advocacy groups, civil society and the general public. A series of best practices bear witness of the efforts of cities to achieve a balanced approach to preserving, yet enhancing, marine and coastal biological diversity.

Marine spatial planning and integrated coastal management create a most suitable framework for addressing human impacts on marine and coastal biological diversity and for promoting sustainable use. Crucial components include industrial activities such as construction and mining in coastal areas, mariculture, mangrove management, tourism, recreation, fishing practices and watershed management. Destruction and degradation of vital habitats should be prevented and deteriorated habitats, including spawning areas and nurseries of stocks of living marine resources, should be restored.

Urban strategic plans have to reconcile thematic and territorial policy objectives and be analysed and discussed, co-decided, co-implemented and co-evaluated. Integrated urban coastal management should empower, enable and invite all stakeholders to have more control over the natural resources upon which their livelihoods depend. Improved understanding of the various marine ecosystems surrounding a city, their functions and the ways that human activities impact them is a first critical step. The participation of all individuals involved helps develop multiple management approaches to the use of coastal ecosystems and resources, and the broader identification of the actions which allow meeting of objectives without adversely affecting the natural ecosystems sustaining their activities. The process can incorporate a framework for arbitration and conflict resolution and shared vision building and develop awareness on the threats and opportunities. Every coastal city should invest in developing the ability of those with responsibility for urban policy to manage natural, including marine, resources in a sustainable way.

Many cities envision ambitious agendas for 2030. The PlaNYC was released in 2007 as an unprecedented effort to prepare the city for welcoming one million more residents, strengthen the economy, address climate change, and enhance the quality of life for all present and future New Yorkers. The Plan brought together over 25 City agencies to work toward the vision of a "Greener, Greater New York". The updated plan included 132 initiatives and more than 400 specific milestones (New York City Mayor's Office of Long-Term Planning and Sustainability 2011).

By 2030, the population of New York City is expected to increase to more than nine million, including newcomers and present citizens along with the next generations. This intergenerational multicultural city can offer tremendous opportunities, enrich communities, enhance green environments and optimise well-being for all. Urban planning processes have to enhance the capacity of public infrastructures to improve quality of life.

Accountability is a sine qua non condition in the journey to sustainability. Progress against goals should always be closely monitored and publicly disclosed. In New York City, 64,000 units of housing have been created and 20 new neighbourhoods became accessible by public transport, during the years 2007–2011. The city's first bus rapid transit system has been launched and $1.5 billion was committed for green and blue infrastructures. Nearly half a million trees are being planted and investments focus on the drinking water supply network. Over 30 % of the yellow taxi fleet is already green and regulations are expected to phase out polluting fuels. The process to remediate and revitalise brownfields is being streamlined and the clean-up of the most degraded lands is progressing. Last but not least, public plazas have been created for pedestrians, including on Times Square, the "crossroads of the world", attracting residents and visitors.

In Singapore, the Concept Plan, initiated in 1971 and regularly updated, is a strategic land use and transportation plan to direct development in the longer term. The Concept Plan focuses on spatial resources and dynamics to meet long-term population needs and aspirations. The first Concept Plan laid the foundation for Singapore's growth for a better quality of life with new towns, transport infrastructure and access to recreation. The Concept Plan was subsequently reviewed in 1991, in 2001 and 2011 to factor in changes in local and global trends, and ensure that plans remain relevant to address challenges and meet future needs. The plan has played a vital role in helping balance the diverse land use needs, such as port, industry, commerce, housing, parks and recreation, transport, culture and community facilities.

Singapore could have a population of between 6.5 and 6.9 million by 2030. The plan outlines the strategies to support population and economic growth, with the strategic intent of ensuring a high quality living environment for all Singaporeans. The reviews benefited much from extensive public consultations carried out through various channels, such as surveys, focus group discussions, and public forums. Through engaging all stakeholders and the public at large, the city can better understand the concerns and aspirations of the various communities. The most recent 2011 review took into account the opinions gathered by the National Population and Talent Division (Urban Redevelopment Authority of Singapore 2011).

Tokyo, a labyrinth of cities, enjoys limited atmospheric pollution, low unemployment, an efficient public transport system and high healthy life expectancy for its citizens. Super-ageing is a key challenge for the future of the city, which has also to address many natural threats. Risk governance has often been the source for new concepts and processes. The Kobé earthquake in 1995 provoked a whole range of innovative managerial responses. The plans reconstruction includes cardinal innovations for the disaster-proof city, born out of the eternal urge to create something eternal.

The WBCSD Urban Infrastructure Initiative proposed several solutions that will help Kobe to address economic, environmental and social challenges, including an ageing population, and achieve its sustainability ambitions. The current 5-year master plan emphasises creativity and innovation, local economic and sustainable community development, population and livelihoods. Major companies and the WBCSD worked with officials from several city departments to identify key sustainability issues and propose policy options. They identified priorities for action, including energy efficiency and renewable energy production, sustainable mobility, knowledge networks and leadership in disaster resilience.

A transformation study led to the formulation of 14 practical solutions including energy efficiency improvements, sustainable mobility measures and knowledge-based economic development. The process demonstrated the value of providing multi-sector input early in the city's strategic development. This enabled city officials to consider a variety of ideas and engage with the private sector collectively, rather than only in relation to specific projects. Such holistic thinking is increasingly important for inclusive and sustainable urban development and of special value for coastal cities (WBCSD 2013).

The population of London is expected to rise as the city's capacity to attract people from all over the world, from wealthy expats to asylum seekers, is expected to continue. This could be reflected on all plans. In 2030, London could be a city that is over 40 % overseas-born communities and may be constituted 40 % from non-white population. The city could be more cosmopolitan and tolerant, and more attractive to talented creative mobile citizens of the world.

Integrated management programmes have already demonstrated their potential as an effective tool in developed and emerging countries around the world. Xiamen, one of the world's top 20 ports, has sustained one of the fastest rates of economic growth in China. The rapid growth and the increased intensity of sea use, coupled with the lack of adequate regulations, coordination and enforcement, led to conflicts among various marine and maritime activities and provoked the degradation of native species and natural habitats. Largely sector-oriented legislation and operational mechanisms to harmonise development across sectors were in general weak. In 1994, the Prevention and Management of Marine Pollution in the East Asian Sea introduced integrated coastal management. The aim was the integration of various coastal and marine activities and coastal environmental management for holistic and sustainable development. The cornerstones of this approach included an inter-agency coordinating mechanism, a multi-disciplinary experts group, an integrated profile of the coastal sectors and a strategic management plan. The results included reduced conflicts among various sea operators, better enforced marine protected areas, rehabilitation of threatened species and degraded habitats and a better environment for all to enjoy.

Sustainability debates insist on density and intensity and call for limiting urban sprawl. Urban sprawl can disproportionately increase the ecological footprint of cities and aesthetically deform their peripheries, the "lost cities", as thay are called in Latin America. The threats are very high for cities that are surrounded by trademark landscapes or very fragile water and marine ecosystems. The benefits of maritime spatial planning include generation and enhancement of synergies between different activities, encouragement of investment, by instilling predictability, transparency and clearer rules, reduced conflicts and a better ownership of the plan by local stakeholders.

Strategic plans can provide legitimacy for activities to support sustainable development, such as development of renewable energy sources and grids, and establishment of Marine Protected Areas. They should also increase coordination and help balance the development of a range of marine and maritime activities. The early identification of impacts and opportunities for multiple use of space and the coordination of land-sea interactions of coastal activities is a major advantage. A holistic approach and multi-stakeholder dialogue are always the most critical parts of the process.

Climate change, in particular the rise of sea levels, acidification, increasing water temperatures, and frequency of extreme weather events, is likely to cause a shift in economic activities in maritime areas and alter marine ecosystems. Strategic planning can play an important role in prevention and mitigation, by promoting the efficient use of maritime space and renewable energy, and a cost-efficient adaptation to the impact of climate change. Strategic plans for sustainability focus upon bringing abandoned urban land into mixed-use development and seek to restrain peripheral growth to key nodes near public transport stations. Cooperation beyond administrative and sectoral borders is a major issue and has often to overcome local divisions and invest in bridging territories, people and opinions. Integrated risk management has to take into account the perception of risks and the tolerable levels by all stakeholders.

As all cities, coastal cities are chronotopes, with interconnected spatial and temporal dimensions and interrelated historical and geographical aspects. Port cities and cities which support large flows of passing passengers and goods are particular chronotopes. Like space and water, time is a scarce resource for cities. The time dimension does greatly matter in advancing towards sustainable development. It introduces concerns about inter-generation distribution of capital and serves as a litmus test for the well-being of individuals and societies. Time management has a potential for extending the possibilities of spatial planning. Some governments and cities have been pioneers in promoting time plans that strengthen places (INU-Politecnico di Milano 1997). Special 4 seasons or 24 h plans for territories like harbours could be very beneficial for all actors spending a significant part of their life on the coast.

Seasonal planning can integrate temporary structures for cultural or emergency reasons. London has commissioned the world's most renowned architects to design temporary structures to host summer events. The concepts of seasonal planning and policy could also be most interesting for tourist resorts or coastal cities which serve as gateways for important tourism flows or host different populations during various seasons. The ecosystem-based principle could be extended to embrace place and time-based approaches to resource management. Sound understanding and strong citizen participation to address time management issues are crucial for managing human uses and impacts for the well-being of all.

Strong sustainability actions ask for the enhancement of every possible space or resource and especially the ones traditionally seen as waste which should be seen just as a by-product or as another product. A combined rubbish incinerator and power plant with a ski slope on its roof by the architect B. Ingels is an inspiring example in Copenhagen. A mountain created on a multi-storey garage can help achieve functional, aesthetic and health objectives and address Denmark's lack of mountains. The same architect is also the creator of the "8" house on the edge of Copenhagen, an apartment complex that wraps around itself. The rising and falling roof creates a continuous platform and park and cycle track and inhabitants can bike to their elevated front door.

Crowd-sourcing urban design and crowd funding can lead to joint ownership of urban commons to be collectively enjoyed. In Copenhagen, the Superkilen, an extraordinary one-mile long urban corridor that runs through the cosmopolitan waterfront Nørrebro quarter, reflects the diverse cultures of its residents originating from 50 different countries. The creation of the park, also by architect B. Ingels, has been a laboratory of ideas and ideals with the participation of all foreign citizens expressing the wish to find in Superkilen some essential elements of their original culture. The urban corridor was enriched with a Moroccan fountain, Chinese palm trees and benches, cutting down barriers and creating bridges with the origins of the inhabitants.

Increasing population and employment density in metropolitan areas could reduce vehicle travel, energy use, air pollution and CO_2 emissions. Urban sprawl, largely made possible due to private cars and extensive highways, still reflects the preferences of many citizens for living in single-family homes. Dispersed, car-dependent development patterns, however, involve high economic and ecological costs, including the use of vast quantities of land, increased reliance on private cars and greenhouse gas emissions. In the US, compact development is focused on new housing, as converting existing housing to higher densities could be prohibitively expensive. Compact, mixed-use neighbourhoods could reduce trip lengths, and make walking, biking, and public transit more viable alternatives to driving. The key precondition is for jobs, schools and shops to be equitably integrated in the urban fabric (TRB 2009a, b).

Compact coastal cities manage resources at remarkably lower levels of material and energy consumption, compared to diffuse settlements and dispersed populations and tentacle-shaped low-rise suburbs across the coastline. The interrelated issues of density and compactness are critical indicators for sustainability. Compact settlements encourage functional diversification and integration of land uses at the neighbourhood level and prevent urban sprawl and resource overconsumption. The Danish model of decentralised concentration highlights the importance of all these components, while the Dutch compact city policy is based on the principle of spatial multi-functionality. Creating multi-purpose urban cells is a guiding principle for many city plans (World Bank 1995; OECD 2012a).

Amsterdam considers functional mix as a valuable attribute of an inner city heritage and tries to strike a balance among spaces for housing, offices, commerce, services, tourism and leisure. The compact city policy, introduced already in 1985, aimed at enhancing scarce space as efficiently as possible, introducing more scrupulous resource management which created multi-functional residential environments and curbed the overall ecological burden. Diversity and mixed-land uses are linked to enhance the city's unrivalled character as a cultural melting pot.

Sustainable coastal cities must reflect a true urban, marine and social intermixture. The integration of urban functions should reinforce identity, but also improve communication and openness to the world. Many cities experience a need for functional mix, a community desire for "a real urban gene, an organic part of the city, everywhere in the city". The "open block" proposed by the architect Ch. de Portzamparc advocates for urban blocks which open citizens to the city and the world. The concept could favour the interactions and social life at the scale of each neighbourhood in the world of the city (De Portzamparc 2007).

Sustainable regeneration and consolidation of urban areas in decaying waterfronts has been a key instrument by many cities in their efforts to attract new business and residents. Sustainable renewal has to address the unrealised potential of land-sea interactions and involve all stakeholders willing to invest and promote life and work in healthy marine environments. Revitalising an urban area entails recreating its economic diversification, its social heterogeneity and cultural diversity of the city. Successful schemes address both the hardware and software of the areas and try to reconcile environmentally sound revival of physical structures with socio-economic and cultural enhancement.

Some harbours went a long way from mono-functional areas towards multi-purpose mixed-use neighbourhoods with integrated housing, employment and educational infrastructures. Previously deserted and dangerous in the evening harbour and commercial streets have become the thoroughfares of vibrant neighbourhoods. Further inspiration could be offered from the "Living above the Shop" project in Dublin, a prime example encouraging and assisting shop owners to convert their upper floors into residential spaces. Similar concepts were developed for many port areas recreated as lively neighbourhoods, providing for instance housing for immigrant workers.

Managing the dual space port-city is of the highest importance. The port of Antwerp invested much in reviewing its sustainability performance in the context of the city. With this benchmarking exercise, the Antwerp port community gave form to its ambition of being a sustainability leader. It was the first time that the many interlinked sustainability aspects, of both of the port and of the hinterland and foreland, were reviewed by the entire array of port community. Transparent, open and systematic consultation was seen by the various stakeholders as particularly useful and concluded with recommendations for improvement.

Particular attention has been paid to safety, with the emphasis on accident prevention. The modern port is a very busy environment with many risks, and safety is a priority for the port community. Much attention was given to local nuisances affecting either the environment or people, or both. But the scientific data collected to monitor and prevent the nuisances does not always match citizen's perception of the negative aspects of living near a port. A comprehensive survey of the perceptions of nuisances in and around the Antwerp port provided particularly useful complementary information about the negative effects of port activities and the ways that are experienced by the local population (Port of Antwerp 2010).

As regards energy, the first concern is to make more rational use of energy and increase efficiency. In an industrial environment such as the port of Antwerp with one of the world's largest chemical clusters, there is a huge demand for energy. The port community therefore strives to maximise the use of green energy. The port authority tries to lead by example through purchasing 100 % green electricity. Significant expansion of the wind power capacity within the port is also planned, and the possibilities of bio-based energy sources are being examined. Further research in energy efficiency is continuing, including on a heat distribution network.

The study of the feasibility and desirability of urban-maritime combinations along a river paid particular attention to multiple factors such as maritime value, housing demands, metropolitan accessibility, landscaping, investment costs and overall sustainability prospects. From the viewpoint of sustainability, the growth potential and land demand of both city and port are in balance and an impulse is given to better connectivity through urban public transport.

The future of blue urbanism may include floating cities or other forms of permanent or semi-permanent habitation of the ocean environments, either on the surface or underwater. The American designer J. Fresco, already before the second war, invented the trend home, one of the first glass aluminium structures, and proposed model ocean cities. These cities are unique to their regions. By maximising efficiency with space and context they also can be interlocked and formed into constellations. French visionary architect J. Rougerie proposed many concepts to realise the tremendous potential of inhabiting the oceans, thus raising awareness about the beauty and the fundamental role of the sea in humanity. He proposed underwater laboratories, a subaquatic archaeological museum for a journey to sunken cities, a floating scientific city and an underwater habitat-observatory.

Many more urban utopias have been proposed such as self-contained cities in the shape of a lily pad or a lotus flower and fertile cities such as the "Recycled City", of half a million residents, that would be created from recycled plastic from the North Pacific Garbage Patch. Other concepts include amphibious tourist resorts, as included in the Greek pavilion during the 2014 biennale of Architecture in Venice (Aesopos 2014) and the extension and expansion of cities in vertical and submarine spaces. Human populations already living on ships and multi-purpose marine platforms, in a close connection to oceans, could be the precursors of the citizens of the floating cities of the future (Beatley 2011).

8.3 The Hallmarks of Urban Ports and Noble Waterfronts

Advantageous port conditions, adequate defence infrastructure and good inland connections are the principal factors that have guided the evolution of harbours and their complex relationships with the rest of the world. Ancient harbours were usually more exposed to the sea than the treasured agoras, at the very heart of the city. Many ancient port cities still host civil and military harbours protected by towers and fortifications. Gates ensured the entrance into the town and the ancient agoras. Many harbours throughout history were later turned into commercial ports enhancing valuable port conditions, defensive infrastructure and inland connections.

Security from the sea has always been extremely important for coastal cities. In Dubrovnik, the city walls are considered to be one of the most grandiose fortification monuments in Europe and a fine example of fort architecture. The first fortifications were built already in the eighth century, but the most intense construction took place from the mid-fifteenth to the end of the sixteenth century. The ramparts encompass the city in an irregularly shaped polygon form, with Fort Minčeta at the highest Northwestern landward corner of the city and with Fort St. Johns on the South-eastern seaside. Other strong forts are also part of this defensive urban infrastructure, even though they are separated from the ramparts. The ramparts are 22 m high in places, and from 4 to 6 m thick on the landward side and from 1 to 3 m thick on the seaside. In front of the central wall on the mainland side is an outer wall with 10 semi-circular bastions, in front of which used to be a moat. The main wall has 14 quadrangular and 2 circular towers, 2 angular fortifications and 4 bastions. This imposing structure has become a precious legacy.

Much of the monumental historic parts of harbours can be shared and enjoyed by all. Noble public spaces on the seaside or with specific sea views may foster citizen participation and promote exchanges and interactions. Open-air infrastructures, like ancient fortifications, but also cultural places like theatres can play an important role as shared civic spaces. Urban leaders and citizen associations should always take care for public places to be accessible to all and include everybody.

The Agora, the focus of civic life in the archetypal City-State, constitutes a powerful public model place embracing cultural and commercial spaces and orchestrating all urban functions. Citizenship, justice, culture, and exchanges were well anchored in these spaces, where the assembly, the theatre, the stadium, the market place and the sanctuaries had the noble aim of promoting the physical and mental well-being of citizens. Environmental and cultural landscaping of public spaces can help to forge urban identity. Qualitative recommendations for the functional and aesthetic character of squares, waterfronts, seaside plantations and public lighting have been developed and implemented in many cities.

Historic public spaces can invite travellers to endless journeys into ancient times. The Sacred Street in Athens goes deep into place and time. The city which invented theatre, philosophy and democracy reorganised its ancient heritage surrounding the Acropolis, crowned by the Parthenon, into an urban archaeological park to be best enjoyed by citizens and visitors as a backdrop of everyday environments. The park incorporates also green spaces and micro-squares and pedestrian paths and bicycle routes providing a particular essence distilled from the magic of the ages.

Amphitheatric planning of coastal cities provides visual access to the sea for everyone. Often the built environment skilfully couples with the physical landscape. Some Mediterranean cities provide extraordinary lessons. Taormina occupies the site of an ancient town, on a lofty hill. The site is about 250 m above the sea, while a very steep isolated rock, crowned by a castle, rises about 150 m higher, indicating the place of the ancient citadel, the inaccessible position of which is mentioned by ancient writers. Numerous fragments of ancient constructions are scattered in the landscape, including extensive reservoirs of water, sepulchres, and pavements.

The most remarkable monument in Taormina is the ancient theatre, one of the most celebrated in Sicily, due to both its outstanding location and preservation state. Rebuilt upon the foundations of an older theatre of the Greek period, this theatre is still used for theatrical performances and music concerts. The greater part of the original seats have disappeared, but the wall which surrounded the amphitheatre and the proscenium with the back wall of the scene are preserved in singular integrity, and contribute much to the majesty of the site. Parts of a temple are also visible, converted into the church of San Pancrazio.

Catania offers tangible and intangible assets which bear witness of a stunning relationship between the city, the nature and the sea. Founded in the eighth century BC, colony of the Greek metropolis Chalcis, the city counts with three theatres which couple with the landscape. They include the Teatro Romano, built on the site of a Greek theatre in the second century AD, the Odeon, a smaller theatre used for more intimate occasions, such as musical performances, and the Amphitheatre dating from the Greco-Roman period. The city lies on the Ionian Sea, under the shadow of ever-present Mount Etna. On several occasions, volcanic eruptions destroyed the city. In 1669, an eruption covered the city with lava and, in 1693 an earthquake shook it down to its foundations. The entire old urban part was rebuilt in Baroque style, with large, wide open squares and avenues. The most amazing aspect was the building material, lava, which gave the city its particular colour!

Catania is also a lively city with a colourful and bustling fish market and an atmosphere that has remained virtually unchanged for hundreds of years. The old fortress of Castello Ursino is an imposing structure on a rocky cliff overlooking the sea. However, the massive lava spill of the 1669 eruption pushed the sea back, creating a new coastline and stripping Castello Ursino of its strategic position.

Many modern cities try to enhance elements of their coexistence with the sea. Contemporary cultural spaces often draw vigour from their connection to the sea and endow citizens with the legacies of the future. The Kursaal auditorium and congress hall in San Sebastian, conceived as two gigantic rocks stranded at the mouth of the Urumea River, has been conceived as an organic part of the landscape. All other facilities, including the exhibition and music halls, meeting rooms, offices, and supporting spaces are located in the platform at the base of these cubes, where the cultural centre meets the city, and ensure an open public access.

The auditorium celebrates its character of almost a geographical accident with a slight inclination towards the sea. Its volume, integrated asymmetrically inside the glass prism, seems to float within it. The glass surfaces protect against sea winds, making the volume a dense, opaque mass by day and a source of light by night. The orientation makes every visitor entering the foyer to unconsciously walk towards the highest level where Mount Urgull and the sea can be contemplated in all their splendour from a singular window. Similar design and structural criteria have been used in planning the smaller congress hall, also conceived as an inclined prism. The rectangular congress hall adheres to the best acoustical and functional requirements. The asymmetry is less evident, but the view from the foyer of Mount Ulía and the sea in the background is just as almighty.

Architecture inspired by the sea and the forms of waves have provided many outstanding examples. The hegemony of the sea is often reflected on public spaces and exceptional buildings designed in coastal cities and strengthening visual and spatial connections between the ocean and the city. The Oslo Opera House features a dramatic granite roof sloping into the city's fjord and creates an awe-inspiring public plaza, a symbolic springboard of a coastal city to the world.

The project has been designed as the first element in the transformative urban agenda of the Norwegian capital. The marble-clad roof forms a large public space in the urban landscape and the fjord. Endowed with an imposing size and compelling aesthetics, it powers a dynamic osmosis among three elements, the wave wall, the factory and the carpet. The wave wall divides but also unites a dynamic surface, represented as an awakening wave. The factory is a symbol for the production facilities of the opera house, organised as a self-contained, functional and flexible space. Finally, the carpet epitomises the horizontal monumentality of the city, the web of shared ownership, providing access for all to the sea of art and culture.

Reykjavik, the world's Northernmost capital of a sovereign state, has a unique relationship with the sea. The whole city is turned to the sea, with its multicolour houses and its lake surrounded by green spaces in the city centre. Sustainable architecture by the sea endowed Reykjavik with Harpa, the Concert Hall and Conference Centre also inspired by the boreal light. Situated on the waterfront, Harpa stands out as a radiant construction with a clear view of the enormous sea and the surrounding mountains reflecting both sky and sea space, as well as the vibrant life of the city. Harpa features a welcome foyer area in the front of the building, four halls in the core space and a backstage area with offices, administration, rehearsal hall and supporting services. At the heart of the rock, the largest main concert hall reveals a powerful interior. Seen from the foyer, the halls form a mountain-like massif, similar to the basalt rock on the coast, in a stark contrast to the impressive facade.

Waterfronts are spaces of expectation and wonder for a very diverse range of populations. Being there is already being somewhere else. They play a cardinal role in the integrated management of the dual city-harbour resource. The recreation of the Belfast waterfront has been driven by the strong will to create a new face for the city with spaces of hope superseding places marked by violence. In Galway, the renewal of a derelict central area, past hub of economic activities, led to a harmonious marriage of rehabilitated buildings and new constructions. The regeneration respected Galway's unique character and atmosphere and promoted a functional mix, essential for the vitality of the city centre. A balance has been struck among residential, commercial, cultural and tourism functions (BURA 1997).

In Boston harbour, a symbol of the city's birth right, civic activism triggered a dramatic regeneration and an outstanding new waterfront. The Boston vision for 2030 charts the way to a resilient city reinventing the innovation economy and promoting a civic agenda. Disused dock infrastructures have been transformed into exhibition halls, shops, craft workshops and centres for ecological, leisure, civic and cultural activities. Business incubators bring new capital into the areas and help create local enterprises and services. Functional diversity is increasingly important and public access to the waterfront is considered to be decisive. Waterside promenades gradually replace industrial docks and welcome citizens.

The Eastern Dock Island at the east of Amsterdam's Central Station is a component of the large-scale project to develop the South bank of the bay. The move of harbour functions has presented new opportunities to transform the area into an intensive mixed urban neighbourhood with residential, commercial, recreational and cultural functions, including the new Public Library and the Conservatorium. The Library, designed by architect J. Coenen, after an exploratory complex quest, includes a central underground system for long term energy storage delivering air conditioning and heat.

In Estonia, the waterfront of Tallinn has been enriched with the Seaplane Harbour, a maritime museum inaugurated in 2012, in a building originally constructed as a hangar for seaplanes. The hall, in the historical seaplane hangar, out of service during the Soviet era, has been thoroughly renovated to host the exhibition bearing witness of the maritime past. The museum presents the history of the old maritime Estonia in a modern visual language. The exhibits are located in the air, on the sea and below the sea. The museum has submarine and flight simulators, and a pool to sail aquatic animals and miniature ships. A 1936 submarine has been renovated to its original state. The museum also displays a yellow submarine and a full-scale replica of a World War I seaplane, the wreck of the wooden ship Maasilinn, and the icebreaker Suur Tõll, conquered from the Russians near Helsinki in 1918 and finally donated to Estonia.

Genoa's harbour, the most important in Italy and one of the most eminent in the Mediterranean, has become the centre of cultural, political, tourist and commercial activities. In 1992, on the occasion of the Columbian Celebrations, the waterfront was thoroughly redeveloped and the ancient harbour zone rehabilitated and opened to the sea. Native architect R. Piano restored the historical buildings, including the cotton warehouses, and created new landmarks like the Aquarium, the Bigo trademark of the port activity, and a Sphere made of metal and glass, installed in the water, and unveiled in 2001 at the occasion of a decisive G8 Summit.

In Thessaloniki, the creation of the new seafront has been an ambitious project involving the development of 80,000 square metres, pedestrian and green areas, athletic and recreational facilities, public sculptures, an outdoor amphitheatre and playgrounds, next to water commons. Almost 1000 trees were planted and cycling paths have been expanded. Upgrading the relationship between the city and the sea has been a major goal of the project, which also provided a green lung and 1 km pedestrian walkway along the seafront.

In Finland, Turku addressed creatively the decline of the harbour industry and infrastructure on the river Aurajoki. The majority of Finnish maritime industry companies are based in the region. In 1987, the municipal council organised an architectural competition for a new master plan and its phantom spaces of closed down factories and warehouses. The winning entry "Sigyn" introduced a resplendent mix of old and new structures in brick, steel and glass, and proposed magnificent functional spaces for educational, economic and cultural purposes. Two massive former shipbuilding halls and a former rope factory, once voted as the ugliest building in town, composed a major fine arts complex, including a conservatory, the Turku School of Art and Communication and the School of Fine Arts.

Shaping the places to anchor a sustainable future is a critical issue for most cities. The Rotterdam innovation dock campus, at the very heart of the Rotterdam City Ports area, has been established on the former premises of the Rotterdam Dry Dock Company. Founded in 1902, the former booming shipyard closed down in 1983 and started a new trajectory in 2007. Building on the ideas of the Rotterdam Climate Initiative, bringing together all stakeholders to achieve decisive reduction in CO_2 emissions, the area was redeveloped into a low-carbon place and an incubator for creative and innovative businesses. Educational institutions and companies work together on sustainable options in the fields of building, mobility and energy. Far more cities have initiated beacon initiatives for the transformation of former naval and industrial infrastructures into campuses for educational activities, research institutes, and innovation businesses.

Many urban waterfront regeneration projects focus on areas which are far larger than the immediate waterfront and often have the ambition to improve the whole city and its access to the sea. The renewal of the London docklands is more than emblematic of transformative waterside projects. During the nineteenth century, London's harbour was one of the busiest in the world, but, by the end of the 1950s, decline of the port industries and manufacturing, containerisation asking for docks equipped with large cranes and an increase in ship size made many docks derelict. These developments led to a spiral of decline, marked by decrease of population and employment, inadequate services, lack of open space and recreation facilities, and poor access to the rest of London with heavily congested narrow roads and a lack of public transport.

The London Docklands Development Corporation, set up in 1981 to lead the project, has worked for 17 years to bring a new face and significance to the place in quest of a new future. Other organisations involved in the redevelopment process included the national government offering incentives to encourage private investment, property developers, responsible for building large office blocks like the Canary Wharf, local housing associations which obtained home improvement grants and civil society organisations and advocacy groups (Hall 1998).

Ecological improvements included a network of pedestrian and cycle routes through the area with access to the river, the creation of bridges and new open public spaces, a water ecological park and London's first bird sanctuary at East India Dock Basin, and the planting of 200,000 trees. The economic regeneration led to the doubling in employment and numbers of businesses, the conversion of old warehouses into new homes, the opening of the Docklands Light Railway and the City Airport in the former Royal Docks, attraction of high-tech firms, and media headquarters.

The rapid rail connection to central London which greatly improved the accessibility of the docklands has been decisive for local businesses. Benefits included a wide range of economic, environmental and social advances, including thousands of new housing units and quality jobs. Most of the criticisms related to the lack of benefits for the local population, as the original "East enders" including many old dockers were unable to afford the high costs of the new expensive houses and had not the skills required by the new economy. The community spirit of the area has also been altered as the newcomers did not truly mix with the previous local society.

Creating a true Porto Maravilha is a great plan for Rio de Janeiro. Its waterfront, one of the city's oldest strategic assets, played a fundamental part in the past economic and social development but declined during the last decades. The revitalisation of the waterfront is expected to reintegrate it with the city centre, respecting principles of integrity, sustainable urban development and social inclusion. Enhancing the sustainable value of the place and the creation of new cultural facilities will make it one of the most attractive areas of the city. The municipal government also initiated the construction of the Rio Museum of Art.

The Porto Maravilha Urban Operation aims at providing Rio with a vibrant city centre, a sustainable urbanised area and a multimodal transportation system blending with the region's historical and architectural heritage. The project is also based on real estate development with emphasis on socio-economic development and inclusion, in order to mitigate the gentrification process. The infrastructure improvements in the city would enable the creation of an innovative and high technology area, the High Technology Park of the Rio de Janeiro Federal University. The Porto Maravilha Operation also established an innovative financial model of Public-Private Partnership to ensure the project's implementation in partnership with the private sector.

The project covers a total area of five million square metres and hosts 32,000 people. It plans to host 100,000 citizens in 10 years. The first phase focused on infrastructure works, such as renewed sewage, drinking-water networks, telecommunications and street lighting public networks. In order to solve flooding problems, new water galleries were installed taking into account rising sea levels in the near future. The second phase focused on a new urban mobility model that prioritizes pedestrians, cyclists and sustainable urban transport. Ecological restructuring measures are expected to increase the green spaces from 2.5 % to 11 % of the city's surface, improving the soil permeability. Furthermore, water, air and noise pollution are expected to drastically diminish.

Outstanding waterfronts have often been the dynamic legacies of unique events bringing special opportunities for cities, regions and nations. They address the two-fold multi-challenge of creating an infrastructure and an environment with the occasion of a short-term event but planning it to serve long-term purposes. Global attractions based on once-a-lifetime events, can put a city on the map and a sporting or cultural event can act as a magnet to draw public attention and propel the city on the world stage. However a city should rely on more sustainable assets and attractions or use once-a lifetime events as milestones in a much longer journey.

As F. Pessoa had put it "the ways of change also change". Lisbon seized the opportunity offered by two catalysing major events, the Cultural Capital of Europe in 1994 and the Universal Exhibition in 1998 on "The Oceans" to re-invent and transform itself. The public works related to this exceptional event redesigned the map of Lisbon and helped emerge a new quarter to the East, the Oriente. Oceanic vocation, one of the city's essential attributes has offered a vital thread. Expo 1998 invested in an abandoned former industrial area transformed into a vibrant innovation and creativity site. The project was not confined to the exhibition precinct of 50 ha but created a whole new resourceful city of 330 ha. Investments in bioclimatic architecture, quality design and advanced energy concepts have led to the aesthetic and functional metamorphosis of the area. An eco-efficient distribution network for thermal energy, heat and cold air, was set up, together with an observation and monitoring system. The adopted standards were higher than the ones required by the national regulations and the performance of the area has been exemplary. The Vasco de Gama Tower is among the emblematic buildings which form the legacy of Expo 1998.

The 1992 Olympics was a key catalyst for the renaissance of Barcelona. The city, which had long lived with its back to the Mediterranean, invented a new waterfront and was definitely reconciled with the sea. Enriched with new and better places and functions, including a public beach, and integrating the Olympic village, the city was enhanced with thorough restoration projects and noble public places. The rehabilitation of the Ciutat Vella has been an unparalleled event, in terms of investment, effort and civic spirit.

Ciutat Vella is nestled between the sea and the Eixample district, a strict grid pattern crossed by wide avenues designed by the visionary urban planner I. Cerdà. Running down the heart of the district, Las Ramblas offer the delights of a vibrant public artery and seal the link to the sea. The four historic quarters of Ciutat Vella, including the gothic quarter, have been thoroughly transformed through selective renovation, rehabilitation, eco-retrofitting and eco-constructions, civic centres, pedestrian precincts and green public spaces. Thoroughly designed small interventions acted as stem cells that injected in the body of the city led to a positive metastasis and overall renewal. The strong neighbourhood groups cooperated with the authorities and played a pioneering role in the allocation of housing and services and the enhancement of public life (Ajuntament de Barcelona 1995).

The Sydney 2000 Olympics, promoted as the first "Green Olympics", endowed the city with a sustainable new organic part by the sea. A 640-ha industrial wasteland in Homebush Bay, previously intended for an urban renewal project after 100 years of industrial and military uses, was fully reinvented for the Games and transformed into the Sydney Olympic Park. After the end of the 2000 Olympics, the Park was converted to a multi-purpose facility which continues to host sporting events. The Sydney Olympic Park Master Plan encourages a broad range of residential, commercial, recreational and cultural activities that attract new assets. Urban design and landscaping principles adopted in the Master Plan emphasise excellence and efficiency (Mega 2010).

The 2004 Olympic Games have endowed Athens with an integrated Olympic Public Transport System, after the radical restructuring of the network throughout the greater Athens region. The legacy of the Olympics to the Greek capital includes 120 km of new roads, 90 km of upgraded arteries, 40 km of suburban railway, 7.7 km new metro lines, a 23.7 km tram network, modern train stations and a new state-of-the-art traffic management centre. The Olympic village has been transformed into high quality homes for low-income working families, which have been selected by draw.

The city of Cape Town owes much to the organisation of the 2010 World Cup. Improvements to the city centre, the downtown area known as the city bowl, and the public transport system attracted citizens and visitors. Concerns about safety and the haunting legacy of apartheid have been successfully addressed and sports were used as a means to overcome racial fears and tensions. The short distance to the Cape of Good Hope was truly filled with hope. This coastal city can hope a better future.

A triple Olympic city, London has long been a multi-faceted city and a great commercial and financial centre in its way to become a global centre for the arts, culture and entertainment. The city has been enriched with new landmarks. The new iconic tallest building in Europe, the Shard, has been praised for its creative and bold design and criticised for disrupting the historic views of the London skyline. Already competing on the London skyline with the Gherkin, it is entering the world emulation of the tallest world marvels. Since the San Gimignano towers, this competition never stopped and leaves the Shard far behind the Burj Khalifa in Dubai, 828 m high.

Sustainability was a prominent criterion in the organisation of the 2012 London Olympics. The 2012 Olympics brought a revival to London's East End and in particular on a brownfield site, previously isolated from the city centre. Adjacent districts with warehouses have also benefitted and transformed into artist homes and workplaces. The Olympic stadium, hosting up to 80,000 participants, features flooring made from recycled tennis balls, while the wave-shaped Aquatics centre designed by Z. Hadid demonstrates sea-inspired design. The 115 m tall Orbit, a unique fusion of art, architecture and engineering, by A. Kapoor, is the landmark of the Olympic park which initiated its new life in the aftermaths of the Olympics. The Games acted as a catalyst for growth through the legacy of world-class infrastructures, business hubs and new vibrant neighbourhoods.

Functionality and aesthetics of the waterfronts are critical for quality of life, work and travel. Harbours, airports and railway and bus stations are gateways to cities, temples of welcome and farewell. The design of the Kansai International Airport illustrates human ingenuity and cooperative effort. Created in a typhoon zone three miles offshore, it is a good example of the integration of structure, function and environment with a sense of purpose. Protection of its coastal and marine environments was extensively studied and carefully addressed. The construction and administration of the airport were conducted by a special corporation established through joint investment by the national government, local authorities, and the private sector. The airport, inaugurated in 1994 as Japan's first international airport to be operational round-the-clock, serves the entire Kansai region, the historical, economic and political heart of the country.

8.4 Composing Better Policies for Better Lives by the Sea

Better strategies and policies should lead to even better policies and ultimately better lives in all cities and regions. Many organisations and think tanks have tried to measure and appraise the quality of life with the use of various quantitative and qualitative indicators. These issues raised questions about the inability of traditional measures of national and urban welfare built around macro-economic statistics to capture the reality of citizens' lives in a complex world and serve as significant yardsticks and compasses towards sustainable development.

Gross domestic product, a measure of the past, served to encapsulate as fully as possible the state of national welfare during the twentieth century. Already in 1962, Nobel Prize winner S. Kuznets suggested that "Distinctions must be kept in mind between quantity and quality of growth, between costs and returns, and between the short and long run. Goals for more growth should specify more growth of what and for what". In 1968, R. Kennedy suggested ,with reference to the GDP, that "…it measures everything in short, except that which makes life worthwhile". These considerations are more valid than ever in search of sustainability metrics and become more complex in intense spaces like coastal cities (EC 2009).

Through indicators, cities try to measure and represent in concise form what they most value. Efforts for developing sustainability indicators were often linked to the greening of national accounts, aggregating and comparing data in monetary forms. Genuine saving indicators or Gross welfare indices attempt to broaden the usual measure of saving to account for the cost of environmental depletion and degradation and investment in human capital or social welfare.

Indicators should capture critical features of a city, such as the quality of coastal and marine environments, and contribute to making them more visible and transparent, help structure and harmonise data banks, enrich decision-making with relevant and timely information, assist appraisals, comparison and prediction, stimulate communication and promote citizen empowerment and participation. They should embrace all sectors and neighbourhoods contributing to the co-evolutionary process of sustainable development. An indicators assessment board should validate the set of indicators, and ensure that the framework is regularly updated and assessed.

A holistic policy framework has to be accompanied by a number of urban thematic indicators monitoring a city's performance in all fields contributing to sustainable development and according to its specific policy objectives. Aggregate indexes, like the human development indicator, the genuine savings or the ecological footprint and happy planet index, can shed light on the overall performance of coastal cities. The development of a sustainability index after thematic policy indicators is a complex task, since indicators have to be weighted according to their contribution to sustainability levels and all the levels of aggregation have to be taken into due account. Finally, it is important to highlight that no indicator can inform if a city integrates socio-economic and environmental policy objectives which however is at the heart of sustainable development practices (OECD 1996; EFILWC 1998).

Many cities have tested and introduced frameworks of indicators during the recent years. Seattle is often quoted as a classic example of a dynamic city, breeding ground of successful businesses such as Boeing and Microsoft, with a coherent set of award-winning indicators. The Seattle framework demonstrates that indicators can enhance local attributes like the health of the marine environment, reinforce communication processes and promote common understanding and political accountability (Sustainable Seattle 2008).

Qualitative indicators and state of the city and the environment reports are also very important. The City of Amsterdam initiated biannual "State of the City" reports, which proved to be inspiring monitoring tools. The reports draw information from municipal statistics and data banks, and other sources, including a survey after a questionnaire sent to residents. Results are compared to the national situation as well as to other Dutch cities. Such "State of the City" reports could offer an assessment of policies and a compass for the future in an accessible citizen-friendly form.

The "State of the City of Amsterdam" report offers a broad understanding of the city and its residents. The first 2000 edition focused on public participation. It then developed into a monitor to take the pulse of the economy, quality of life, education and culture, minorities and safety. The monitor highlighted the strong connection between education, employment and participation, provided insights into spatial and ethnic patterns and shed light on the city's congested housing market (City of Amsterdam 2004).

Boston is a city in constant transformation. The Boston Indicators Project, initiated in 2000 with the goal of assessing and presenting progress through 2030, Boston's 400th anniversary, offered new ways to understand the city and its neighbourhoods in a broader context. Simple yet fine-grained benchmarks help democratise access to information, foster insightful public debate, evaluate progress on shared civic goals, and report on achievements in the essential sectors of Civic Vitality, Cultural Life and the Arts, Economy, Education, Environment, Health, and Housing, and the cross-cutting topics of Children and Youth, Competitive edge, Fiscal Health, Race / Ethnicity and Sustainable development.

On the basis of these indicators, for more than 10 years, a series of imaginative biannual reports underlined the dynamics and identified emerging prospects for Boston. A first report on "The Wisdom of Our Choices: Measures of Progress, Change and Sustainability" introduced the framework of indicators through a rigorous process involving more than 300 experts and stakeholders. The report noted that the booming knowledge economy had creating an "education divide". A second report highlighted Boston's institutional, physical and cultural assets, but remarked a worrying brain drain, due to the move of young people away from Boston and Massachusetts, mainly because of the high cost of living (Boston Foundation 2001, 2003).

True progress should be widely shared and also problems should be jointly addressed. All following reports revealed particular aspects of the city and the region. The third report "Thinking Globally, Acting Locally: A Regional Wake Up-Call" noted that the region was suddenly competing for jobs and talent not only with other US regions, but with China, India and other emerging economies. It called for a coherent, collaborative response and a civic agenda. The fourth biennial report suggested remarkable civic progress during the years in which the local and regional economy strengthened considerably (Boston Foundation 2005, 2007).

Boston could be described as a resilient twenty-first century city striving to develop a more robust, sustainable local economy. The vibrant Innovation District tries to capture synergies and embrace collaboration. The city wants to reach up, and, in order to achieve it, tries to reach out. Greater Boston has ridden out the economic downturn better than most US cities. The 2012 report, celebrating a city of ideas, suggests that engagement is the key to prosperity and to sustainability. It signals that equality is a major issue, as the region's economic sectors create new wealth, but leave poorer parts of the city lagging behind (Boston Foundation 2009, 2012).

Boston has been among the top cities of the SustainLane US City Rankings of the 50 largest cities, an inspiring benchmark exercise on the unfolding efforts of cities towards sustainable development. A San Francisco green media company, SustainLane, went through an examination of sustainability initiatives in US cities, including indicators of quality of life, such as local food availability, air and water quality, pedestrian and park space and road congestion. SustainLane also recognised the growth of clean technologies, developments in renewable energy, waste management, advanced transport services, alternative fuels and green buildings. Since the first SustainLane rankings in 2005, hurricanes and crises underlined vulnerability and the need to foster urban resilience.

Coastal US cities topped the SustainLane City Rankings focusing on urban practices which differ across the country and can reveal their distinctive figures. The US winner of the 2005 and 2008 ranking exercises was Portland, recognised as the most sustainable city, followed by San Francisco, Seattle, Chicago, New York, and Boston. Portland particularly excelled in clean technology and green building development, overall quality of life, and sustainability planning and management. Citizens pinpointed their high quality of life and engaged involvement in urban policy-making, boards, projects and practices that impact sustainability.

The vision of Portland promises a sustainable future that meets today's needs without compromising the ability of future generations to meet their needs. The city accepts responsibility to support a stable, diverse and equitable economy, protect the quality of the air, water, land and other natural resources, preserve native vegetation, fish, wildlife habitat and other ecosystems, and minimise human impacts on local and worldwide ecosystems.

International comparative analyses are usually constrained by culturally-dependent definitions and statistics. Until recently, there was no harmonised definition of what is a city for the European Union and OECD countries. This weakened the comparability of cross-country analysis of cities. To improve the situation, the OECD and the European Commission developed a new definition of a city and its commuting zone in 2011. This new definition identified 828 cities with an urban core of at least 50,000 inhabitants in the EU, Switzerland, Iceland and Norway. Half of these European cities are relatively small with a centre between 50,000 and 100,000 inhabitants. Each city is part of its own commuting zone or a polycentric commuting zone covering multiple cities. These cities host about 40 % of the EU population without their suburbs, which together with other towns cover another 30 % of the EU population. In addition, this methodology identified a further 492 cities in Canada, Mexico, Japan, Korea and the United States. The cities and commuting zones together, called Larger Urban Zones, account for 60 % of the EU population (OECD 2012b).

Cities use indicators to measure their performance in service delivery and sustainable development, but there has not been a comparable standard for use by local authorities. Existing indicators are often not standardised, consistent, or comparable over time or across cities. Indicators are traditionally limited by national differences in data definitions and collection and composition methods. Systematic territorial indicators are necessary complements to national indicators serving for international comparisons. Regularly reporting on territorial progress towards international targets and commitments can promote policy coherence and accountability of national decision-makers at the local and international level.

In 2014, the first ISO standard for city indicators was launched, providing city managers, politicians and planners with the opportunity to objectively evaluate their progress and compare their achievements against other cities. The ISO 37120:2014 has been developed as part of an integrated suite of standards for sustainable development in communities. It covers numerous themes, including education, energy, transportation, governance, finance, shelter, safety, sanitation and recreation. The standard is designed for use by any city, municipality or local government irrespective of size and location or level of development. Recognising that urban conditions are very diverse in terms of resources and capacity, the indicators have been divided into core measures, which must be followed, and supporting measures, which should be followed.

Profile indicators also provide basic statistics and information to help comparison purposes. This first edition is based on the Global Cities Indicators, supported, amongst others, by the World Bank, UNEP and UN-HABITAT. The Global Cities Indicators Programme suggested that indicators should be standardised, consistent and comparable over time or across cities. A single comprehensive system for measuring and monitoring city performance and urban quality of life should facilitate programme monitoring over time and enhance governance accountability.

Further experiences with indicators for sustainable urban development include the EU Reference Framework for Sustainable Cities (RFSC). The RFSC is an online toolkit for the integrated approach of all urban issues contributing to smart, sustainable and inclusive cities. Designed for European local authorities, the RFSC provides a common space for very diverse cities to share their experiences and enrich their approaches. The initiative helped many local authorities to shape their sustainability policies and set up the monitoring frameworks. By joining the RFSC community, cities get access to various services, including peer learning, dedicated training sessions and advice from urban governance experts.

References

Aesopos Y (ed) (2014) Tourism landscapes. Remaking Greece (in the framework of the Greek participation in the International Exhibition of Architecture, Venice 2014)

Ajuntament de Barcelona (1995) Barcelona: un modelo de transformación urbana 1980 -1995 Gestion Urbana, Vol. 4, Barcelona

Beatley T (2011) Blue urbanism: the city and the ocean. Islands Press, Washington, DC

Boston Foundation (2001) The wisdom of our choices: measures of progress, change and sustainability. Boston Foundation, Boston

Boston Foundation (2003) Creativity and innovation: a bridge to the future. Boston Foundation, Boston

Boston Foundation (2005) Thinking globally, acting locally: a regional wake up-call. Boston Foundation, Boston

Boston Foundation (2007) A time like no other. Chartering the course of the new revolution. A summary of Boston indicators 2004–2006. Boston Foundation, Boston

Boston Foundation (2009) A great reckoning: healing a growing divide. Boston Foundation, Boston
Boston Foundation (2012) City of ideas: reinventing Boston's innovation economy. Boston Foundation, Boston
British Urban Regeneration Association (BURA) (1997) The future of cities. Conference report. BURA, Belfast
City of Amsterdam (2004) The state of the city of Amsterdam III. Developments in participation and life situation. O+S, Department of Research and Statistics, City of Amsterdam, Amsterdam
De Portzamparc C (2007) Rêver la ville. Le Moniteur, Paris
Elmqvist T et al (2013) Urbanization, biodiversity and ecosystem services: challenges and opportunities. A global assessment. Springer, New York
European Commission (EC) (2009) GDP and beyond measuring progress in a changing world. European Commission, Brussels
European Foundation for the Improvement of Living and Working Conditions (EFILWC) (1998) Challenges for urban infrastructures in the EU. EFILWC, Dublin
EC (2014) Directive of the European Parliament and of the Council establishing a framework for maritime spatial planning and integrated coastal management SWD (2013) 64 final. European Commission, Brussels
European Council of Spatial Planners (ECSP) (2013) The charter of European planning. ECSP, Brussels
Hall P (1998) Cities in civilisation. Weidenfeld & Nicholson, London
INU-Politecnico di Milano (1997) Il Tempo e la città tra natura e storia. Atlante di progetti sui tempi della città. INU-Politecnico di Milano, Milan
Mega V (2010) Sustainable cities for the third millennium. The odyssey of urban excellence. Springer, New York
Metropolis (1996) Metropolis for the people. V world congress documents. Metropolis, Tokyo
New York City Mayor's Office of Long-Term Planning and Sustainability (2011) Plan NY: a greener, greater New York (Update 2011). PlaNYC, New York
Organisation for Economic Co-operation and Development (OECD) (1996) The ecological city. OECD, Paris
OECD (2012a) Compact city policies. A comparative assessment. OECD, Paris
OECD (2012b) Redefining "Urban": a new Way to measure metropolitan. OECD, Paris
Pistor R et al (1994) A city in progress. Physical planning in Amsterdam. Gemeente Amsterdam, Amsterdam
Port of Amsterdam (2009) Port-City. Port of Amsterdam, Amsterdam
Port of Antwerp (2010) Sustainability report. Port of Antwerp, Antwerp
Sesini M (2011) The garbage patch in the oceans: the problem and possible solutions. Columbia University, New York
Sustainable Seattle (2008) Indicators of sustainable community. Sustainable Seattle, Seattle
Transportation Research Board (TRB) (2009a) Driving and the built environment: the effects of compact development on motorized travel, energy Use, and CO_2 emissions. The National Academies, Washington, DC
TRB (2009b) Critical issues. The National Academies, Washington, DC
UN/Tokyo Metropolitan Government (1998) Eco-partnerships Tokyo. Cultivating an eco-society. UN/Tokyo Metropolitan Government, Tokyo
Urban Redevelopment Authority of Singapore (URAS) (2011) Concept plan. URA, Singapore
World Business Council for Sustainable Development (WBCSD) (2013) A solutions landscape for Kobé. WBCSD, Geneva
World Bank (1995) The human face of the urban environment. World Bank, Washington, DC

Selected Internet Links

www.stockholmroyalseaport.com
www.glasgow.gov.uk
www.bcn.es
www.harbourbusinessforum.com
www.urban-villages-forum.org.uk
www.urbantaskforce.org
www.biennialtownplanning.org
www.smb.museum/smb/standorte/index.php
www.adottaunaguglia.duomomilano.it
www.big.dk
www.superkilen.dk
www.nyc.gov/html/planyc2030/html/home/home.shtml
www.ec.europa.eu/regional_policy/urban2/urban/audit
www.beyond-gdp.eu
www.sustainlane.com/us-city-rankings
www.bostonindicators.org
www.metrofuture.org
www.sustainableseattle.org
www.veolia.fr
www.tempomat.it
www.neweconomics.org/files/The_Happy_Planet_Index_2.0
www.mori-m-foundation.or.jp
www.rfsc.eu
www.cityindicators.org

Watercolour 9
Singapore at the Crossroads of the Seas

Chapter 9
Coastal Cities for and with Local and World Citizens

Abstract Present and future citizens are the political stakeholders of cities in an increasingly multipolar and interconnected world. They have the right to consultation on investments of the future and the duty to exercise democratic scrutiny of policies. Climate change and the path to sustainable development have encouraged the emergence of multiple forms of active citizenship. New governance architectures seek to enhance the potential of all invisible hands of urban societies and economies and to build a social consensus on a future vision and the steps and means to make it become true on land and the sea. Citizen empowerment is a mobilising force and public-private partnerships are recognised drivers of transformational change.

This chapter examines the emergence of new models of citizen participation in responsible coastal cities and links with world citizens. Institutional alliances, enriched with a variety of participatory leadership schemes and innovative partnerships, can maximise the potential of synergies, enhance the content and the methods of cooperation and serve as catalysts of change. A global solidarity bond involving coalitions and networks of cities, both from the emerging and the developed world, can play a major role in jointly addressing global common challenges. Initiatives such as the Compact of Mayors, a ground-breaking agreement created during the 2014 Climate Week, can raise the bar of excellence and enhance resilience to climate change.

9.1 Urban Governance 2.0: Co-leading Open Coastal Cities

The governance of cities has experienced many ebbs and flows since the Athenian laboratory of urban democracy and the development of city-states, which often expanded as maritime powers and created colonies beyond seas. Cities have evolved much as other prominent political entities, but citizens remain the key political stakeholders and their participation is the common denominator of mobilising transformative projects. The array of stakeholders has been enlarged and private local actors and civil society organisations play an expanding role. The ways and methods

V.P. Mega, *Conscious Coastal Cities*, DOI 10.1007/978-3-319-20218-1_9

of active involvement have been multiplied and the joint actions by interconnected local partnerships are limited only by human imagination.

Urban democracies depend on interactions among the many diverse stakeholders and actors of the cities and multi-level governance. In an increasingly multipolar and interconnected world, with a progressive diffusion of power, urban leaders often compete with national leaders in proposing innovative policies, as well as initiating and scaling progress towards sustainable development. The role of businesses and NGOs is paramount, especially in coastal cities with maritime industry and advocacy groups striving for a healthier marine environment. Public multi-stakeholder partnerships can play a decisive role for better urban politics.

Urban governance can provide the keys to the city (Storper 2013). It can be defined as the science and art of co-governing societies with the participation of all actors having a stake in the future of a city. Thinking ahead and together with citizens can spectacularly open the spectrum of optimal futures and help co-evaluate the drivers, the barriers and the conditions for change. A transition from government to governance implies the recognition that visions, strategies and policy options and decisions should actively involve all those concerned. In many coastal cities, new civic bonds have been sought with maritime business, environmental advocacy groups and civil society, expected to invigorate the debate between governments and the constituencies they represent, increase public transparency and accountability and enhance collective capacity for reflection, judgement and decision.

Virtuous and respectable governance is essential for cities to become global democratic leaders for sustainable development. Democracy is always reputed to be a less imperfect form of government for open societies. An urban governance framework to enhance the marine capital should embrace vertical governance between different levels of government, as well as horizontal governance across multiple sectors, including engagement with non-governmental actors, and governance across and between cities or coastal territories. A key issue for national policies is the empowerment of cities and citizens to become more effective in the design and implementation of policies, also including actions driven from the bottom-up and innovations that could subsequently be scaled up. A hybrid of the two frameworks provides opportunities for participative leadership towards a mutual trust paradigm.

Urban democracy, representative and direct, is vital for the permanence of cities and their capacity for continuous renewal. Representative democracy has to address the challenge of the duly constituted authorities, linked to the proportional representation of all legitimate local groups. Participative democracy can lead to truly publicly-owned policies and facilitate their respectful implementation. Many cities have deployed exemplary efforts to make the best out of the participation of all stakeholders. Citizens have been gradually invited to support broadening visions about the future of cities and act as strategic partners rather than reactive protesters. Empowerment has become ethically and politically correct and is now recognised as a driver of transformative change.

At the dawn of the civilisation of sustainability, and with the advent of participative and interactive 2.0 models, there has been a clear shift from direct representational systems of democracy to a more interactive and direct democracy. Normally associated with web applications that favour interactivity, interoperability and stakeholder-centred design, the label 2.0 has been grafted into diverse forms of urban governance for a continuous reinforcement of the civic bond and the incessant confirmation of civic values.

Anticipatory democracy can increase considerably the political capital of a city and become enriched with instant direct democracy, enabled by web 2.0 processes. Much depends on the quality and commitment of human and social capital and political leadership. Governments should give unlimited opportunities to the concepts and ideas proposed by stakeholders, including the ones usually without a voice. Horizon scanning and anticipation may be decisive and mediation at an early stage is critical to building trust. Leadership at all levels is a sine qua non condition and the exchange of selected practices a most inspiring tool.

A preferred shared future implies a common vision and consensual actions to make a vision come true. Civic Agreements signal an engagement by cities conceiving and forging new partnerships with their multiple and diverse stakeholders, far beyond the exchange of information and views. Citizens should become more than voters or customers, they should be engaged as co-designers and co-deciders in the policy-making process. The transient population should also commit and contribute in a spirit of openness and shared awareness. Manifold partnerships with civil society organisations and citizen consultation on the gestation of visions and plans have expanded as a precondition for civic bonds and bold collective action (Putnam et al. 2002).

Leadership at all levels can bring great impetus. Participatory and shared leadership has become a subject of attention among political analysts, social experts and behavioural scientists. Shared leadership, occurring when the role of leader is actively and intentionally shifted, depending on the issues or the context, can raise ownership and dynamism in many projects. It empowers all citizens to experience the responsibilities of the leaders, and enables a gradual building of consensual mobilising visions.

Democracy has always to provide real and virtual forums to exert sound judgement and help citizens to be transformed from mere consumers and users of urban infrastructures and services into responsible city actors, sharing values, embodying visions and performing actions. Their representation and participation can be extended through action planning and schemes, citizen and stakeholder forums, dialogue and consensus workshops, bringing together often opposed actors on neutral grounds and on equal terms (Abbott 1996).

Formal and informal mechanisms of horizontal and vertical cooperation between government bodies and partnerships with non-governmental actors, mainly industry and civil society organisations, are necessary for weaving the democratic web of a city. Governments have, in some cases, begun to promote the formation of new spatial structures, such as inter-communal frameworks, regional platforms, territorial pacts, and sustainability contract areas. These processes, beyond traditional administrative boundaries, promise more coordinated policies and more coherent allocation of public resources, as well as greater transparency, visibility and accountability (Barber 2013).

Coastal regional and local authorities should be endowed with the resources to manage the multiple functions leading to marine and coastal sustainability, fulfil new mandates and ensure policy coordination, implementation and evaluation. Fiscal federalism, based on the search for a balance between distribution of powers and allocation of resources, may be instrumental for sustainability. Responsible monitoring and evaluation may also be facilitated by 2.0 schemes and crowdsourcing schemes can promote marine stewardship.

National leadership is necessary to provide a coherent conceptual framework and coordinate the thematic policies of different national ministries, enable territorial authorities to set priorities and commit resources, promote public-private partnerships and share the risk and cost of innovations. Leadership at the local and regional levels is necessary to define territorial needs, propose visions, coordinate the implementation of programmes, mobilise public and private funds to invest in marine resources and develop a permanent dialogue with a broad array of marine stakeholders and other territorial levels of governance.

Leaders have to interpret the needs, values and preferences of citizens in relation to the sea, elaborate plans towards sustainable development, facilitate and stimulate interactions of different actors and partners and guarantee coherence among options, means and decisions. Stockholders, owners of the physical assets, stakeholders, having particular interests in local coastal life, and outside partners, such as shipping lines and offshore industries, could invest great energy in developing coastal and marine resources in an integrated and responsible way.

Local involvement to address national and global challenges is crucial. The engagement of local and regional authorities is critical for the achievement of national and international objectives. In the EU, the Committee of the Regions, representing local and regional authorities, warned that the execution of the strategy Europe 2020 was being undermined by a clear lack of involvement of local government. A political declaration, the Athens declaration on the mid-term review of Europe 2020, approved by the Committee during the sixth European Summit of Regions and Cities, argues that the European Union must reconsider its approach and actively involve all governance levels, if it is to achieve a smart, sustainable and inclusive economy by 2020 (CoR 2014).

Whilst some coastal EU regions are progressing, others are lagging behind, also impacted by national and regional dynamics. The Committee of the Regions' seven-point plan for Europe 2020 suggested giving the strategy a clear territorial dimension, managing the National Reform Programmes in partnership, making multi-level governance the standard approach, aligning the European Semester with genuine long-term investment, enhancing the Europe 2020 Flagship Initiatives for strengthened policy coordination, mobilising funding for long-term high-quality investment, and strengthening administrative capacity for more effective implementation. A specific Public Sector Innovation Platform could support and coordinate public sector innovation in cities in direct interference with marine resources.

The marine pillar of the Europe 2020 strategy, the Marine Strategy Framework Directive (MSFD), introduced in 2008 to promote sustainable use of Europe's seas and marine ecosystems, relies much on active stakeholder involvement. The main goal of the directive is to achieve Good Environmental Status in Europe's seas by 2020. It is a cyclical process which requires EU countries to adopt the ecosystem approach in their marine strategies, set objectives and targets, identify policy options, and engage in continuous evaluation and adjustment. The directive defines Good Environmental Status as "marine waters that provide ecologically diverse and dynamic oceans and seas which are clean, healthy and productive". Eleven descriptors have been selected to guide evaluation of Good Environmental Status, including biodiversity, eutrophication, seafloor integrity, non-indigenous species, population of commercial fish/shellfish, elements of marine food webs, hydrological conditions, contaminants for seafood, marine litter, and underwater noise.

European legislation has been a major driver for active citizenship and participation of organised civil society. Implementation of the Marine Strategy Framework Directive gave rise to multiple forms of public participation in coastal environments. Non-governmental organisations, including the general public have become very active despite a frequent lack of knowledge about European legislation and the descriptors. Access to information is the first step of any consultation process and active participation the most advanced step involving empowerment and co-ownership. Cities can gain much by participating in the public debate and joining the consultation in marine issues which are largely deconcentrated processes.

A joint NGO paper brought together Seas At Risk, the Marine Conservation Society, the Coalition Clean Baltic, the Swedish Society for Nature Conservation, Oceana, The World Wide Fund, BirdLife Europe, the Mediterranean Information Office for Environment, Culture and Sustainable Development, The Fisheries Secretariat, Black Sea NGO Network, France Nature Environnement, the North Sea Foundation, the Surfrider Foundation Europe, the Coastwatch Europe, the Sustainable Water Network, and the Mediterranean SOS Network. This joint paper provided directions for discussion with many local partners and the civil society and brought forward the positions of organisations that were very sensitive to the quality of maritime environment. The participatory process concluded with priorities for policy action (Seas At Risk et al. 2014).

9.2 Reinventing Citizenship, Trust and Accountability to the Future

Citizens are the political stakeholders and society is the ultimate frontier for all policies. If governments wish their policies to be owned by citizens, they should allow as many voices as possible to be heard, and as many values as possible to be represented during consideration of possible futures, shaping of the vision of desired futures and the elaboration of consensual plans for concretisation of that vision. Residents, users of public infrastructures and services, port and coastal communities have to be properly informed on important emerging issues, and duly involved in the formulation of possible policies. Decision-makers should invest in a better understanding of public opinion and the preferred options. Making the community, especially the under-represented social groups, better informed and more aware and willing to take part in a shared future is a noble challenge. Projects must not only be scientifically robust, but also socially acceptable and ecologically unbeatable.

Partnerships are linked to the shift in public policies from direct interference to indirect or conditional policies, such as incubation and mediation. They should enhance the capacity, contribution and commitment of the public, private and community sectors and improve the ability of an urban coastal society to act proactively and drive change. Public-private partnerships should work like an orchestra under a public conductor for the overall improvement of urban functions executed by private actors on the coast and the sea.

Sometimes the role of orchestra conductor can be played by other citizen groups. The Committee for Sydney is an independent leadership group bringing together all sectors that share a passion for Sydney. In 2014, the Committee created a Task Force to prepare the Sydney 2054 strategy in collaboration with government. The Committee favours a "One Sydney approach" and enhanced coordination across the city's numerous councils. It also invited inputs from citizens and especially young leaders through the Sydney's Future Leaders sounding board.

Strategic public-private multi-stakeholder partnerships have a great potential in balancing objectives of competitive strength, social justice and environmental well-being. They have the potential to take more risks, reduce the social costs of projects and lead to enhanced outcome from public and private investment. They offer ample grounds for coalitions to overcome sector and institutional dissonance and play a critical role in the implementation of sustainable development policies. A clear vision and structure, a consistent and dynamic, strategic and tactic approach, a critical mass, assertive leadership, flexible adjustments and continued monitoring and assessment are usually suggested as the success factors of the partnerships.

Citizens, co-builders of visions and initiators of actions, may contribute decisively in creating a collective momentum for better public policies. Structured collaborative events, which unbind creative individuals and articulate a sense of vision, can create a thrust for the future. Processes like the charette are being used to bring together the richness of diverse opinions and ideas and build consent on possibly controversial projects at the earlier possible stage. Citizen platforms can provide more permanent and effective interfaces among experts, policy-makers and citizens (EFILWC 1998).

The health of coastal ecosystems has been at the heart of some citizen consultation schemes. Amsterdam introduced referenda to elicit the residents' opinion on the extension of the city on artificial land and supporting infrastructures on sensitive coastal ecosystems. Sustainable development public debates offer an excellent opportunity for raising awareness of threats to unique marine and land interactions and strengthening the urban social fabric.

The City Mayors Foundation, an international think tank for local government, proposed the World Mayor Project and every 2 years recognises a mayor with an outstanding sustainability vision for urban communities. Good and honest local government is the foundation of any nation that strives to provide its citizens with happiness, security and prosperity. Incompetence, corruption and misconduct in local government threaten the fundamental decency in a society and undermine the essentials of democracy.

The Code of Ethics underlines the Mayors' pledge to execute their office respecting the laws for the common good of their communities while refraining from actions that discriminate against individuals or groups because of their politics, communities, race, religion, gender, disability or sexual orientation. It states their full responsibility for any acts performed by themselves or their administrations and emphasises that mayors should manage public resources for the public benefit of their communities and not for privileges or advantages for themselves and their close circle.

According to the code of ethics, mayors have to prove independence of judgment and demand the same degree of impartiality from their administrations. Mayors are subject to public scrutiny and should report any improper actions they may witness. They should work to strengthen civil society by raising public awareness of, and trust in, their city government's activities. Last but not least, Mayors must be committed to using their influence to promote cooperation and good will between cities, nationally and internationally.

The World Mayor Prize, awarded since 2004, honours mayors with vision, passion and skills to make their cities outstanding places to live, work in and visit. The project aims to highlight the results of city leaders with knowledge, enthusiasm and integrity, social and economic awareness, ability to provide security and to protect the environment as well as capacity to foster good relations between communities from different cultural, racial and social backgrounds.

Citizens of the future should be given particular attention, as the well-being of children is a critical test for the present and future of society. It seems that in coastal cities, children tend to develop a particular relationship with the sea. They love blue horizons, the ever-changing weather patterns and watching boats rising and falling on the waves. The Irish Marine Institute has organised poetry competitions on marine life and the ocean. The marine environment can stimulate imagination and civic awareness on issues of sustainability and engage children as citizens of the future.

Marine subjects can greatly expand children's imagination and nurture plans for the future. In Ireland, the Marine Explorers' Education Programme has been running for many years in Galway, Mayo, Dublin and Cork. The Programme aims to raise awareness and interest in the marine environment and teaches subjects in primary school that relate to marine species, ocean technology, and the shared heritage of the oceans.

The citizens of the future should be given the place that they deserve in all forums for Sustainable Development. In preparing for the Rio+20, three NGOs, Oikos, Terre des hommes and the World Future Council, launched an online petition for signatures demanding world leaders to support the establishment of Ombudspersons for Future Generations. Such initiatives and the Future Policies Awards by the World Future Council can inspire further action for sharing and protecting the interests of future generations. In 2012, the Future Policy Award celebrated exemplary policies for Oceans and Coasts (WFC 2012).

Ombudspersons for Future Generations could act as a driving force behind the sustainable development agenda while ensuring accountability and accessibility. They can bring long-term interests to the heart of decision making at all levels. This offers an opportunity to break with immediate concerns and stand up for future generations.

9.3 Transparent Stakeholder and Community Engagement

City authorities are in a unique position to effectively engage resident and transient stakeholders and design and implement locally tailored urban coastal policies. Reinvigorating stakeholder participation and engagement in responsible cities is a sine qua non condition for sustainability and blue green growth. Transparent partnerships can maximise the potential of all synergies, enrich the content and the methods of cooperation and serve as catalysts of change.

Transparency of public administrations, citizen participation and cooperation processes form a triad bottom line of requirements supported also by social media and eGovernment. In Italy, the compass of transparency, an online iplatform offers citizens the possibility to monitor, in real time, the implementation of all data and information requirements imposed by the Italian legal transparency framework. Moreover, users can express their comments and opinions about the quality of the published information. These opinions are immediately made available online and can be accessed by everybody. The aim is to foster direct involvement of citizens in the continuous improvement of the quality of online and digital services. The initiative focuses on continuous improvement of transparency compliance in more than 20,000 Italian public administrations, with returns in terms of increased efficiency, reduction of corruption and lower costs across the whole public sector. The heart of the system is a validation mechanism, composed of software sensors and mathematical algorithms, with the ability to analyse public administration websites in both real-time and at significant intervals.

Voluntary approaches with the active participation of citizens and industry are most important for sustainable development. In Rio+20, the plethora of diverse voluntary pledges has been one of the most encouraging signs. Many cities develop particular approaches to attract voluntary commitment on various fronts of a sustainable development agenda, especially on climate change. Cities like Helsinki extend the city's reach through voluntary agreements with the private sector to further their climate protection goals (CDP 2012).

From voluntary schemes to volunteering for the city, the leap is not without meaning. Citizens regularly offering their time, energy and skills pro bono in their city constitute a precious urban human resource which has to be appreciated and enhanced. Exceptional events often served as catalysts of commitment. In preparing for example for the 1992 Olympic Games, Barcelona created and trained a body of 40,000 volunteers. After the Olympic Games, this body was considered a living asset and offered new opportunities for local action. The municipality helped to create the association "Volunteers 2000" and availed of its assistance in policies and projects.

In Athens, the Olympics of the Sea were a great part of the 2004 Olympic Games which attracted 55,000 volunteers, Greek and foreign, carefully selected and trained in order to offer their services during the 2004 Olympic Games. The municipality of Athens organised a dedicated body of volunteers to help visitors discover another face of the Greek capital city and the coastal landscapes beyond tourist stereotypes. Volunteers are also active participants.

Volunteers are an invaluable resource for beach clean-ups in coastal cities. The CSO Ocean Conservancy has been leading the way since 1986 with a vision of trash free seas. The International Coastal Clean-up is the world's largest volunteer effort to clean up beaches and waterways, with partners around the globe. During Ocean Conservancy's 2014 International Coastal Clean-up, 648,015 volunteers collected more than 12 million pounds of trash around the shores of the world. The organisation publicises data that help the public better understand the problem of ocean trash, cooperates on innovative solutions to the ocean trash issue with leaders from the private, public and academic world and tries to build a movement to prevent the creation of trash.

Committed citizens can act as ambassadors of their relationships with the seas. Many citizens have often acted as ambassadors of their cities and contributed to city twinning schemes. The earliest examples of twinning cities include the treaties between ancient city-states designed to protect each other's interests in times of hostilities. Most recent twinning schemes in Europe have their origins in the hope of peace and the unprecedented involvement of the citizenry in world conflicts.

Citizen diplomacy is vital for open coastal cities. The twinning activity became much more intense after World War II and brought together many former enemies. In 1947, Bristol, for instance, sent citizens as ambassadors to Hanover and Edinburgh and signed a twinning agreement with Nice. And on the other side of the Atlantic, the US Sister City programme in 1956 became a citizen diplomacy network strengthening partnerships for global cooperation at the municipal level, cultural understanding and socio-economic development. The programme created a movement for volunteer action and local community development, by motivating and empowering citizens, municipal officials and business leaders to engage in citizen diplomacy.

The choice of twin cities may be based on various geographical, industrial or cultural characteristics and factors, growing from long-standing traditions rooted in past or recent links prompted by political solidarity. Old port or maritime cities have close links, as for instance Marseille twinned with Piraeus. Sustainability ethics create new forms of global solidarity among local communities increasingly aware of their common destiny.

Over recent decades, city twinning became frequent and active in Europe, including not only cities from countries with long-standing cooperative experience but also from the post-socialist states. Twinning among coastal cities, especially ports or tourist resorts, is viewed by many municipalities as an instrument for both addressing local problems and ensuring sustainable development (Joenniemi and Sergunin 2011).

9.4 Local Shared Leadership and Democratic Global Governance

On the global scene, city twinning schemes have planted seeds towards the creation of a world coalition for cities. The social compact among emerging countries, responsible for their development, and developed countries, supporting them through aid, debt relief and trade access, has been beneficial but further commitments are needed. Progress is also required on an international agenda for trade, knowledge sharing and technology transfer, access to essential care, and promotion of youth employment. Cities could play a great role.

A strong global partnership of cities could yield sufficient collective power and prod further up the sustainable development global agenda. Cities often have more freedom than nation states to put into place progressive strategies that are changing citizen's lives. The C40 Cities, the global network of large and engaged world cities, is committed to implementing locally meaningful actions to help address climate change. The C40 alliance was formed in 2005, when representatives from 18 mega-cities came together to cooperate on reducing climate emissions through decisive and immediate action on procurement policies and the uptake of climate-friendly technologies and market dynamics. This was the start of the C40 Climate Leadership Group, comprised of cities in diverse stages of the development arc. The initiative invested in meaningful partnerships with the Clinton Climate Initiative (CCI) and Siemens, with which organised the City Climate Leadership Awards (C40 and Siemens 2013, 2014).

The Clinton Climate Initiative aims to address the core drivers of climate change at the invitation of city and national governments and with businesses around the world. The Initiative supports green growth programmes to reduce greenhouse gas emissions, such as increasing accessibility and deployment of clean energy, reversing deforestation, and reducing carbon emissions in cities and communities. The initiative embraced a holistic approach to tackle the major sources of greenhouse gas emissions and influence the people, policies, and practices that impact them.

Municipal governments can have a major direct impact on the global market for blue green technologies. Cities purchase goods and services for thousands of activities, such as schools, hospitals, administrative offices and police stations, port functions and waterfront lighting. Cities also buy and operate municipal fleets of vehicles and run their water and waste systems. Through the C40–CCI partnership, cities are able to pioneer energy-efficient and clean-energy products and technologies. This collective effort can significantly reduce greenhouse gas emissions and mitigate climate change on a large and measurable scale.

Cities and citizens and networks of cities joined the People's climate march on 21 September 2014 and made cities echo with hope. From Buenos Aires to Naples and Jakarta, 2646 solidarity marches in 162 countries asked for a global breakthrough in view of the new agreement to save the world from run-away climate change.

The leadership role of cities on the world scene continues to become more solid. In 2014, the C40 welcomed member cities and partners in New York for the Climate

Week, which helped demonstrate the growing momentum towards a global agreement on climate action. The City Climate Leadership Awards 2014 by Siemens and the C40 Cities Climate Leadership Group honoured ten world cities for excellence in urban sustainability and leadership in the fight against climate change. They are all coastal cities and include Amsterdam (Finance & Economic Development), Barcelona (Intelligent City Infrastructure), Buenos Aires (Solid Waste Management), London (Carbon Measurement and Planning and Air Quality), Melbourne (Adaptation and Resilience), New York City (Energy Efficient Built Environment), Portland (Sustainable Communities), Seoul (Green Energy) and Shenzhen (Urban Transportation). Last but not least, Taipei was awarded the Citizens Choice, for which all were free to cast one vote, and all 31 finalists were eligible to win.

Emerging world coastal cities could be the gates for the cleaner technologies to upgrade the world marine infrastructure. Significant penetration of cleaner technologies in emerging cities would need access to the best available technology, finance for the necessary investments and an adequate institutional framework. Cities, in cooperation with business coalitions, could facilitate smooth international exchanges, access to effective markets and services and transfer of state-of-the-art technology for coastal management and the emerging offshore economy.

The penetration of renewable technologies in the emerging coastal megacities of the world can yield multiple benefits for the environment, security of energy supply and the global economy. Offshore generation presents a special opportunity and can complement the grids which presently leave a large share of the population outside, and facilitate access of the population to cleaner and emission-free systems fuelled from the exploitation of offshore renewable resources.

City networks can have a great impact on mobilising the cities of the emerging world. Twinnings and schemes of cooperation among coastal cities should help facilitate the transition to clean energy options also from offshore. Cities in the EU, the world's largest donor of development aid, have a critical role to play, particularly for the sharing of knowledge and the transfer of cleaner energy technologies. Advanced marine technology projects, carried out in cooperation with emerging cities, could greatly expand the development of urban services based for example on offshore renewable sources. Furthermore European cities could help developing cities to strengthen their democratic governance structures and balance and integrate their pro-poor, pro-jobs, pro-growth and pro-environment agendas.

The 2014 Climate Week has been instrumental for catalysing global leadership action. The Action statement by the Compact of Mayors is a ground-breaking agreement by city networks to undertake a transparent and encouraging approach to reduce city-level emissions, and to enhance resilience to climate change, in a consistent and complimentary manner with national efforts. Not only have local governments cooperated more closely, but their organisations and networks created multiplier partnerships in cooperation with business and scientific organisations. The Compact of Mayors builds on the ongoing efforts of world and European networks including C40, ICLEI, United Cities and Local Governments, World Resources Institute, Energy Cities, Eurocities, The Climate Group, the Chinese Global Mayors Forum and Citynet.

The Compact of Mayors represents an overarching commitment of cities which set ambitious, voluntary climate targets for reduction of greenhouse gas emissions to address climate risk, report on progress against targets, strive for rigorous and consistent reporting standards and make that information publically available on a recognised city platform. The Compact enables recognition of new and city-level commitments and promotes publication of annual reporting data on local climate action. It helps to establish robust and transparent data collection standards, and commit to common, consistent and reliable processes for local climate action.

The Compact demonstrates the commitment of city governments to contribute to more ambitious, transparent and trustworthy climate targets and encourage national governments to actively support additional city action by recognising local commitments, establishing more enabling policy environments and directing resources to local climate action for both mitigation and adaptation.

The adoption of minimum standard and transparent reporting is considered a way of measuring progress towards achieving mitigation targets and adaptation commitments. The Global Protocol on Community-scale GHG Emissions (GPC) is recognised as a new global standard for community emissions. For mitigation purposes, compliance means reporting that is sufficiently robust and rigorous to allow for reliable sectoral-level reporting of GHG emissions. For adaptation, compliance, assessed on an annual basis, means reporting on climate change adaptation commitments such as plans to reduce vulnerability or enhance resilience to specific hazards, such as storm surge and the rise of sea-levels.

The carbon Climate Registry has been designated as the single platform to serve as a unique publically available repository. A common risk framework is to be jointly developed by the cooperating city networks, building on existing frameworks, and in line with relevant international processes including sustainable development and disaster risk reduction. The GPC 2.0 has been released at the COP 20 in Lima, as the new global accounting and reporting standard for city-level GHG emissions. This is a joint activity by the World Resources Institute, C40 and ICLEI, and is supported by a number of international organisations (WRI et al. 2014).

The 2014 Climate Week has been a tipping point, a critical threshold on the way to COP 20. The week brought together 400,000 people in New York and millions of citizens in 162 countries in the largest climate change demonstration in history to keep the pressure on, and tip governments into action. During Climate Week, citizens and the world's largest investors made their voices loud in that tackling climate change is an ethical and an economic imperative. Nearly 350 investors managing $24 trillion urged governments to put a price on carbon, phase out fossil fuel subsidies, and forge a strong global agreement on climate change by 2015. Some leading investors announced new commitments to act on climate change and invest in clean energy. These actions included an initiative to decarbonise investment portfolios, a commitment by commercial banks to issue $30 billion of green bonds by 2015, and an announcement by three major investors declaring their will to accelerate their investments in low-carbon assets to a combined $31 billion by 2020.

The commitments announced during Climate Week are significant and can open a new avenue. Forty companies, including Kellogg's and Nestlé, announced significant new commitments to reduce and eventually eliminate tropical deforestation from unsustainable palm oil production. In order to limit global warming to 2 °C, the world needs to invest $44 trillion in clean energy by 2050, an average of more than $1 trillion per year for the next 36 years, coined by CERES as the Clean Trillion (CERES 2014).

The year leading up to the COP 21 in Paris in December 2015 is crucial. The Lima Accord reached at the conclusion of the UN Framework Convention on Climate Change's 20th meeting, marked the first time that all nations joined their voices to fight climate change. Although non-binding, it is an encouraging development that sets the stage for further action. Major achievements include the pledges made by both developed and developing countries prior to and during the COP 20 that took the capitalisation of the new Green Climate Fund past an initial $10 billion target. Transparency and confidence-building reached new heights as several industrialised countries questioned their emission targets under a new process, the Multilateral Assessment. Among the inspirational initiatives launched in Lima, the "Change initiative" proposes to evaluate the risks of stranded assets for economies and their sovereign debt. Last but not least, the Lima ministerial declaration on education and awareness raising called on governments to put climate change awareness into school curricula and into national development plans.

Cities, citizens, investors, and businesses should ask governments to commit to stronger climate policy objectives. Government policy is critical to unleashing clean investment at the speed and scale needed. The great forward movement towards this direction is the landmark USA–China climate deal, which created momentum heading into COP 20. It overcame a political taboo and created a promising foundation on which to build. This represents a strong recognition of the need to advance towards decarbonisation in the not too far away future by the two countries with the largest CO_2 emissions. The US and China commitments to reduce greenhouse gas emissions and promote renewable energy by 2025 and 2030 respectively send a promising signal to the world community on the path to the Paris COP 21. According to the White House, the new goals would keep the United States on the trajectory to achieve deep economy-wide carbon emission reductions of the order of 80 % by 2050. China has a target of 20 % energy consumption coming from zero-emission sources by 2030.

References

Abbott J (1996) Sharing the city: community participation in urban management. Earthscan, London

Barber BR (2013) If mayors ruled the world: dysfunctional nations, rising cities. Yale University Press, New Heaven

C40 Cities and Siemens (2013) City climate leadership awards. C40 Cities and Siemens, London

C40 Cities and Siemens (2014) City climate leadership awards. C40 Cities and Siemens, New York

Carbon Disclosure Project (2012) Seven climate change lessons from the cities of Europe. CDP, London

Coalition of Environmentally Responsible Economies and Societies (CERES) (2014) Investing in the clean trillion: closing the clean energy investment gap. CERES, Boston

Committee of the Regions (CoR) (2014) The Athens declaration. Documents for the 6th European summit of regions and cities. CoR, Brussels

European Foundation for the Improvement of Living and Working Conditions (EFILWC) (1998) Challenges for urban governance in the EU. EFILWC, Dublin

Joenniemi P, Sergunin A (2011) Another face of integration: city twinning in Europe. Res J Int Studies 22:120–131

Putnam RD et al (2002) Democracies in flux: the evolution of social capital in contemporary society. Oxford University Press, New York

Seas At Risk et al (2014) Priorities for MSFD programmes of measures. Seas at Risk, Brussels

Storper M (2013) Keys to the city: how economics, institutions, social interaction, and politics shape development. Princeton University Press, Princeton

World Future Council (WFC) (2012) The 2012 future policy award. WFC, Hambourg

World Resources Institute (WRI) (2014) Global protocol on community-scale GHG emission inventories. An accounting and reporting standard for cities. World Resources Institute/C40/ICLEI, Washington, DC

Selected Internet Links

www.ec.europa.eu/citizenship
www.arau.org
www.coe.int/clrae
www.c40cities.org
www.clintonfoundation.com
www.oceanconservancy.org
www.wiserearth.org
www.unhabitat.org
www.worldfuturecouncil.org
www.twinning.org
www.worldmayor.com/

Watercolour 10
Naples, a Mediterranean City at Risk?

Chapter 10
Interconnected World Anthologies of Future Blue Cities

Abstract The European Union, the world's first maritime power, is the continent with the densest network of conscious cities. Many coastal cities have developed a myriad of innovative flagship actions which can inspire world cities. But the path to sustainable development will be long and arduous. This final chapter builds on messages from promising cities and pioneer networks for increasing the resilience of vulnerable coastal cities and their potential to grasp opportunities offered by the blue green growth.

Global strategic partnerships and alliances, such as the C40, the Asian Cities Climate Change Resilience Network and the 100 Resilient Cities, can greatly improve the capacity of coastal cities to withstand threats and seize opportunities for sustainable development. Promising emerging cities, such as Jakarta and Manila, can benefit from good cases and integrate no-regret multi-beneficial practices in their transformative agenda to become more resilient and thrive together with their citizens.

10.1 Towards Sustainable Cities: Messages from the European Rainbow

The European Union is first and foremost urban, a constellation of cities and urban agglomerations, shaped by legendary cultures and surrounded by majestic seas. The first maritime power of the planet includes the most polycentric urban network. London and Paris are the only global cities. Smaller world cities, like Amsterdam or Barcelona, try to enhance their human, social, economic, political and cultural assets. Vibrant urban conglomerations have been developed in many countries in symbiosis with the sea. The Randstad surrounding the Dutch "green heart", sixth-largest metropolitan area in Europe, is a noteworthy example of an urban region in balance with its water environment. Intangible bonds and networking are crucial for cities willing to build upon each other's experiences and achieve quantum leaps (Hall 2013).

Europe is a polycentric urban network of small and medium-sized cities boosting the dynamism of European regions. "Small large" cities are usually on a more human scale. Coastal medium-sized cities usually offer easier access to multiple

---

V.P. Mega, *Conscious Coastal Cities*, DOI 10.1007/978-3-319-20218-1_10

physical and social resources, including the sea. Intermediate cities are often more open to the countryside and may act as an interface between larger cities and coastal regions. Medium-sized cities, on the orbit of metropolitan areas, face special opportunities and threats. Their geographic position may provide advantages of access to international links, but also disadvantages in developing autonomous identities and functions (Bellet et al. 1998).

European cities mobilised for sustainable development even before the 1992 Rio Summit. The wave of conferences on European Sustainable Cities and Towns (ESCT) brought progress in the move towards urban sustainability. The first conference (Aalborg, May 1994) focused on the discussion and signature of the "Charter of European Cities and Towns: Towards Sustainability", starting point for the European Campaign of Sustainable Cities and Towns. This generated a movement of conscious cities in Europe and an important pillar in the pantheon of world networks and movements (ICLEI 1995).

The Charter of European Cities and Towns Towards Sustainability, seen as the European version of Local Agenda 21, states that cities and towns should base their living standards on the carrying capacity of nature and advance towards social justice, prosperous economies and environmental improvements. Social equity is considered to be a precondition to the achievement of sustainability, as the inequitable distribution of wealth causes both unsustainable results and increases resistance to change. The Charter embraces an ecosystem approach to urban policies and declares the responsibility of cities for many global problems. Land use and transport models, production and consumption patterns, values, culture and leisure have myriads of consequences. Sustainable development cannot be achieved without governments, local communities and citizens rising to the purpose of transmitting their preserved urban capital to the next generations (ESCTC 1994).

Six years before the turn of the century, the ECSC Charter engaged with cities to improve their environmental condition and not export problems into the broader environment, including the marine environment, or the future. High priority has to be given to the reduction of greenhouse gas emissions, the enhancement of biodiversity and preservation of the ecosystems. Finally, the Charter advocated the development of urban sustainability indicators as yardsticks of progress and compasses for the way forward.

Coastal cities and their associations, especially in the Mediterranean Sea and the Baltic Sea, were among the first signatories of the charter in 1994 and the partners of the European campaign of sustainable cities and towns which also brought together important constellations of local authorities wishing to benefit from the collective momentum. The Association of Cities and Regions for Recycling, the Climate Alliance, the Council of European Municipalities and Regions, the Énergie-Cités, Eurocities, ICLEI, and the World Health Organisation Healthy Cities were among the partners. The campaign introduced the Sustainable City Award which recognised best practice and promoted noble emulation.

In many cities, sustainability became an innovative, creative and pro-active process gradually embedded in the institutional culture and practices, and the subject of a structured, and meaningful dialogue with citizens and other stakeholders. The second conference on European Sustainable Cities and Towns (which took place in Lisbon in 1996) urged cities to move from charter to action (ICLEI 1997), whereas the third conference (in Hanover in 2000) marked a leap to the new millennium with the Mayors convention declaring local sustainability as their highest political priority. The Charter of European Cities and Towns celebrated 10 years of interconnected efforts for Sustainable Development with the "Aalborg+10" Conference in 2004. This fourth conference was marked by discussions of the Aalborg Commitments, conceived to assist cities and towns in engaging with sustainability. The Aalborg Commitments provided an instrument to help local governments to shape robust policies and set clear qualitative and quantitative targets to implement the urban sustainability principles of the Charter.

And the move from words into deeds continued. The coastal city of Dunkirk hosted the sixth conference of the European sustainable cities in 2010, which concluded with two political declarations. The Local Sustainability Declaration advocates for the transition to a sustainable, green and inclusive economy. It demands the strategic prioritisation of investments and a more efficient and ecosystem-based management to enhance citizens' quality of life. European cities that signed the declaration insisted that change can be made only with the support of national and international governments and institutions. The second political declaration was the Call on Climate Action which asked for more ambitious reduction targets to achieve a meaningful longer-term climate agreement. This call also suggests that to achieve a low carbon, climate resilient and green economy in Europe, local and regional governments must be a fully integrated party in negotiations and decisions and work in concert with national governments and supranational bodies.

In the EU governance architecture, cities were long considered to be the level closest to citizens. In the 1990s, the European Commission started a series of actions to promote cooperation and exchanges among cities, notably on environmental issues. The European Urban Initiative tried to address problems and enhance opportunities in cities. "Urban I" (1994–1999) assisted 118 cities of the European Union to improve their living and working conditions, whereas "Urban II" (2000–2006) assisted 70 sensitive urban districts in undertaking socio-economic and environmental regeneration. The Urbact programme capitalised on the experiences of exchanges that were promoted between cities through thematic networks, and the dissemination of exemplary practices. The programme expanded its training of the elected representatives to address the challenges of future leadership (European Commission, Urbact II 2011).

European Union Member States and the Council of Ministers have also been active in urban affairs. National governments started cooperating on urban strategies in the 1990s. During the 2007 German presidency, ministers in charge of urban affairs signed the Leipzig Charter on the sustainable European city (Leipzig, May 2007), a statement of shared principles for urban development policy in Europe. The charter recognised European cities as organic parts of a common heritage. Reinforcement of distressed areas, youth employment and sustainable architecture and planning were presented as key objectives. Recommendations addressed all aspects of urban policy, including sustainable infrastructures, energy efficiency, innovation, education and training, culture and heritage (Council of Ministers 2007).

They apply to all sizes of cities, coastal or not, but have always to be adapted to local circumstances.

The EU territorial agenda provided a springboard for European governments to promote a polycentric cooperation between cities and create new forms of partnerships between the cities and the countryside, to cement regional clusters for increasing competitiveness to promote the trans-European management of risks, such as the effects of climate change and sea level rise, and to strengthen ecological and cultural resources. In 2008, a Declaration of the ministers in charge of Urban Development (Marseille, November 2008) brought forward the main challenges for cities, including environment and climate change, competitiveness, social cohesion and citizenship. It recognised that cities are key actors for reducing emissions, waste and risks (Council of Ministers 2008).

Exchanges between responsible cities and regions, and open to citizens, Europe and the world, have greatly intensified since the start of the millennium. Since 2003, the open days of the European week of the regions and cities, organised by the European Commission and the Committee of the Regions, bring together in Brussels every October thousands of participants of all regions of Europe. In 2014, the twelfth edition brought a wave of events throughout Europe. Sharing ideas and insights on a wide range of issues, from boosting innovation to employment generation, and addressing short, medium and long-term societal challenges, were at the heart of the debates. Platforms for exchanges with public and private investors serve as laboratories of ideas and actions.

To promote research and innovation and address urban challenges, some Member States of the European Union joined forces as they had done in the Joint Programming Initiative "Urban Europe", an initiative of 15 European countries dedicated to making better use of public research opportunities across Europe. Research issues at various levels include urban scenarios and foresight activities for policy-oriented roadmaps and planning concepts, demonstration and pilot projects, monitoring and benchmarking. The initiative identified vital urgent issues to be addressed in order to create attractive, sustainable and economically viable urban areas in which European citizens and communities can thrive. Building on existing research and policy initiatives, Urban Europe brought together multiple and diverse strands of thought and action to implement a strategic research agenda. To make cities attractive places to live and work in a global village, governments agreed to enhance the advantages of their cities and help realise their potential for innovation and competitiveness (European Commission, VU University Amsterdam 2011). This initiative can inspire joint action by other integrated regions such as the Association of Southeast Asian Nations.

Global sustainability depends much on the multiplication of exemplary action and events at the urban/region level. Some inspiring models come from the annual European Green Capitals, launched in the EU towards the end of the first decade of the millennium. The candidate cities have presented a wealth of proposals for a year of green technological and social innovation. Twelve indicators help to benchmark the cities according to their contribution to GHG emissions, local transport and green areas, nature and biodiversity, quality of air and the acoustic environment, water consumption and waste management, waste water treatment, eco-innovation and sustainable employment, environmental and energy performance. All candidates must prove that sustainability ethos is high on their agendas and leadership for sustainability is exercised at all levels. The award-winning cities must be committed to excellence in environmental performance, act as role models and inspire other cities in their uphill battle for sustainable development (EC 2013a).

The majority of cities which have already been crowned are coastal ones. They all aimed at igniting the debate and driving decisive action towards urban sustainability. Stockholm, first European Green Capital in 2010, was followed by Hamburg (Green Capital 2011), Vitoria Gasteiz (Green Capital 2012) and Nantes (Green Capital 2013), which honoured their respective years with great distinction. The 2014 award celebrated Copenhagen, a city with a record of consistently high environmental standards, profoundly engaged in pursuing ambitious green goals. The city opted for the most sustainability oriented actions in everyday life and all urban functions. Bristol took over the torch as the 2015 European Green Capital.

The combined Green Capitals present a rich record of ideas and actions towards urban sustainability. Stockholm has been the epitome of green urban growth efforts in the European Union, and became a world model (OECD 2013). Hamburg boosted environmental awareness and invested considerably in renewable energy and the creation of HafenCity. Vitoria-Gasteiz praised its semi-natural green area that had been partially reclaimed from degraded areas, surrounding the urban heart and bringing nature into the city. The city wishes to ensure that the entire population lives within 300 m of an open green space and implements many tangible measures to assist and increase biodiversity and ecosystems services. Nantes has developed a sustainable transport policy with a focus on public transport and optimal conditions for pedestrians and cyclists.

The Green Capital movement invited exemplary concepts and solutions for the future of European cities. Innovative partnerships with citizens, businesses, local universities and the civil society were also forged and provide lessons and insights. To enhance the exchange of ideas among European cities, Hamburg put on rails a "train of Ideas", an interactive exhibition on a moving train, with critical information about major projects such as the reintroduction of a tram system and the creation of low/zero carbon districts. "Ships of Ideas" could probably be interesting for coastal cities and their partners.

Next to the European Green Capitals, the European Green Leaf initiative focused on smaller cities with a population between 50,000 and 100,000 inhabitants. It is an award that recognises commitment to environmental improvement, with a particular accent on efforts that generate green growth and new jobs, while developing citizens' environmental awareness, culture and involvement. The initiative tries to identify cities able to act as a "green ambassador" and to encourage other cities to progress towards a better sustainability future. "Blue ambassador" schemes could also be interesting for coastal cities but also other cities engaging with the ocean.

Despite the many achievements, the journey of European cities towards sustainability is a long and challenging one. The European Environment—State and Outlook 2015 Report by the European Environment Agency, the largest integrated assessment of Europe's environment that includes data at global, regional and country levels, as well as cross country comparisons, has alerted us to the risks of environmental deterioration and its impacts on human well-being and prosperity. The report concludes with recommendations to recalibrate policies in line with the 2050 vision of "living well within the limits of the planet".

The EU environment and climate policies have delivered considerable benefits for functioning of ecosystems and for the health and well-being of citizens in Europe, and also has created economic opportunities. The environment industry sector grew by more than 50 % between 2000 and 2011. The quality of drinking and bathing water has improved in recent decades and hazardous pollutants have been reduced. European greenhouse gas emissions have decreased by 19 % since 1990, despite a 45 % increase in economic output. Total resource use has declined by 19 % since 2007, less waste is being generated and recycling rates have improved (EEA 2015a).

However, Europe's natural capital is still being degraded by socio-economic activities such as agriculture, fisheries, transport, industry, tourism and urban sprawl. Air and noise pollution continue to cause serious health impacts, particularly in urban areas. In 2011, about 430,000 premature deaths in the EU were attributed to fine particulate matter (PM2.5). Exposure to environmental noise is estimated to contribute to at least 10,000 premature deaths each year. Perhaps the most difficult challenges for European environmental governance arise from the fact that environmental drivers, trends and impacts are increasingly subject to global processes and complex systems. A variety of long-term megatrends affect Europe's environment, consumption patterns and living standards (EEA 2015b).

Land degradation and climate change remain major concerns, threatening the flows of environmental goods and services that underpin Europe's economic output and well-being. A high proportion of protected species (60 %) and habitat types (77 %) are considered to be in unfavourable conservation status, and Europe is not on track to meet its overall target of halting biodiversity loss by 2020, despite success in relation to specific targets. Fresh water quality has improved over recent years. However, around half of Europe's freshwater bodies are unlikely to attain "good ecological status" in 2015. Marine and coastal biodiversity is of particular concern. Across all of Europe's regional seas, marine biodiversity is in poor condition, with only 7 % of marine species assessed as "favourable conservation status". Seafloor damage, pollution, invasive alien species and acidification are particularly threatening. Overfishing has decreased in the Atlantic and Baltic, but the Mediterranean shows a more negative picture, with 91 % of assessed stocks overfished in 2014 (EEA 2015a).

The way to sustainable development is long and arduous. The level of ambition of existing environmental policy may be inadequate to achieve Europe's long-term environmental goals. Projected greenhouse gas emissions reductions are currently insufficient to bring the EU onto a pathway towards its 2050 target of reducing emissions by 80–95 %. Recalibrating existing policy approaches can make an essential contribution to such transitions. In the environment and climate policy domain, four established and complementary approaches could enhance progress to long-term transitions if considered together and implemented coherently: Mitigation, Adaptation/Resilience, Precaution/Prevention, and Restoration.

With the "EU Cities Adapt" and "Mayors Adapt" Initiative, European cities signalled the world that adaptation to climate change is no longer a matter of policy option. Even if GHG emissions are radically curbed, natural ecosystems have already been perturbed and may take centuries to recover. The journey to resilience is a journey to responsibility (EC 2013b; ICLEI 2013).

10.2 The Journey to Urban Responsibility: A Bird's Eye View on Never-To-Regret Actions

Asia is particularly at risk of apocalyptic disasters and has major cities in vulnerable areas which are insufficiently prepared to face extreme natural disaster and violent shocks. Inadequate citizen education and awareness, congested infrastructure, poverty and deficient public services and chronic tensions are aggravating their conditions. Coastal Asian cities located on deltas or along eroded coastlines, under typhoon belts and on earthquake-prone zones, have to prepare to address the additional pressures associated with climate change and extreme events. Emerging Asian megacities under explosive urbanisation, such as Jakarta and Manila, have been pointed out as having a great potential as emerging cities (Kearney 2014) and should in urgently try to overcome their vulnerabilities.

Strategically positioned in the Indonesian archipelago, Jakarta, the capital and only megacity of Indonesia, is ranked by A.T. Kearney at the top of emerging cities. Located on the Northwest coast of Java, Jakarta is an economic, cultural and political hub, the most populous city in Southeast Asia and one of the fast growing cities in Asia. A global monitor of metropolitan areas highlights that in a 2011 ranking according to economic growth, Jakarta occupied the 17th place among the world's 200 largest cities, a jump from a 2007 ranking of 171th place. Jakarta has grown more rapidly than Kuala Lumpur, Beijing and Bangkok (Brookings 2012).

The history of Jakarta in a nutshell encapsulates the evolution of many emerging cities. Jakarta is located at the mouth of the Ciliwung River on Jakarta Bay. Established in the fourth century, it became a strategic and thriving trading port for the Kingdom of Sunda within the sphere of influence of Srivijaya maritime empire. The first European fleet involved four Portuguese ships from Malacca, which arrived in 1513 looking for spices. Dutch ships landed in 1596 and, once victory over the English was consolidated in 1619, the capital of the Dutch East Indies was renamed Batavia, which still brings to life the city's historic roots. The city became the capital of Indonesia after World War II and was renamed Jakarta.

Indonesia's founding president envisioned Jakarta as a great international city, and instigated most-ambitious, government-funded projects with nationalistic and modernist architecture. Flagship projects included a clover-leaf highway, monuments such as the National Monument, and a new parliament building. In 1965, Jakarta was the scene of an abortive coup attempt which precipitated a forceful anti-communist purge and the beginning of Suharto's New Order. In 1966, Jakarta was declared a "special capital city district", and gained a status approximately equivalent to that of a state. The capital city was endowed with health and education services, equipped with transport infrastructure, encouraged in their culture and the arts, and invested in modernisation. Slum dwellers were moved to new development projects and control of migration to the city tried to reduce overcrowding and poverty. Foreign investment contributed to a real estate boom which changed the face of the capital.

The boom ended with the 1997–1998 East Asian Economic crises which placed Jakarta at the centre of violence, protest, and political tensions. However, the capital remained the focal point of democratic change in the dominant ASEAN economy. In 2007, Jakarta held its first ever direct elections to choose a governor. The city has a two-tier governance system and is divided into five cities, formerly municipalities, each headed by a mayor, and one regency headed by a regent.

Jakarta leads a country with more than 17,000 islands and a coastline of about 81,000 km and has historically been a most important trading port in the region. Its economy depends heavily on financial services. The main problems for sustainability include the extreme population growth, high levels of poverty, and inequitable distribution of wealth. This megacity is highly vulnerable to climate change and must reduce its greenhouse gas emissions. This is critical to prevent future climate impacts and ensure adequate adaptation action to withstand stress. Improved urban planning and infrastructure, such as water and waste services, and green infrastructures are multi-beneficial actions which not only minimise climate impacts but are long-term avenues to sustainable development. Green growth policies started to expand in the capital cities of the region (OECD 2015).

The topography of Jakarta is critical for vulnerability. The capital city includes areas with very diverse morphology. The Northern part is low-lying, just some meters above the average sea level and frequently floods, while the Southern part is hilly. The Ciliwung River is the most significant river and divides the city from West to East. In addition, about 12 other rivers drain the hilly southern part of the city into the sea. Central Jakarta is the capital's administrative, cultural and political heart. It has large parks, colonial buildings and landmarks such as the Jakarta Cathedral. West Jakarta has the highest concentration of small-scale industries, the Chinatown district and colonial landmarks. West Jakarta contains part of Jakarta Old Town, while South Jakarta, originally planned as a satellite city, hosts affluent residential areas. South Jakarta functions as Jakarta's ground water buffer, but the green belt areas are threatened by new housing developments. East Jakarta hosts several industrial sectors, while North Jakarta is bounded by the Java Sea and includes the Tanjung Priok Port. Thousand Islands, formerly a sub-district of North Jakarta, constitute a unique ecosystem of 105 small islands in Java Sea. Marine tourism, such as scuba diving and wind surfing, is the most important activity in this territory which has a high conservation value.

As two-thirds of Indonesia's territory consists of marine and inland waters with an abundance of natural resources, the development of aquaculture was promoted to encourage community-based rural economic activities and ensure food security. About 90 % of the country's total fish production is consumed domestically. Shrimp from both capture fisheries and aquaculture played an important role in exports.

The development of aquaculture according to sustainability principles and the FAO "Code of Conduct for Responsible Fisheries" should enhance the capability of communities in applying environment friendly technologies.

The creation of new green open spaces, improvement of waste management, education and awareness raising among the population, especially the more disadvantaged groups are key priorities. Conservation of mangroves should protect against coastal erosion and strong waves. Furthermore, adaptation action to address sea-level rise, extreme weather phenomena, and threats to ecosystems and biodiversity should be a high priority. Education and awareness should also be the first priority investment and this is also to be shared with the other dynamic emerging ASEAN megacity of Manila.

Manila, capital of the Philippines, was ranked second promising emerging city (Kearney 2014). Metropolitan Manila is the political, economic, social, educational, cultural and recreational hub of a vast archipelago of 7107 islands. Metro Manila is a confederation of 17 different cities and municipalities, creating around one third of the country's domestic product. Manila is an extreme city, marked by huge economic disparities. Reportedly, 97 % of the national GDP is controlled by 15 % of the population.

A large proportion of the population of Philippines lives on archipelagos and in low-lying river deltas, which are particularly susceptible to sea-level rise and flooding. The metropolis is a pulsating city offering a glimpse in the country's multifaceted culture and striking contrasts. In old parts of Manila, like the walled city and Chinatown, life continues like in the previous centuries. In the old port Sunda Kelape many of the scenes on the quayside have been unchanged for centuries. Glass and concrete towers co-exist side-by-side with Spanish colonial houses and neo-classical government buildings. Dragon arches, gold-domed mosques and elaborately-styled mansions define the enclaves of the Chinese, the Moslems, and the rich. Vibrant marketplaces are only steps away from peaceful parks and silent churches. Outside the rush of modern air-conditioned shopping centres, crowds bargain with sidewalk vendors.

In a typhoon-prone region and near exceptional global marine resources, Manila is a vulnerable megacity. The average elevation of metropolitan Manila is a mere 10 m. The Metro Manila is built on alluvial deposits of the Pasig River, Manila's primary waterway, which meanders through the metropolis before draining into Manila Bay. Building resilience and protecting marine resources are key priorities. Manila's protected harbour is also the main seaport for the Philippines and a major manufacturing centre, for chemicals, steel, textiles, clothing, and electronic goods. Manila's passenger and container ports bustle with the activities of local and international shipping lines. Cruise ships, as well as private vessels, also can find a berth in Manila Bay. Shipping and tourism have to develop the most sustainable way to provide lasting benefits (WWF 2009).

The Philippine archipelago has vast marine and inland resources, lakes, rivers, reservoirs and estuarine areas. The fisheries sector plays a vital role in the national economy. Aquaculture has a long history and involves many species and farming practices in diverse ecosystems. It contributes significantly to the country's food security, employment and foreign exchange earnings. Farmed seaweeds by far remain the top export commodity. The Philippine aquaculture industry needs to embrace a global sustainability perspective to become part of a healthy local blue economy.

The no-regret options: ten essential engagements to both respond to urgencies and advance to sustainable development

All the above can help garner some advice for fundamental investments which will be beneficial, anyway, even if extreme events do not occur. Given the medium-term risks and uncertainties and the potential long-term benefits from sustainable development, it is important to concentrate investments on the most urgent and important actions.

Engage with All citizens and stakeholders

Education, awareness raising and citizen empowerment, well before extreme events occur, is essential. Action for resilience is still essentially disaster-driven. Emergency services often help repair damages, but proactive longer-term action can be far more beneficial.

Engage with the sea and the water

Coastal cities have to invest in the health of their marine resources. They should drive for sustainable fisheries and aquaculture, close their ports to illegally harvested seafood and take action for emission reduction, raise awareness on ocean acidification and strive to reduce marine litter.

Engage with blue green infrastructures

Responses to sea-level rise have to start with green infrastructure and wetland restoration. The enhancement of natural processes should become an essential part of flood planning. Urban retention areas and ocean parks have multiple benefits.

Engage with win-only energy options

Investments in energy efficiency and renewable energy sources can only be beneficial. Transforming marine waste into renewable energy and exploring ocean, tidal and offshore wind energy options are responsible choices for coastal cities.

Engage with inclusive resilience

Every city, in cooperation with all stakeholders, has to carefully evaluate its vulnerability, and the social tolerance to risks. The optimal adaptation levels should be co-decided. To foster their immunity and better withstand threats, coastal cities have to integrate threats and climate-related mitigation and adaptation in daily action.

Engage with Democracy and Trust

Democratic governance, active citizenship, leadership and accountability are of the highest importance. All cities have to reinforce institutional capacity and strive for the well-being of present and future citizens, in an increasingly multipolar interconnected world.

Engage with blue green inclusive growth

Sustainable development of all blue activities, from fisheries and shipping to cruise tourism is the fundamental precondition for the blue economy to thrive. Multiple-dividend innovations are essential for creating new assets, often out of liabilities, and capturing blue green synergies. Eco-responsible businesses have a key role in leading action.

Engage with integrated ecosystem-based policy and structural reform

Transformative agendas have to be integrated, strategic, holistic, transparent, multi-stakeholder, ecosystem-based. Urban coastal planning has to provide a framework for the optimisation of urban functions on land and the sea and address synergies and conflicts.

Engage with the world

Openness to the global world of ideas and actions is decisive. World partnerships and networks are a great source of peer-learning and a driver of change through sharing of knowledge and exchange. They also constitute an advocacy means at the global level.

Engage with complexity and uncertainty

Decision-making processes should incorporate futures thinking. Anticipatory democracy has to help build trust and a common vision for the desirable future, after discussion of all possible futures.

10.3 World Partnerships and Networks for Urban Resilience and Sustainability

World partnerships and alliances are a great driver and catalyst of progress. Sharing experience with other cities in the framework of structured world networks is highlighted as very beneficial by most partners. Vulnerable and poor cities have a special interest to learn from their peers, even if they always have to find their own way to resilience and sustainable development. Innovative initiatives originate from UN agencies and especially the UN-HABITAT, but also the C40 Cities Climate Leadership Group, ICLEI, the WBCSD and foundations such as the Rockefeller Foundation. Partnerships are extremely useful for bringing together evidence and also creating common tools and adaptable models and landscape solutions. This is particularly interesting for smaller and poorer cities and local governments which lack expertise and resources. Furthermore, the advocacy power of global networks and coalitions is of the highest importance.

Examples abound. For accounting and reporting city-wide greenhouse gas emissions, cities can use the Global Protocol for Community-Scale Greenhouse Gas Emission Inventories (GPC) presented during the COP 20 in Lima. It is a GHG Protocol standard for cities, developed jointly by the World Resources Institute, C40 Cities Climate Leadership Group and ICLEI. The GPC helps cities develop a comprehensive and robust greenhouse gas inventory, set reduction targets, and monitor their performance, ensure consistent and transparent measurement and reporting of greenhouse gas emissions according to internationally recognised greenhouse gas accounting and reporting principles, and gain insights through benchmarking and possible aggregation of comparable data (WRI et al. 2014).

In 2013, the Rockefeller Foundation launched the 100 Resilient Cities Centennial Challenge to stimulate 100 cities to better address the increasing shocks and stresses of the twenty-first century, by adopting already known and shared methodologies and solutions. The initiative enables cities to evaluate their exposure and risks to threats, to develop a concerted integrated plan to address them, and to respond more proactively and effectively. It suggests that enhancing resilience is about making cities better, for both the short and long-term, for all and everyone (Rockfeller Foundation 2014).

The two first waves of cities were selected in 2013 and 2014. Each city was asked to present a clear and compelling description of its approaches and planning to decrease vulnerabilities and reinforce resilience. The selected cities demonstrated a dedicated commitment to building capacities to prepare for, withstand, and bounce back rapidly from shocks and stresses. The city resilience framework, developed by Arup, provides a lens and a common language for cities to understand their complexity and the drivers that contribute to their resilience. It assists cities to assess the extent of their strengths, identify critical weakness factors, and design actions to address their vulnerable areas of concern.

Other worthy of mention initiatives includes the "AsianCitiesAdapt" partnership, involving local governments in eight cities in India and the Philippines. The partnership carried out a cyclical adaptation management process, which integrated local adaptation needs into local government daily operations to address vulnerability. Scientists from the Indian Institute of Technology in Delhi and the Potsdam Institute for Climate Impact Research worked together and suggest that vulnerability assessment and resilience can be extremely important for development policies. There is no blueprint for adaptation; it is up to local stakeholders to incorporate the adaptation principles into their policies (ICLEI 2013; ICLEI et al. 2015).

The Asian Cities Climate Change Resilience Network (ACCCRN), launched in 2008, aims to catalyse attention, funding, and action on building climate change resilience for vulnerable communities. The network brings together over 20 small and medium-sized cities in India, Vietnam, Thailand, Bangladesh, the Philippines and Indonesia, and generates practical examples for climate change adaptation and mitigation, and building urban resilience in rapidly urbanising, low and middle income countries.

The vision statement of the network suggests that ACCCRN partners collaborate to build a robust knowledge basis of realistic and actionable know-how to build resilience, which will ultimately improve the lives of the poor. The ACCCRN programme, funded by the Rockefeller Foundation, incorporates knowledge and project management and network capacity building.

The CityLinks programme facilitates connections and relations to promote climate-related mitigation and adaptation as a component of all local planning and sectoral policies. Through CityLinks, US cities with effective mitigation and adaptation strategies are partnered with cities in developing and emerging countries to help adapt well-tested policies into local conditions. The initiative facilitates exchange visits that allow partners, for instance Fort Lauderdale and Durban, to address the impacts of climate change and shape strategies for adaptation. A pilot initiative promoted a city-to-city partnership of La Ceiba and Choloma in Honduras with Los Angeles, in order to strengthen violence prevention through environmental design.

Risk reduction is certainly an investment in the way of sustainable development. Some initiatives focus on risk governance and propose toolkits and handbooks for local government leaders striving to make cities more resilient The UNISDR handbook discusses the benefits of disaster resilience, the range of possible required actions and the ways to organise the tasks. It offers practical guidance to understand and take action on the "Ten Essentials for Making Cities Resilient" as set out in the global campaign "Making Cities Resilient: My City is Getting Ready!" (UNISDR 2012).

The UN-HABITAT launched the City Resilience Profiling Programme (CRPP) in 2013. CRPP aims to develop a comprehensive, integrated urban management and planning approach to monitor urban risk resilience, including climate change and multi-hazard catastrophic events. The programme aims to achieve an adaptable urban systems model, and indicators, profiles and standards to support cities in strengthening urban systems. The UN-HABITAT is testing and refining the Programme's guidelines and tools in ten partner cities, including Balangoda, in Sri Lanka, Barcelona, Beirut, Dagupan in the Philippines, Dar es Salaam, Lokoja in Nigeria, Portmore in Jamaica, Concepcion in Chile, Tehran and Wellington. Many more, coastal or non-coastal cities, will be able to benefit. History is in the making.

References

Bellet C et al (1998) Ciudades Intermedias. Urbanizacion y sostenibilidad, Milenio, Lleida

Brookings Institution (2012) Global metro monitor 2011. Volatility, growth and recovery. The Brookings Institution, Washington, DC

Council of Ministers (2007) Leipzig charter on sustainable European cities. Council of Ministers, Leipzig

Council of Ministers (2008) Déclaration finale des ministres en charge du développement urbain. Council of Ministers, Marseilles

European Commission (EC) (2013a) European Green Capital. Urban environmental good practice and benchmarking report. EC, Brussels

EC (2013b) EU adaptation strategy. EC, Brussels

EC, Urbact II (2011) Urbact projects results. EC/Urbact II, Brussels

EC, VU University Amsterdam (2011) Joint programming initiative on urban Europe. Global challenges and local responses in the urban century. EC/VU University Amsterdam, Amsterdam

European Environment Agency (EEA) (2015a) The European environment. State and outlook 2015. Synthesis report. EEA, Copenhagen

EEA (2015b) The European environment. Assessment of global megatrends. EEA, Copenhagen

European Sustainable Cities and Towns Campaign (ESCTC) (1994) Charter of European cities and towns: towards sustainability. ESCTC, Brussels

Hall P (2013) Good cities, better lives: how Europe discovered the lost art of urbanism. Routledge, London

International Council for Local Environmental Initiatives (ICLEI) (1995) Towards sustainable cities and towns. Report of the first European conference on sustainable cities and towns. Aalborg, May 1994. Freiburg

ICLEI (1997) From charter to action. Report of the second European conference on sustainable cities and towns. Lisbon, 1996. Freiburg

ICLEI (2013) Local governments championing adaptation to climate change. A primer for decision-makers in Asian cities. ICLEI, Freiburg
ICLEI, WMCCC, Bangkok Metropolitan Administration (2015) Resilient cities Asia-Pacific 2015. ICLEI/WMCCC/Bangkok Metropolitan Administration, Bangkok
Kearney AT (2014) Global cities index and emerging cities outlook. Global Cities, Present and Future
Organisation for Economic Co-operation and Development (OECD) (2013) Green growth in Stockholm, Sweden. OECD, Paris
OECD (2015) Green growth in Bangkok, Thailand. OECD, Paris
Rockfeller Foundation (The) (2014) Insights. An annual publication exploring emerging issues of consequence. Rockfeller Foundation (The), New York
United Nations International Strategy for Disaster Reduction (UNISDR) (2012) How to make cities more resilient. A handbook for local government leaders. UNISDR, Geneva
World Resources Institute (WRI) (2014) Global protocol on community-scale GHG emission inventories. An accounting and reporting standard for cities. World Resources Institute/C40/ICLEI, Washington, DC
WWF (2009) Mega-stress for mega-cities: a climate vulnerability ranking of major coastal cities in Asia

Selected Internet Links

www.ec.europa.eu/regional_policy/urban
www.cor.europa.eu
www.jpi-urbaneurope.eu
www.ec.europa.eu/energy/index_en.htm
www.oecd.org/greengrowth
www.epa.gov/jius
www.100resilientcities.org/resilience
http://unhabitat.org/initiatives-programmes/city-resilience-profiling-programme/
http://icma.org/en/cl/home
www.iclei.org
http://www.asian-cities-adapt.org
http://www.arup.com/Projects/Asian_Cities_Climate_Change_Resilience_Network.aspx

Watercolour 11
Sydney, a Coastal City with High Liveability

Watercolour 12
Helsinki facing the Baltic Sea

About the Author

Dr. Voula Mega graduated as an engineer from the National Technical University of Athens and completed her DEA Diploma at the National Geographical Institute in Paris. She continued with a DEA at the French Institute of City Planning. She was conferred with a Ph.D. on City and Regional Policy and Planning at the Institut Français d'Urbanisme and the École Nationale des Ponts et des Chausséees in Paris.

Her post-Doctorate studies include policy-oriented research on Regional Policy at the School of Planning, Oxford Brooks University, and Environmental Economics and Policy Analysis at the Harvard Institute for International Development, Harvard Kennedy School of Government. She also received a scholarship from MIT and attended a summer programme on innovation and the dynamics of technology and organisations.

She started her career as Special Adviser to the Greek Government and held important positions in various European and International organisations. She has been Research Manager at the European Foundation for Living and Working Conditions, an EU agency in Dublin, and worked as consultant at the Organisation of Economic Cooperation and Development in Paris. In 2001, she joined the European Commission in Brussels, where she works as senior policy analyst at the Directorate-General for Research and Innovation. She contributed to important policies, including, *inter alia*, energy research policy and social and ethical implications, strategic planning and programming, the European Research Area, the European Partnership for Researchers, foresight studies in relation to global societal challenges and the European Forum on Forward Looking Activities and the Blue economy and marine resources.

V.P. Mega, *Conscious Coastal Cities*, DOI 10.1007/978-3-319-20218-1

Her academic activities include invited Master Lectures at the University of California at Berkeley, Tokyo Institute of Technology, Harvard Chapters, Columbia University, Prince of Wales Institute of Architecture, Institut Français d'Architecture, the European University of the Environment, University College of London, Catalan Institute of Technology, and the Universities of Athens, Bologna, Trieste and Parma. She has also been Associate Professor at the National School of Public Administration, in Athens.

Citizen of the world and passionate European, Voula was nurtured by many cultures, including Spanish, Latin American and Thai. As a strategist, she strives for leadership in far-reaching ideas and concepts and injects open futures thinking in policy analysis and negotiation processes. Her policy analysis and advice, research management and publications address the scientific foundations, values, culture and innovations for sustainable development policies and the cities of the future. She firmly believes that Science and the Arts can herald a better world and that strategic foresight, visions and plans for desired futures can far extend the limits of engagement and action.

Always in search of excellence, she published and co-published many books and hundreds of articles with the European Union, OECD, UN, UNESCO and international publishers and magazines, in Greek, English, French, Spanish and Italian. Most recent titles include "Quintessential Cities, Accountable to the Future. Sustainability, Innovation and Citizenship" (Springer, New York, 2013), "Bio-diver-cités" (Saarbrucken, EUE, 2012), "The Desirable Future of Innovative Cities" (Saarbrucken, LAP, 2011), "Sustainable Cities for the Third Millennium. The Odyssey of Urban Excellence" (New York, Springer, 2010), "Modèles pour les villes d' avenir" (Paris, L'Harmattan, 2008) and "Sustainable Development, Energy and the City" (Springer Science+Business Media, New York, 2005).

Furthermore, she tries to actively promote the dialogue between Science and the Arts and among various forms of arts, especially poetry, design and painting. She published the poetry collections "Siren Cities" (in Greek, English and French) accompanied with drawings (Athens, Exantas, 1997), "Dawns and Souls for Europe" (in English and French by Persée, Paris 2008), official selection for the European Book Prize 2008, and "Héliopolis. Sourires de Villes solaires" (Paris, HD, 2013). An exhibition of watercolours, drawings, diptychs (poetry/design) and photos at the European Parliament (16–21 February 2004) brought together her artistic work on cities under the title "The Song of Siren Cities".

Watercolour 13
Copenhagen: The 2014 European Green Capital

Index

V.P. Mega, *Conscious Coastal Cities*, DOI 10.1007/978-3-319-20218-1

D

MIX
Papier aus verantwortungsvollen Quellen
Paper from responsible sources
FSC® C105338

If you have any concerns about our products,
you can contact us on
ProductSafety@springernature.com

In case Publisher is established outside the EU,
the EU authorized representative is:
Springer Nature Customer Service Center GmbH
Europaplatz 3, 69115 Heidelberg, Germany

Printed by Libri Plureos GmbH
in Hamburg, Germany